T0172087

PLANETARY CRUSTS: THEIR COMPOSITION, ORIGIN AND EVOLUTION

Planetary Crusts is the first book to explain how and why solid planets and satellites develop crusts. This extensively referenced and annotated volume presents a geochemical and geological survey of the crusts of the Moon, Mercury, Venus, the Earth and Mars, as well as the distinct crusts of the asteroid Vesta and the satellites Io, Europa, Ganymede, Callisto, Titan and Triton.

Spanning a much wider compass than mere descriptions of the diverse crusts encountered throughout the Solar System, the book begins with a discussion of the nature of Solar System bodies and their formation. The authors then adopt a comparative approach to investigate the many current controversies surrounding the development and evolution of planetary crusts. These include the origin of the Moon and Mercury, the nature of the Mercurian plains, the exotic chemistry of Mars, differences in the geological histories of Venus and Earth, the significance of the rare earth element europium, the primitive crusts on the Earth, the onset of plate tectonics, the composition of the mantle, the origin of granites, why Ganymede differs from Callisto, and many other debated topics. The authors conclude that stochastic processes dominate crustal development, and the book ends with a discussion of the likelihood of Earth-like planets and plate tectonics existing elsewhere in the cosmos.

Written by two of the world's leading authorities on the subject, this book presents an up-to-date survey of the numerous scientific problems surrounding crustal development. It is a key reference for researchers and students in geology, geochemistry, planetary science, astrobiology, and astronomy.

STUART ROSS TAYLOR was born in New Zealand and is now an Emeritus Professor at the Australian National University. He is a trace element geochemist and carried out the initial analysis of the first lunar sample returned to Earth at NASA, Houston in 1969. He has a D. Sc. from the University of Oxford, is a Foreign Member of the US National Academy of Sciences, and has received the Goldschmidt Medal of the Geochemical Society, the Leonard Medal of the Meteoritical Society, and the Bucher Medal of the American Geophysical Union. He is the author of 6 other books including *Solar System Evolution,* Second edition (Cambridge University Press, 2001). Asteroid 5670 is named Rosstaylor in his honour.

SCOTT M. MCLENNAN is Professor of Geochemistry at the State University of New York at Stony Brook. He conducts research into the geochemistry of sedimentary rocks, and has published 140 papers in the fields of geochemistry, planetary science and sedimentology. Since 1998, he has applied laboratory experiments and data returned from missions to Mars to understand the sedimentary processes of that planet, and is on the science teams of the 2003 Mars Exploration Rover and 2001 Mars Odyssey missions. He received a Presidential Young Investigator Award from the National Science Foundation in 1989 and a NASA Group Achievement Award as part of the Mars Exploration Rover Science Operations Team in 2004.

Professors Taylor and McLennan are also the authors of *The Continental Crust: Its Composition and Evolution* (1985).

PLANETARY CRUSTS: THEIR COMPOSITION, ORIGIN AND EVOLUTION

STUART ROSS TAYLOR

Department of Earth and Marine Sciences
Australian National University
Canberra, Australia

AND

SCOTT M. McLENNAN

Department of Geosciences
State University of New York at Stony Brook,
Stony Brook, NY, USA

CAMBRIDGE
UNIVERSITY PRESS

CAMBRIDGE UNIVERSITY PRESS
Cambridge, New York, Melbourne, Madrid, Cape Town, Singapore,
São Paulo, Delhi, Dubai, Tokyo

Cambridge University Press
The Edinburgh Building, Cambridge CB2 8RU, UK

Published in the United States of America by Cambridge University Press, New York

www.cambridge.org
Information on this title: www.cambridge.org/9780521142014

© S. R. Taylor, S. M. McLennan 2009

This publication is in copyright. Subject to statutory exception
and to the provisions of relevant collective licensing agreements,
no reproduction of any part may take place without the written
permission of Cambridge University Press.

First published 2009
This digitally printed version 2010

A catalogue record for this publication is available from the British Library

Library of Congress Cataloguing in Publication data
Taylor, Stuart Ross, 1925–
Planetary crusts : their composition, origin and
evolution / Stuart Ross Taylor and Scott M. McLennan.
p.   cm.
ISBN 978-0-521-84186-3
1. Planets – Crust.   2. Planets – Origin.   I. McLennan, Scott M.   II. Title.
QB603.C78T39   2009
551.1′3–dc22
2008036966

ISBN 978-0-521-84186-3 Hardback
ISBN 978-0-521-14201-4 Paperback

Cambridge University Press has no responsibility for the persistence or
accuracy of URLs for external or third-party internet websites referred to in
this publication, and does not guarantee that any content on such websites is,
or will remain, accurate or appropriate.

Ross dedicates this book to Angelo in the hope that he will find the planets just as interesting as has his grandfather.

Scott dedicates the book to his wife Fiona and daughter Kate.

Contents

Preface

This work is not intended as a textbook, or as a review, but represents an enquiry into the problem of how and why solid planets produce crusts. As this seems to have happened at many different scales throughout the Solar System, we were curious to see whether some general principles might emerge from the detail. The formation of the planets themselves is the outcome of essentially random processes, constrained mainly by the history of the inner nebula and by the cosmochemical abundances of the chemical elements. But perhaps the production of crusts might be a simpler or more uniform process, a notion supported by the frequent appearance of basaltic lavas of assorted types on the surfaces of rocky bodies.

This book is also written from geochemical and geological perspectives, the areas with which the authors are most familiar. We were immediately faced with the problems of ordering the discussion in a logical sequence because "good reasons could be found for placing every chapter before every other chapter" [1]. Although one might reasonably expect to begin such a book with a discussion of the continental crust on which we are standing, this useful feature, like the Earth itself in a wider planetary context, is one of the least enlightening places from which to discover how planets form crusts. For this reason, our familiar continental crust appears late in the discussion. We decided instead to begin with simpler examples.

The formation of the two types of crust on the Moon, the highlands and the maria, indeed form the clearest and best understood examples of the complex processes that lead solid bodies in the Solar System to form crusts. So after considering the formation of the Solar System, we open the debate with the interesting and well-resolved example of primary and secondary crusts on the Moon. Next we describe what little is known about the crust of Mercury, that forms at present the closest analogue to the well-studied lunar example.

The recent investigations of the martian crust provide an excellent example of the complexities of crustal development on a more evolved planet. Mars is one of the few planetary bodies for which samples, in the form of martian meteorites, are available for study. The amount of information now available for Mars is substantial enough to warrant two chapters. The strange case of Venus, our "twin planet", discussed next, shows bizarre contrasts to the familiar geology of the Earth and provides a sobering insight into how similar planets can develop in different directions throughout geological time.

The five chapters on terrestrial crusts that follow illustrate the unique importance of water in crustal development. Beginning with the oceanic crust, another example of a secondary crust, we discuss in the following four chapters, the development of our useful continental crust. These latter chapters illustrate the complexities of terrestrial crustal development, the onset of plate tectonics and the slow evolution of the present continental crust from its enigmatic beginning in the Hadean.

In the penultimate chapter, some readers may be surprised to find a discussion about the crust of an asteroid and of some of the major satellites of the giant planets. However, these provide further examples of crustal diversity, as well as providing insights into early planetary evolution. Their crusts likewise provide interesting contrasts with those in the familiar inner Solar System. We conclude with some speculations on the likelihood of extra-solar examples of geological processes such as plate tectonics and of the possible presence of Earth-like planets with benign crusts elsewhere. We close each of the major chapters with a synopsis of the salient points.

We hope that the broader perspective adopted in this book will encourage our more terrestrially oriented colleagues to consider the interesting and unusual scientific problems presented by other bodies in the Solar System. Here we are mainly concerned with why and how crusts develop on planets in the Solar System rather than with local geological details that are unique to each planet. While interesting in their own right, the details on the Earth, for example of the formation of mountain ranges, or the location of subduction zones, belong in another treatise. Thus while continent–continent collisions may be broadly similar, the resulting geology exposed for our observation may be strikingly different. For example the European Alps, derived from the collision between Africa and Europe, differ in much tectonic detail from the Himalaya and the Tibetan Plateau, that formed from the collision of India with Asia during the Tertiary.

The operations of these stochastic processes mimic, on a smaller scale, the larger-scale sequence of random events that led to the accretion of the terrestrial planets. Thus readers seeking enlightenment on topics such as the tectonic evolution of the Archean crust of the Earth or the manner of the accretion of terranes to cratons during the assembly of continents on the Earth will need to look elsewhere [2].

The proliferation of planetary missions over the past decade, especially to Mars, has resulted in a tidal wave of spectral data and truly spectacular images. Although some of these results are of direct interest to our enquiry, most shed light more on surface processes and geological phenomena. Accordingly, we have resisted the temptation to focus too much of our attention on these pictures, stunning as they may be. We use planetary images sparingly, mostly where they illustrate specific points being discussed.

Although we try to refer to much of the relevant literature, the book is not intended as a review, but reflects our assessment of the evidence. Here we have chosen to take a broad overview so that many may become irritated by a perceived cavalier treatment of their specialty. However, specialist treatises often suffer from the Law of Diminishing Returns (witness the proliferation of multi-author volumes on the Archean or on the early Solar System) and we make no apology for this attempt to widen horizons.

Here we have attempted to give references to all sources of fact, information, opinion and interpretation other than our own. The rapidly increasing number of papers on all topics has made it inevitable that we have missed some significant contributions. We apologize for such omissions and have found the existence of "invisible colleges", within which authors circulate their preprints, to be a partial solution in dealing with the deluge.

We list references by number in order of appearance in each chapter, rather than break up the continuity of the text with lists of names and dates. Except for two-author papers, first authors only followed by *et al.* are given, as it is pointless in a work such as this to provide lists of authors that on occasion extend to 50 or more names [3]. Most material is easily accessible on the Smithsonian/NASA ADS Abstract Service or Georef (the premier database from the American Geological Institute).

We have referenced mostly more recent works, except where older papers have acquired classic status, remain the sole source of information or provide cautionary tales. The average lifetime of a scientific paper rarely exceeds five years. After that period, the results are either incorporated into the general corpus or, if erroneous, are mostly ignored. We have excluded all references to conference abstracts and other grey literature except in a few cases where the information is not available elsewhere. We rarely include references to websites and the like, regarding them as too ephemeral for incorporation here. We have attempted to survey the literature through 2007.

In attempting such a broad synthesis, we are conscious of the risks either of offending the specialist or of boring the general reader. The individual chapters and sections could be the subject of books in their own right (and in many cases have become so). In trying to solve this problem, extensive use has been made of notes and comments that attempt to steer a course between the Scylla of minutiae and the Charybdis of lack of precision. However, the huge amount of literature that we feel requires comment means that the material in the notes sometimes exceeds the length of the discussion in the main text.

We have endeavoured, not always successfully in such a seriously overloaded discipline, to avoid jargon and acronyms, which continue to spread like a virus. Among much bad usage, it seems impossible to read anything about Precambrian geology without encountering those terrible twins, autochthonous and allochthonous. These have displaced the basic English terms, native and foreign, with words that demonstrate the erudition of the writer and send the less erudite searching for their dictionaries. However, like much jargon (delamination is another example), such usage casts a veneer of understanding that serves to obscure the underlying complexities. Sophistication should not be confused with explanation.

Elemental abundances, as in the vast majority of the geochemical literature, are given in wt%, ppm, ppb or ppt rather than as wt%, mg/g, ng/g, µg/g or pg/g. We regard the latter convention as likely to lead to error and confusion, as well as having elements of scientific pretension. As wt% (parts per 100) has been retained in this latter usage, it even lacks the pedantic excuse of consistency. Density is expressed in units of g/cm^3 rather than kg/m^3 because most of the geological and geophysical literature continues to use these units. We list units of pressure as kilobars (kbar) or gigapascals (GPa) as appropriate [4]. We apologize that in the absence of better alternatives for these stupendous periods of time, millions of years are abbreviated to Myr and billions of years to Gyr.

Books should reflect the opinions of the authors. It is no service to readers to provide a list of ongoing controversies or of problems without making some assessment of a likely resolution or outcome. This is indeed not without hazard. Here we have attempted to give our own judgment on such controversial matters as the existence of plate tectonics in the Archean, the nature of the Hadean crust of the Earth, the value of the decay constant for ^{176}Lu, the stratigraphic record on Venus, the evolution of Mars, the age of martian meteorites, the composition of the continental crust and of the Earth and many other difficult topics. On occasion, as in the discussion of mantle plumes, the origin of eucrites or the source of the bodies responsible for the lunar "cataclysm", we have preferred to wait for new evidence.

Meanwhile, as Francis Bacon remarked, we remain conscious of "the subtilty (sic) of Nature, the secret recesses of truth, the obscurity of things, the difficulty of experiment, the implication of causes and the infirmity of man's discerning power" [5].

Notes and references

1. Kerridge J. F. (1988) *Meteorites and the Early Solar System* (eds. J. F. Kerridge and M. S. Matthews), University of Arizona Press, p. xvi.
2. Among the masses of literature on these subjects, excellent expositions can be found in Cloud, P. (1988) *Oasis in Space: Earth History from the Beginning*, W.W. Norton; and (2002) in *The Early Earth: Physical, Chemical and Biological Development* (eds.

C. M. R. Fowler *et al.*, 2002), Geological Society of London Special Publication 199. The assembly of the terrestrial continents, is well treated in Rogers, J. J. W. and Santosh, M. (2004) *Continents and Supercontinents*, Oxford University Press; and in Bleeker, W. (2003) The late Archean record: a puzzle in *c.* 35 pieces. *Lithos* **71**, 99–134. A good example of the problems involved in such reconstructions can be found in Windley, B. F. *et al.* (2007) Tectonic models for accretion of the Central Asian orogenic belt. *J. Geol. Soc. London* **164**, 31–47.

3. Workers in particle physics are even less (or more) fortunate with lists of authors often exceeding 500.

4. As 10 kbar equals one GPa, we wonder about the necessity for this change, apart from the need to celebrate the life of that distinguished French scientist, Blaise Pascal (1623–1662). Bars bear an obvious relationship to atmospheric pressure, just as cycles per second seem more understandable than Hertz. We deplore the attempts of committees to force-feed the use of "international units". It is interesting that that useful unit, the Angstrom (10^{-8} cm, about the size of an atom) has survived the attempts of such committees to order the universe in multiples of 1000 from an essentially arbitrary base.

5. This translation from Latin of aphorism 92 from *"Novum organum"* by Francis Bacon (1620) is from Peter Medawar (1979) in *Advice to a Young Scientist*, Pan Books, p. 6.

Acknowledgments

First we wish to acknowledge the continued support and patience of our families during the writing of this book. Ross thanks Richard Arculus and David Ellis of the Department of Earth and Marine Sciences at the Australian National University for much encouragement and advice. Ross also wishes to thank Brian Harrold (Department of Earth and Marine Sciences, Australian National University) for crucial assistance with computing and the Australian National University for the award of a Visiting Fellowship that has enabled this book to be written. We are highly appreciative of the assistance of Dr Judith Caton (Department of Earth and Marine Sciences, Australian National University) who has transformed many of our rough drafts of the figures and tables.

Scott wishes to thank his graduate students, past and present, for helping to create a rich and exciting environment for studying crustal processes on both Earth and Mars. He is also grateful to the entire Mars Exploration Rover and Mars Odyssey science teams who, over the past four years, have brought to life in exquisite detail the geology of another planet.

We are grateful to Simon Mitton and particularly to Susan Francis of Cambridge University Press for their encouragement to write this book, for the assistance of Diya Gupta and Eleanor Collins during production, to Zoë Lewin for meticulous copy-editing and all for their patience in dealing with contrary authors.

We have discussed the topics in this book for many years with too many colleagues to thank them all individually although special thanks are always due to Robin Brett. However, we are indebted to the following scientists who have reviewed the various chapters in this book.

Chapter 2 Lunar highlands and Chapter 3 Lunar maria: Marc Norman (Australian National University).

Chapter 4 Mercury: Bob Strom (University of Arizona).

Chapter 5 Mars: planetary composition: Jeff Taylor (University of Hawaii).

Chapter 6 Mars: crustal composition: Hap McSween (University of Tennessee) and Jeff Taylor (University of Hawaii).

Chapter 7 Venus: Ellen Stofan (Proxemy Research, VA).

Chapter 8 Oceanic crust: Mike Perfit (University of Florida).

Chapter 9 The Hadean: Balz Kamber (Laurentian University, Sudbury, Canada).

Chapter 10 The Archean: Stephen Moorbath (University of Oxford).

Chapter 11 The Post-Archean and Chapter 12 The continental crust: Richard Arculus (Australian National University).

Chapter 13 Minor bodies: Brett Gladman (University of British Columbia) and Bill McKinnon (Washington University, St Louis).

They have performed an invaluable service, saving us from various misinterpretations and errors. We remain completely responsible for any mistakes and omissions.

Abbreviations

Here we list those that appear commonly throughout the text and that may not be familiar to some readers.

Å:	Ångstrom unit (10^{-8} cm)
AGU:	American Geophysical Union
AU:	Astronomical unit ($149\,597\,871$ km)
CI:	Type 1 carbonaceous chondrites
EPSL:	Earth and Planetary Science Letters
GCA:	Geochimica et Cosmochimica Acta
GRL:	Geophysical Research Letters
GSA:	Geological Society of America
HREE:	Rare earth elements Gd through to Lu
JGR:	Journal of Geophysical Research
KREEP:	Acronym from potassium, rare earth elements and phosphorus
LIL:	Large ion lithophile
LPI:	Lunar and Planetary Institute, Houston, Texas
LPSC:	Lunar and Planetary Science Conference
LREE:	Rare earth elements La through to Sm
MORB:	Mid-ocean ridge basalt
MPS:	Meteoritics and Planetary Science
OIB:	Oceanic island basalt
PEPI:	Physics of the Earth and Planetary Interiors
ppb:	Parts per billion (10^9, hg/g)
ppm:	Parts per million (10^6, μg/g)
ppt:	Parts per trillion (10^{12}, pg/g)
REE:	Rare earth elements

Prologue

We are apt to judge the great operations of Nature on too confined a plan.

(Sir William Hamilton) [1]

It seems inevitable that rocky planets, like bakers, cannot resist making crusts, heat being the prime cause in both cases. Although trivial in volume relative to their parent planets, crusts often contain a major fraction of the planetary budget of elements such as the heat-producing elements potassium, uranium and thorium as well as many other rare elements while the familiar continental crust of the Earth on which most of us live is of unique importance to *Homo sapiens*. It was on this platform that the later stages of evolution occurred and so has enabled this enquiry to proceed.

Planetary crusts in the Solar System indeed have undeniable advantages for scientists: they are accessible. Unlike the other regions of planets that we wish to study, such as cores and mantles, you can walk on crusts, land spacecraft on them, collect samples from them, measure their surface compositions remotely, study photographs, or use radar to penetrate obscuring atmospheres. Despite this accessibility, the problems both of sampling or observing crusts are non-trivial: most of our confusion in deciphering the history of crusts ultimately turns on our ability to sample them in an adequate fashion. We discuss these diverse problems in the appropriate chapters.

This advantage of relatively easy access to crusts is also offset by the distressing tendency for crusts to be complex, so that one may easily become lost in the detail, failing to see the forest for the trees. This is particularly true of the continental crust of the Earth that is sometimes heterogeneous on a scale of meters. One consequence of this myopia is that one sometimes encounters claims that extrapolate from a small region to produce a world-embracing model. The furore over whether there was an early granitic continental crust during the Hadean is a familiar example of the perils of extrapolation from a handful of zircon grains preserved in younger sedimentary rocks. As Charles Gillispie has remarked "the inherent difficulties of the science, Lyell thought, had rendered it peculiarly susceptible to the interpretations of ancient miraclemongers and their modern successors" [2]. Moreover the fundamental lesson from comparative planetology is that each rocky planet and satellite has some significant variation from the geological insights gained by studying our own planet.

Another major problem besets attempts to understand the origin and evolution of planetary crusts. Just as it is difficult to trace back the orbit of a near-Earth asteroid that was thrown into an Earth-crossing orbit by Jupiter, or to decipher the oceanic source of an ore deposit that is now outcropping in the middle of a continent, so planetary crusts, that are the final products of extensive planetary differentiation, mostly conceal their previous history.

We usually see only the end product, the classic problem in geology. The upper continental crust of the Earth, that we can investigate so readily, is the product of intra-crustal melting within a crust derived by three stages of remelting of rocks derived from a mantle with a complex history. The other solid bodies in the Solar System display crusts that are often equally complicated, the results of planetary differentiation processes that, although following the laws of physics and chemistry, differ in detail from one body to another.

These are some of the reasons that the development of the geological sciences lagged behind that of most other sciences. Contrary to popular mythology, they are amongst the most difficult and complicated of subjects. This is readily demonstrated by considering the historical development of the various sciences. Thus classical physics was well established by Newton, with the publication of the *Principia* in 1687. Biology was set upon the right track by Darwin in 1859 when he published *The Origin of Species*. The underlying basis of chemistry became understood with the formulation of the Periodic Table of the Elements by Dmitri Ivanovich Mendeleev in 1869. The fundamental nature of atoms was established nearly a century ago in 1911 by Ernest Rutherford. Even the origin of the chemical elements themselves was understood following the work of the Burbidges, Willy Fowler, Fred Hoyle and independently by Al Cameron in 1956.

However, it was only as late as 1963, three centuries after Newton's physical insights, that Fred Vine and Drum Matthews hit upon the fundamental process of plate tectonics. Then geologists finally understood what was going on under their feet. This mechanism explained the architecture of the surface of the Earth that had been painfully established in the previous 150 years following the pioneering works of James Hutton, William Smith and Charles Lyell.

There is a further philosophical problem that bedevils geology, a term that we use here in its broadest sense to encompass the study of the "origin, structure and history" of planets. Planets differ from stars, whose classification and evolution have been understood for nearly a century. Thus the Hertzsprung–Russell diagram, fundamental to astrophysics, dates from 1913. It is nearly a century old, as is the robust OBAFGKM classification of stars [3]. In contrast to stars, the planets, including the terrestrial planets and the Earth and indeed most of the geological record, are essentially the end result of the operation of stochastic processes. Planets are individuals that refuse to be placed into neat pigeonholes, unlike stars.

Thus it is difficult to find geological laws or generalizations of general applicability such as the Hertzsprung–Russell diagram or the Periodic Table of the Elements that enabled the rapid development of astronomy and chemistry. Such problems are responsible for the lengthy development of geology and of the continual appearance of bizarre theories to account for geological phenomena.

So we need to heed the wise advice of Sir William Hamilton that heads this section and that of Al Hofmann, who, in studying the mantle of the Earth, employed "a simple-minded, uniformitarian approach that uses known geological processes and avoids exotic processes wherever possible" [4].

It is only occasionally that the investigation of geological details has led to insights into fundamental processes, examples being the unconformities in Scotland at Jedburgh and more famously at Siccar Point that enabled Hutton to develop the concept of deep time. The K-T boundary outcrop at Gubbio, in Tuscany, Italy is another such that led to the recognition of the catastrophic impact of a 10 km diameter asteroid that ended the Cretaceous Period. But much of the rock record that has been painstakingly assembled over the past two centuries reflects localized events. Standing on the Earth, it is difficult to appreciate the slow process of plate tectonics: it was the data from marine geophysics, not surface outcrops, road-cuts or drill cores, that provided the compelling evidence for sea-floor spreading that was the key to understanding the mobile nature of the surface of the Earth [5].

Early attempts to decipher the geological record were bedevilled by the occurrences of similar-looking rocks that turned out to be of different ages. The study of individual ore deposits that we find so useful for our technical civilization reveals that they form mostly as a consequence of local geological conditions. So they provide only indirect evidence of the processes that have resulted in the concentration of the ore elements by many orders of magnitude from those of the bulk planet. Venus, in contrast to the Earth, has a totally different geological history. Like Mercury and perhaps Mars, all seem unlikely to have much in the way of ore deposits.

Another problem is that geology has had to wait for the development of specialized techniques, from marine magnetometers to mass spectrometers, in order to resolve its problems. As Bill Menard [6] has remarked "geology was moribund during the period from about 1860 to about 1940 because it lacked the techniques to solve its important problems … (and) the geologists … were inevitably doomed to working on trivia until new tools were forged". In the meantime, according to Stephen Brush, "Geologists in the 20th century became accustomed to carrying on interminable controversies about problems that they were unable to solve" [7]. Such debates have often reached levels reminiscent of medieval religious disputes, a classic example, that is worthy of historical study, being the question whether tektites originated from the Moon or the Earth. The wrangle over the reality of mantle plumes forms a current instance.

Fortunately, the advent of sophisticated analytical techniques has helped to resolve many of the problems raised by the field observations and so has enabled us to embark on this discussion.

Notes and references

1. Hamilton, W. (1773) *Observations on Mt. Vesuvius, Mt. Etna and other volcanoes*, T. Cadell, p. 161. Better known to history as the husband of Emma, Lady Hamilton, who was the mistress of Admiral Lord Nelson, this distinguished naturalist and Fellow of the Royal Society was one of the first modern students of volcanoes.
2. Gillispie, C. C. (1951) *Genesis and Geology*, Harvard University Press, p. 127.
3. To which L and T classes have been added for the recently discovered brown dwarfs, thus spoiling the famous mnemonic "Oh be a fine girl kiss me".
4. Hofmann, A. (1997) Mantle geochemistry: the message from oceanic volcanism. *Nature* **385**, 219–22.
5. Earlier attempts to understand the architecture of the surface of the Earth are the classic accounts by Suess, E. (1904) *The Face of the Earth* (trans. H. B. C. Sollas), Clarendon Press (3 volumes) and Umbgrove, J. H. F. (1947) *The Pulse of the Earth* (2 ed.), Nijhoff.
6. Menard, H. (1971) *Science Growth and Change*, Harvard University Press, p. 144. Relevant to this comment is the proliferation of edited volumes on many topics of which the Archean and the early Solar System are favorites. Despite heroic efforts by editors, such collections of papers from disparate authors with varying opinions and styles are rarely successful, while their reference lists extend into the thousands. Much of this material could equally well appear in the normal refereed scientific literature. Indeed attempts to enshrine a topic in a definitive work such as the ten-volume *Treatise on Geochemistry* seem pointless in an active science where any publications rarely survive more than five years. For example, our estimates given in this book, both for the composition of the Earth (Chapter 8) and for the continental crust (Chapter 12), differ significantly from those given in the *Treatise*.
7. Brush, S. G. (1996) *Transmuted Past*, Cambridge University Press, p. 55.

1

The planets: their formation and differentiation

Alphonso, King of Castille, ...was ill seconded by the astronomers whom he had assembled at considerable expence (sic).... Endowed with a correct judgement, Alphonso was shocked at the confusion of the circles, in which the celestial bodies were supposed to move. 'If the Deity' said he, 'had asked my advice, these things would have been better arranged'

(Pierre-Simon Laplace) [1]

1.1 Planetary formation

Although this book is concerned with the crusts of the solid bodies in the Solar System, it is necessary to delve a little deeper into the interiors of the planets, to see how the planets themselves came to be formed and why they differ from one another. It is only possible to understand why and how crusts form on planets if we understand the reasons how these bodies came to be there in the first place and why they are all different from one another. Following 40 years of exploration of our own Solar System, the discovery of over 200 planets orbiting stars other than the Sun has brought the question of planetary origin and evolution into sharp focus. The detailed study of planets is in fact a very late event in science and has required the prior development of many other disciplines.

This highlights a basic problem in dealing with planets, at least in our Solar System, that are all quite different, so that it is difficult to extract some general principles that might be applicable to all of them.

Stars, although they vary in mass, have similar compositions and so are amenable to mathematical and physical laws, a feature that has led to the thriving field of astrophysics. But there is a fundamental difference between stars and planets. Stars form "top-down" by condensation, essentially of hydrogen and helium gas, from dense cores in molecular clouds. Their major differences in mass, luminosity and surface temperature are well displayed on the celebrated Hertzsprung–Russell diagram that is nearly a century old. The success of the Hertzsprung–Russell representation is that the luminosity and surface temperatures of stars are under-pinned by the basic nuclear physics of stellar processes, just as the Periodic Table of the Elements is based on the electronic structure of atoms, something of which the originators of both classifications were unaware.

Planets, in contrast to stars, were assembled randomly, "bottom-up" from left-over material in the nebular disk, at least in our Solar System but likely elsewhere. They are all distinct, forming from a complex mixture of components that can be loosely labeled as gases, ices and rock. From our observations both of our own and extra-solar planets, these bodies may form from any combination of these three components. There is no equivalent of the Hertzsprung–Russell diagram for planets or much sign of one appearing.

It is even difficult to arrive at a satisfactory definition of a planet; witness the furore over the status of Pluto or its larger colleague Eris, that are eccentric dwarfs when placed among the planets, but are the largest icy planetesimals in the Kuiper Belt in their own right [2]. As Confucius remarked "the beginning of wisdom is to call things by their right names".

In our Solar System, we have eight planets, all of them distinct from one another in mass, density, composition, obliquity and rotation rates. Their only common properties are near-circular orbits and low inclinations to the plane of the ecliptic (the Earth–Sun plane), characteristics that enabled Laplace to conclude in 1796 that they had originated from a rotating disk of gas and dust, the solar nebula.

While we still have only one planetary system to examine closely, it includes over 160 satellites [3] but of these, none resemble one another, even among the "regular" satellites. Like the planets, each satellite exhibits some peculiarities of composition or behavior. This tells us that there is no uniformity in the processes of planetary or satellite formation from the gases, ices and rocky components of the primordial nebula. Clones of our planets or our Solar System are consequently expected to be rare.

Our limited sampling of extra-solar planets displays much wider variations from our own system in terms of mass and spacing of planets while, to add additional complexity, many of these newly discovered planets are in highly elliptical orbits. It appears likely that we will find planets forming from Keplerian disks around young stars that will occupy all possible niches available within the limits imposed by the cosmochemical abundances of the elements and the laws of physics and chemistry (Fig. 1.1).

The Earth is the unique planet. No hard-won geological or geophysical truths discovered about our own planet, or even the detailed sequence of geological events, has much applicability elsewhere in the Solar System. Indeed, the sequence of geological processes on Earth has little predictive power. If one had visited the Earth during the Permian, one would not have foreseen the world of the Triassic with its

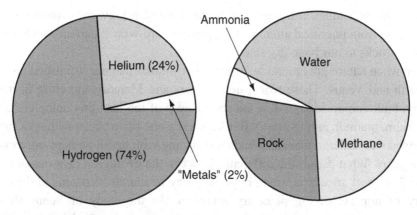

Fig. 1.1 The pie diagram on the left shows the composition of the primordial solar nebula, constituted of 98% gases (H and He) and 2% "metals" (in astronomical jargon). The right-hand pie diagram breaks down the "metals" sector into "ices" (mostly water, ammonia and methane) and "rock" (the remaining elements). Planets may form from any combination of gases, ices and rock. Thus Jupiter and Saturn are dominantly hydrogen and helium ("gas giants"), Uranus and Neptune are "ice giants" while the terrestrial planets are formed from rock.

completely different fauna. To a visitor in the warm Cretaceous, it would have been difficult to imagine the cooling throughout the Tertiary or the onset of the ice ages. Even more unpredictable was the catastrophe that would end that benign period, remove the giant reptiles and lead to the dominance of mammals. That event, that has now resulted in the Earth being overrun by one species, was one consequence of the great K-T boundary collision of the Earth with a 10 km diameter asteroid. Nor could a visitor to Venus a billion years ago have foreseen the total resurfacing of the planet that was to occur shortly thereafter. Planetary history, like planetary formation, is dominated by stochastic and unpredictable events.

The problems of studying planets are well illustrated by the history of attempts to understand the Earth. Often beset by the notions of miracle mongers, the consequence was that geology was a latecomer among the sciences. Even so, it took 150 years following the insights into deep time by James Hutton in 1788 to understand that plate tectonics is the mechanism responsible for the architecture of the Earth's surface. But the Earth is an unusual planet even by the standards of the Solar System. The geological, geochemical and geophysical truths extracted from over 200 years of study are not easily applied to other planets.

Plate tectonics has the useful property both of building continents and of forming ore deposits useful for advanced civilizations and so enabling this discussion to take place. However, this process is unique to the Earth among our planets. The trivial terrestrial water content of a few hundred parts per million, responsible for plate tectonics and the growth of continents, was a late stochastic addition to the planet.

Many of the difficulties in trying to understand the petrology and evolution of the Moon arose from uncritical attempts to apply our hard-won experience with wetter terrestrial rocks to our bone-dry satellite.

Even when nature got around to building two similar planets, it finished up with the Earth and Venus. These twins, unlike Mars and Mercury, are close in mass, density, bulk composition and in the abundances of the heat-producing elements (potassium, uranium and thorium). But Venus is a one-plate planet without a moon and appears to undergo planetary-wide resurfacing with basalt perhaps once every billion years. What causes the difference between the geological histories of these twins? The short answer is water, but much may be due to variations in the early history of impacts during planetary accretion. As the study of Venus shows, similarity is not identity and the Earth resembles Venus much as Dr Jekyll resembled Mr Hyde. As we search for terrestrial-like planets elsewhere, we need to find out the reasons for these differences and the conditions that allow these diverse bodies, or Mercury and Mars for that matter, to form at all. Just as geology arose in the nineteenth century, now the study of planets represents a new area in scientific enquiry.

1.2 The solar nebula and the giant planets

The solar nebula from which the Sun and planets formed had three basic constituents: loosely "gases", "ices" and "rock". The dominant component was gas (98% hydrogen and helium). The heavier elements ("metals" to the astronomers) that amounted to about 2% by mass, had accumulated in the interstellar medium from 10 billion years of nucleosynthesis in previous generations of stars. Abundant elements such as carbon, oxygen and nitrogen were present in the nebula as ices (e.g. as water, methane, carbon monoxide, carbon dioxide and ammonia). The remaining elements, that fill the rest of the periodic table, were present mostly as dust and grains (rock). This rock component had a composition that is given by the CI meteorites, the most primitive stony meteorites (Table 1.1). The rationale for equating their composition to that of the primitive solar nebula is that the composition in this class of meteorites, when ratioed to a common element such as silicon, matches the composition of the solar photosphere. As the Sun contains 99.9% of the mass of the system, their composition is taken to reflect that of the rock fraction of the original solar nebula [4].

Perhaps the most fundamental division in the Solar System is the difference between the giant planets and the small terrestrial planets, although even the giant planets differ significantly among themselves. Jupiter and Saturn, in addition to their massive gaseous envelopes, possess cores of rock and ice that are between 10 and 15 Earth-masses. In contrast, Uranus and Neptune, that are 14 and 17 Earth-masses

Table 1.1 *The composition of the rock fraction of the primordial solar nebula**

Atomic number	Element	Mean CI chondrite		Atomic number	Element	Mean CI chondrite	
		Ref. 1	Ref. 2			Ref. 1	Ref. 2
3	Li (ppm)	1.49	1.50	47	Ag (ppb)	197	199
4	Be (ppb)	24.9	24.9	48	Cd (ppb)	680	686
5	B (ppb)	690	870	49	In (ppb)	78	80
9	F (ppm)	58	61	50	Sn (ppb)	1680	1720
11	Na (ppm)	4982	5000	51	Sb (ppb)	133	142
12	Mg (wt%)	9.61	9.89	52	Te (ppb)	2270	2320
13	Al (ppm)	8490	8680	53	I (ppb)	433	433
14	Si (wt%)	10.68	10.64	55	Cs (ppb)	188	187
15	P (ppm)	926	1220	56	Ba (ppb)	2410	2340
16	S (wt%)	5.41	6.25	57	La (ppb)	245	234.7
17	Cl (ppm)	698	704	58	Ce (ppb)	638	603.2
19	K (ppm)	544	558	59	Pr (ppb)	96.4	89.1
20	Ca (ppm)	9320	9280	60	Nd (ppb)	474	452.4
21	Sc (ppm)	5.90	5.82	62	Sm (ppb)	154	147.1
22	Ti (ppm)	458	436	63	Eu (ppb)	58	56.0
23	V (ppm)	54.3	56.5	64	Gd (ppb)	204	196.9
24	Cr (ppm)	2646	2660	65	Tb (ppb)	37.5	36.3
25	Mn (ppm)	1933	1990	66	Dy (ppb)	254	242.7
26	Fe (wt%)	18.43	19.40	67	Ho (ppb)	56.7	55.6
27	Co (ppm)	506	502	68	Er (ppb)	166	158.9
28	Ni (wt%)	1.08	1.10	69	Tm (ppb)	25.6	24.2
29	Cu (ppm)	131	126	70	Yb (ppb)	165	162.5
30	Zn (ppm)	323	312	71	Lu (ppb)	25.4	24.3
31	Ga (ppm)	9.71	10.0	72	Hf (ppb)	107	104
32	Ge (ppm)	32.6	32.7	73	Ta (ppb)	14.2	14.2
33	As (ppm)	1.81	1.86	74	W (ppb)	90.3	92.6
34	Se (ppm)	21.4	18.6	75	Re (ppb)	39.5	36.5
35	Br (ppm)	3.5	3.57	76	Os (ppb)	506	486
37	Rb (ppm)	2.32	2.30	77	Ir (ppb)	480	481
38	Sr (ppm)	7.26	7.80	78	Pt (ppb)	982	990
39	Y (ppm)	1.56	1.56	79	Au (ppb)	148	140
40	Zr (ppm)	3.86	3.94	80	Hg (ppb)	310	258
41	Nb (ppb)	247	246	81	Tl (ppb)	143	142
42	Mo (ppb)	928	928	82	Pb (ppb)	2530	2470
44	Ru (ppb)	683	712	83	Bi (ppb)	111	114
45	Rh (ppb)	140	134	90	Th (ppb)	29.8	29.4
46	Pd (ppb)	556	560	92	U (ppb)	7.8	8.1

* Two estimates of the composition of type CI carbonaceous chondrites. (1) Mean CI abundances from Palme, H. and Jones, A. (2004) in *Treatise on Geochemistry* (eds. H. D. Holland and K. K. Turekian), Elsevier, vol. 1, Section 1.03, Table 3, p. 49. (2) Mean CI chondrite composition from Anders, E. and Grevesse, N. (1989) *GCA* **53**, Table 1, p. 158. Little significant change has occurred in the 15 year interval between the two estimates.

respectively, contain only 1 or 2 Earth-masses of gas and are mostly composed of ice and rock. These ice giants are analogues for the cores of Jupiter and Saturn. The difference is that Jupiter and Saturn have captured much larger amounts of gas. In addition to the distinction in composition between these giants and the terrestrial planets, there is also a major contrast in mass. Mercury, Venus, Earth, the Moon and Mars contain only a trivial amount (2 Earth-masses of rock) compared with the total of 440 Earth-masses of gases, ices and rock that reside in the giant planets.

It was only after the Sun began the hydrogen to helium nuclear reactions that strong solar winds developed, sweeping out the inner nebula, with the ices condensing at 5 AU at a so-called "snow line". The formation of the planets was thus a very late event in the history of the disk, beginning only after the Sun had entered the T Tauri stage of solar evolution [5]. This enhancement at the snow line of ices and dust, locally increased the density of the nebula by around 5 AU and led to the rapid (10^5 year) runaway growth of bodies of ices and dust of around 10–15 Earth-masses. It is likely that four cores formed of which the ice giants Uranus (14.5 Earth-mass) and Neptune (17.2 Earth-mass) are surviving examples.

The lifetime of the nebula was only a few million years. Disks around stars have lifetimes between three and six million years so that Jupiter and Saturn had to acquire their complement of gases within that period [6]. The early growth of these massive cores enabled them to begin capturing the gases (H and He) before the nebula was dispersed. Perhaps either the core of Jupiter grew faster than the others, or it was closer to the Sun. Whatever the sequence, Jupiter was able to accrete about 300 Earth-masses of gases. This is much less than that present in the original nebula, with the result that Jupiter does not have the composition of the Sun, but is enriched in the ices and rock component, or metals by a factor between 3 and 13 [7]. Saturn, with a similar size core, managed to capture only about 80 Earth-masses of gas and so is more strongly "non-solar" in composition. Uranus and Neptune lost out almost completely and finished up with 1 or 2 Earth-masses of gas.

These non-solar compositions of the giant planets are key evidence for their "bottom-up" or core accretion models of formation from the solar nebula. The core accretion model indeed faces some problems of timing relative to the lifetimes of nebulae, although the times required to form the cores and collapse the gases on to them are not well constrained and probably can be fitted into the few million years of disk lifetimes.

The alternative model for giant planet formation by condensation directly from the gaseous nebula is usually referred to as the disk instability model. Its main attraction is fast formation (a few thousand years), but it also faces theoretical difficulties. Although disks may break up, whether giant planets form from these clumps remains uncertain [8]. Apart from this, there are two fatal flaws. First, the giant planets are predicted to be of solar composition; but Jupiter and Saturn are enriched by several times in ices and rock relative to the solar composition. Secondly the interior of Jupiter is at pressures of 50–70 Mbar with temperatures up to 20 000 K so that the material is present as a plasma of protons and electrons, so-called "degenerate matter". A core cannot "rain out" in a giant gaseous proto-planet in the manner that the iron core of the Earth forms. Our metallic core forms due to the density difference between molten iron and silicate at the much lower temperatures and pressures within the early Earth compared to giant Jupiter. In the center of a giant gaseous protoplanet, such density contrasts do not exist. The core has to be present to begin with, around which the gas can subsequently accrete.

The extra-solar planets provide some evidence in support of the core-accretion model. These are gas giants resembling Jupiter, but they form predominantly around stars with high metal contents. So nature apparently needs metals (i.e., ices and rock) to build gas giants elsewhere. This correlation between the existence of extra-solar planets and metal-rich stars is prima facie evidence that giant gas-rich planets form around ice and rock cores.

Finally, the existence of Uranus and Neptune, 14 and 17 Earth-masses respectively, that are mostly ices and rock with about 1 or 2 Earth-masses of gas, shows that nature indeed managed to make two cores within our own system. Although the proponents of the disk instability model account for these ice giants by evaporating gas from larger bodies due to the intervention of another star, such ad hoc explanations do not explain how the cores came to be, while drastic but unobserved effects on the Edgeworth–Kuiper Belt and the Oort Cloud might be expected.

The differences among our giant planets were probably caused by the earlier growth of Jupiter. As the giant began to dominate the scene, its gravitational field then dispersed the other cores, that had formed around the snow line, outwards into the gas-poor regions of the nebula [9]. In this scenario, the giant planets, that contain so much material far from the Sun, formed before the terrestrial planets. These accreted later from the dry rocky refractory material (about 2 Earth-masses) that was left over in the inner nebula following the dispersal of the gaseous and icy components of the nebula [10].

1.2.1 The depletion of the volatile elements in the inner nebula

One might have expected that the composition of the Earth and the other inner planets would mirror that of the original rock component (represented by the CI chondrites) of the solar nebula. However the moderately volatile elements (that have condensation temperatures in the range 400–1100 K) are strongly depleted both in the Earth, Venus and Mars as well as in many classes of meteorites (Fig. 1.2).

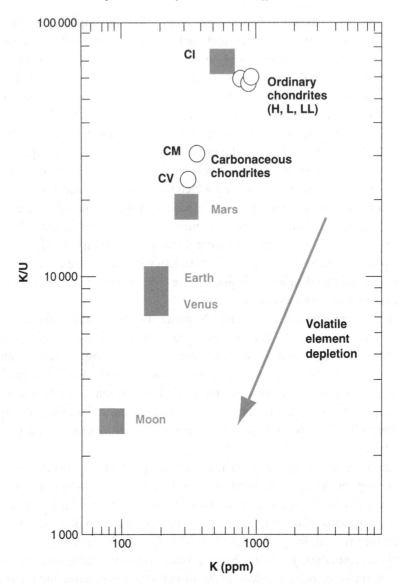

Fig. 1.2 The depletion of potassium, a volatile element, relative to uranium, a refractory element, both relative to CI, in the inner Solar System; CM and CV are classes of carbonaceous chondrite

This depletion in the volatile elements is most readily shown by measurements of the abundances of the gamma-ray emitters, potassium (moderately volatile) and uranium (refractory). The initial solar nebula value of the K/U ratio, as given by the CI meteorites, is near 60 000, but the ratio for the Earth is 10 000 (or 12 500) while that for Venus is similar within rather wide limits. The martian K/U ratio is about 20 000 [11].

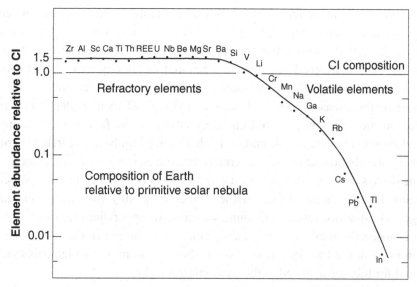

Fig. 1.3 The composition of the Earth relative to CI showing an enrichment in refractory elements and a depletion in volatile elements.

It has been well established that this depletion is a bulk planetary and not a surficial effect. Potassium and uranium, although distinct in ionic radius and valency, are both incompatible (see Section 1.6), being concentrated in residual melts (and crusts) and so remain together during planetary differentiation. Alternative suggestions to account for the depletion of potassium: that it is buried in planetary cores, or is evaporated during planetary accretion, are untenable. As well as potassium, many other elements in the periodic table are also depleted, independently of the size of the planet.

Another useful volatile/refractory pair in this context are rubidium and strontium [12]. The Rb/Sr ratio of the Earth is 0.03, an order of magnitude lower than the primordial solar nebula ratio as given by the CI meteorites. This, coupled with primitive $^{87}Sr/^{86}Sr$ values in lunar samples and meteorites tells us that this volatile-element depletion occurred close to the formation of the Solar System and was not due to later planetary processes such as evaporative loss during accretion.

This depletion is well illustrated by the composition of the Earth, plotted relative to CI (Fig. 1.3). The striking feature of this plot, that resembles the abundances of the elements in the other terrestrial planets as well as in many classes of meteorites other than CI, is that the depletion of the elements correlates with volatility, not with any other chemical parameters. Thus the diagram includes elements of diverse geochemical affinities that all plot on the depletion trend. These include those that enter metal phases (siderophile elements), sulfides (chalcophile elements) or silicates (lithophile elements) during crystallization of a molten planetary body.

In summary, the inner nebula was depleted in volatile elements whose condensation temperatures are less than about 1100 K. This event was probably due to early intense solar activity, so that the Sun was already formed, but occurred before the formation of chondrules. The consequence is that the inner planets do not contain the primordial solar nebular abundances (CI) of the chemical elements.

There are three reasons to raise this topic of volatile-element depletion in planetary compositions in a book about planetary crusts. In the first instance, many of these elements are incompatible and so finish up being highly concentrated in crusts. Secondly, the abundance of the moderately volatile element potassium, whose ^{40}K isotope decays to ^{40}Ar, is a major heat source and so is a driving force in planetary tectonics. Finally, many of the volatile elements (notably carbon, but including nitrogen, phosphorus, sulfur, potassium, sodium and copper) that are essential to life as it occurs on the Earth, are low in abundance in the inner nebula. There is thus a certain irony in the fact that, at least in our Solar System, these elements that are essential for life, are depleted in the habitable zone [13].

1.3 Planetesimals and the accretion of the terrestrial planets

Following the formation of the gas and ice giants, all that was left in the inner nebula was dry rocky rubble, of which the asteroids are analogues. There is little evidence of the presence of either gases or ices during the accretion of the Earth. Rare gases are notably depleted in the Earth; their abundances differ from those in the nebula and the relative depletion of the lighter ones (e.g. ^{20}Ne, ^{36}Ar) suggests that they were derived later from a CI-type source. If the Earth had accreted in a gas-rich nebula, one would expect also that ices would have been present and that the Earth would have accreted much more water ice, as well as methane and ammonia ices. In this case, the planetary budget of water, carbon and nitrogen would be orders of magnitude greater than observed. Thus in contrast to the giant planets, the inner planets accumulated from the dry, rocky debris left after the gaseous and icy components of the nebula had been dissipated. It is worth noting that elements such as potassium and lead, that are much less volatile than water, are depleted in the Earth and the primary minerals of most meteorites are anhydrous. Melting of water ice has formed secondary minerals in CI, but this has not altered their bulk chemical composition, that matches that of the Sun for the non-volatile elements.

The material in the inner nebula, beginning with grains, accreted into bodies that ranged in size from meter-sized lumps to Moon-sized bodies before colliding to form the terrestrial planets. These building blocks are termed planetesimals, a term that originated in a somewhat different sense with T. C. Chamberlin (1843–1928) and F. R. Moulton (1872–1952). The best surviving analogues are the asteroids, along with Phobos and Deimos, the tiny moons of Mars [14].

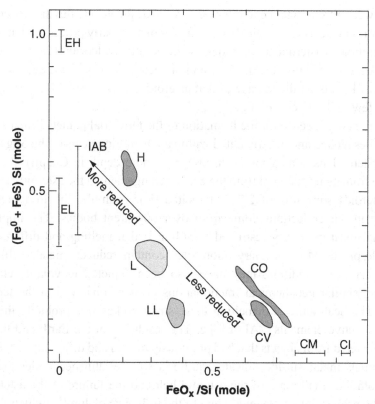

Fig. 1.4 A plot of reduced Fe (Fe and FeS) against oxidized Fe (FeO) using precise data indicates that most meteorite groups do not form linear arrays with slopes of −1. The wider significance is that most chondrites are depleted in total iron relative to CI. As Ni and Co both show similar depletions, this indicates that there was widespread separation of metallic iron (with Ni and Co) from silicate, as well as variations in oxidation state in the nebula before the accretion of chondrites. The abbreviations (EH, IAB, H etc.) refer to various common meteorite groups. Adapted from Larimer, J. W. and Wasson, J. T. (1988) in *Meteorites and the Early Solar System* (eds. J. F. Kerridge and M. S. Matthews), University of Arizona Press, p. 19, Fig. 7.4.1.

Two stages may be distinguished in the process of forming terrestrial planets. The first is a fairly rapid build-up to hundreds of bodies that may approach the Moon, Mercury or even Mars in size from narrow (0.1 AU) zones in the nebula. This occurs on timescales of a few million years [15]. These planetesimals retained many variations in composition such as are displayed by our current samples of meteorites that form analogues for the now-vanished planetesimals. These planetesimals were dry, volatile depleted and had wide variations in the abundance and oxidation state of iron (Fig. 1.4).

Some were differentiated into metallic cores and silicate mantles that are common in the bodies in the asteroid belt [16]. Examples of such early processes that resulted in a differentiated asteroid are provided by the basaltic meteorites (eucrites) derived from 4 Vesta (see Section 13.3), 450 km in diameter. These provide evidence of the eruption of basalts on the surface of that asteroid at about 4560 Myr, a date that is within a few million years of T_{zero} [17].

The chief consequence for the formation of the terrestrial planets is that many of these bodies melted and differentiated within a few million years of the origin of the Solar System. This conclusion is discussed in more depth in Chapter 13 but the evidence from meteorites and from the zoned arrangement in the asteroid belt is that most asteroids sunwards of 2.7 AU melted. For example, the iron meteorites presently in our collections come from over 60 parent bodies. The implication from what must be a very restricted sampling is that melting and differentiation was widespread. Most ordinary chondrites contain reduced metal, sulfide and silicate phases. In addition, they have also been depleted in volatile elements, displaying similar geochemical fractionations to those observed in the terrestrial planets. The heat source for melting these small bodies was probably the short-lived radioactive elements ^{26}Al or ^{60}Fe. The result is that the Earth and the inner planets accreted from objects that had previously melted and differentiated. In these planetesimals metal–sulfide–silicate equilibria were established under low pressures. Both the small mass of the asteroid belt and the failure of the asteroids to collect themselves into a planet are due to the influence of Jupiter, which depleted the belt and pumped up the eccentricities (e) and inclinations (i) of the survivors so that they were unable to collect themselves into a planet.

The larger terrestrial planets, Earth and Venus, took longer to form than their smaller relatives and accreted in a much more violent environment. As the planetary embryos became larger, gravitational effects became dominant and massive collisions between planetesimals became the norm. During the process of collisional accretion of the planets, the intermediary bodies grew to large sizes. Mercury and Mars represent survivors of the final population that accreted to Venus and the Earth. It took somewhere between 10 to 100 Myr for this multitude of bodies to be assembled into the Earth and Venus. The accretion of the terrestrial planets in the planetesimal hypothesis is hierarchical; many of the objects accreting to the Earth were of lunar size. Giant impacts thus were rather common and varied: from head-on collisions of the sort that produced Mercury, to merger of planetesimals, mass loss and break-up of bodies through to glancing collisions. Finally, one body at least the size of Mars, named Theia, that might have survived as a planet in its own right if it had not collided with the Earth, formed the Moon as a result of the glancing collision with the Earth. This Moon-forming event was among the last of the giant collisions [18].

Computer simulations for the inner planets indicate that before the final sweep-up there were probably over 100 objects about the mass of the Moon (1/81 Earth-mass), 10 with masses around that of Mercury (1/18 Earth-mass) while a few exceeded the mass of Mars (1/9 Earth-mass), most of which accreted to Venus and the Earth. In addition there were likely billions of kilometer-size planetesimals.

During the later stages of planetary accretion in an essentially chaotic environment, the large planetesimals were widely scattered, so that during the final accumulation of the Earth and Venus, much radial mixing took place. In contrast to the accretion of the smaller planetesimals from restricted radial zones, the material now in the Earth and Venus probably came from the entire inner Solar System. But whether these two planets or some alternatives were the final outcome, was a matter of chance.

The planetesimal hypothesis thus predicts the occurrence of very large collisions in the final stages of accretion; these account for many of the features of the inner planets and greatly influence early crustal development and evolution. The impact of a Mars-sized object accounts for the origin, angular momentum and composition of the Moon (Chapter 3). The high iron/silicate ratio in Mercury can be accounted for by removing much of its silicate mantle during the collision of Proto-Mercury with an object about 20% of its mass (Chapter 4). The curious siderophile-element composition of the terrestrial mantle, which is parallel to but depleted to about 0.008 times that of CI, may be due to the subsequent addition of material of CI composition that constituted a "late veneer".

These massive collisions have sufficient energy to melt the terrestrial planets, thus facilitating core–mantle separation. When these bodies were assembled into planets, the question of whether re-equilibration between metal and silicate occurred under the higher pressures in planetary interiors depends on the degree of rehomogenization following such events as the Moon-forming collision. The question remains open, as simulations of that event show that the core of the impactor rapidly coalesced with the core of the Earth.

During the differentiation of planetesimals into metallic cores and silicate mantles, separation of tungsten (into iron cores) from hafnium (retained in silicate mantles) occurred. During the accretion of the planets, further melting, perhaps as a consequence of impacts, could have caused rapid and perhaps catastrophic core formation as metal segregates from silicate. This would have rendered it difficult to reach non-unique conclusions about the time of core formation on the Earth from the Hf–W isotopic system. Although a Hf–W model age of 30 Myr (after T_{zero}) is usually quoted, the problem is that the massive cores of lunar-size planetesimals were added to the Earth randomly; the interpretation of the Hf–W isotopic system remains controversial [19].

Such a collisional history also accounts for the variations in composition of the terrestrial planets as the planets accreted from differentiated planetesimals, which had already undergone many collisions. Thus some diversity of composition could be expected. Early planetary atmospheres might also have been removed or added by cataclysmic collisions, accounting for the significant differences among the atmospheres of the inner planets. Thus "in the context of planetary formation, impact is the most fundamental process" [20] while "chaos is a major factor in planetary growth" [21].

The collisions occurring during accretion are quintessential stochastic events. Of course the probability of impacts of bodies of the right mass and at the appropriate angle and velocity to produce the Moon, or remove the mantle of Mercury, is low. However other collisions during the hierarchical accretion of the terrestrial planets, involving different parameters, might produce equally "anomalous" effects; such as a moon for Venus, no moon for the Earth, or different masses, tilts or rotation rates for the inner planets. The variations in composition and later evolution of the terrestrial planets are thus readily attributable to the random accumulation of planetesimals with varying compositions. Computer simulations indeed have difficulty in reproducing the final stages of accretion of the inner planets, commonly producing fewer planets with large eccentricities and wider spacings, that emphasize the importance of stochastic processes in planetary formation [22].

The source of water in the Earth and Venus has remained a contentious topic. The D/H ratio of the Earth's oceans is much higher than that of the solar nebula. The D/H ratio in the only three comets measured is a factor of two higher than oceanic water [23]. Both nebular and cometary origins are thus unlikely as are, in our opinion, wet planetesimals. There is not space here to enter this lengthy debate and we accept the view that the water in the terrestrial planets was derived in a "hit or miss" fashion from later drift-back of icy bodies from the Jupiter region [24].

Three observations support this model. Firstly, the anhydrous composition of most common meteorites is consistent with the snow line model and suggests that the planetesimals forming the Earth sunwards of the asteroid belt were likewise dry. Secondly many elements such as potassium and lead are much less volatile than water but are depleted in the terrestrial planets, so it would be surprising if water somehow escaped the general depletion that affected the nebula out to several AU. Finally if the inner planets had accreted from wet planetesimals, their initial water contents could be expected to be relatively uniform. However, Venus is much less differentiated and degassed than the Earth (see Sections 7.5.2 and 7.6.1) and in that case, the planet might be expected to have retained a wet mantle that would facilitate the development of plate tectonics, contrary to observation.

1.4 The random nature of terrestrial planet formation

An interesting question is that although the four inner planets display much diversity, perhaps if the process was repeated often enough, then clones of the Earth

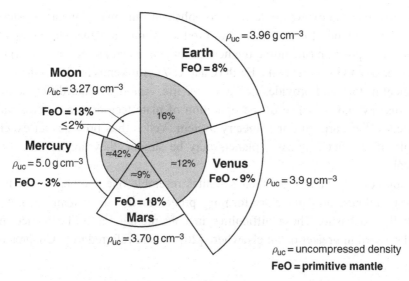

Fig. 1.5 The mantles and cores of the terrestrial planets and the Moon (16% etc = core volume). Adapted from McBride, N. and Gilmour, I. (eds., 2004) *An Introduction to the Solar System*, Cambridge University Press, p. 46, Fig. 2.8.

might appear in other planetary systems. This relates to questions such as the habitability of planets, the possibilities of the development of continental-type crusts on other planets and the emergence of life and technological societies that are addressed in Chapter 14.

Indeed a sober contemplation of our own Solar System, the only example available for close study, reveals just how difficult it is to make planets that resemble our own. But while we have only eight planets, these are accompanied by over 160 satellites. Even disregarding those that have been captured, the regular satellites are often referred to as "miniature Solar Systems". This is not a useful comparison, as the spacing is different and the regular satellite systems of Jupiter, Saturn and Uranus accreted in a different scenario without accretional heating, at much lower temperatures than the terrestrial planets [25]. Even though they are the result of formation in circum-planetary disks, both the systems and the individual satellites are so different that they could well belong to separate planetary systems. Clearly chance events have dominated, an observation that reinforces the random nature of planetary building processes.

Mercury, only 5% of the mass of the Earth, is a body that has survived by reaching a stable orbit, although it is much smaller than the body that formed the Moon following its collision with the Earth. Mars, 11% of the mass of the Earth and about the size of the lunar-forming impactor, is another survivor that might have been swept up into a larger planet. With a lower density than the Earth, it forms another example of a planet assembled at an earlier stage from a distinct collection of rocky planetesimals (Fig. 1.5).

The best that nature was able to accomplish in our own planetary system in constructing a clone of the Earth was to produce Venus, a planet close enough in mass and composition (including the heat-producing elements, K, U and Th) to be considered as a twin planet to the Earth (Chapter 7). But Venus has had a distinctive geological history and provides no haven for life. Amongst other differences, the planet displays no evidence of the operation of plate tectonics, a process that is restricted to the Earth in our planetary system. Venus thus provides an excellent example of the fact that while planets may be similar, they are not necessarily identical.

So none of our planets nor their satellites resemble one another except in the broadest outlines, and planetary-forming processes in our system seem to be essentially stochastic. These difficulties in constructing Earth-like planets have been discussed in greater detail elsewhere [26] and are referred to in Chapter 14.

1.4.1 Meteorites and planetary composition

The most reasonable internal structures for the terrestrial planets involve metallic iron cores overlain by silicate mantles, and it was Victor Goldschmidt who pointed out that the metallic, sulfide and silicate phases in meteorites were analogues [27]. Although this generalization still holds, it has not proven possible to correlate specific classes of meteorites, either alone or in combination, with the composition of the bulk Earth. Neither K/U ratios, volatile element compositions, nor rare-gas abundances in the Earth fit. So although the Earth has a general "chondritic" composition, it cannot be linked either to a specific meteorite class or some mixture of the many groups [28].

The asteroid belt constitutes not much more than 5% of the mass of the Moon, and so it is a poor quarry from which to get material to build the planets. This depleted state of the asteroid belt itself is due to the early formation of Jupiter and dates from the earliest stages of the solar nebula. Thus the belt predates the accretion of the Earth and the inner planets. Oxygen isotope data show that, except for fractionated basaltic meteorites (ruled out on other grounds), no observed class matches the terrestrial data except for the enstatite chondrite class of meteorites.

Because of this similarity in oxygen isotopes and their extremely reduced nature, enstatite chondrites are often thought to be suitable building blocks for the Earth [29]. However their low Al/Si and Mg/Si ratios and their high K/U ratios rule them out as candidates. So it is a coincidence that the Earth and the enstatite chondrites share the same oxygen isotopic composition. As is well known to philosophers, similarity does not imply identity.

1.4.2 Uncompressed density and bulk planetary compositions

The bulk density of planets can be precisely determined from geodetic data. In turn, the best geophysical measure of a planet's bulk composition comes from the average uncompressed (or "zero pressure") density. However, pressure corrections to uncompressed density estimates require detailed knowledge of the internal planetary structure (i.e., details of core, mantle and crust structures), equations of state of the various materials that make up the planet (e.g. bulk moduli and their pressure derivatives) and the thermal structure of the planet.

For Earth, the uncompressed density is well constrained. In addition, the basic materials that make up planets (e.g. metallic phases, silicate minerals) and their equations of state can be reasonably inferred from experiments and from seismological studies on Earth. However, for the other terrestrial planets and moons where there are no seismological data (or very limited data, as for the Moon) the internal physical structures and thermal states are less clear.

For the Earth, Moon and Mars, there are precise moment of inertia factors (I/Mr^2), which provide constraints on internal structure. For Mercury and Venus, no such data are available and accordingly estimates of uncompressed density are much more model dependent. Some reasonable assumptions, such as the planet being fully differentiated, can be made to constrain internal structure; however other important factors, such as the oxidation state of the planet (governing bulk metal/silicate ratios), can only be made with considerably less confidence. In general, the larger the planetary body, the greater and more complex the pressure corrections and consequently the greater the uncertainties. Thus, the uncompressed density of Mercury should be better known than that of Venus.

Stacey [30] reviewed the question of the equations of state of planetary materials and estimated an internally consistent set of uncompressed densities for the terrestrial planets. His values are, in order of increasing uncertainty: Earth: 3.955 g/cm^3; the Moon: 3.269 g/cm^3; Mars: 3.697 g/cm^3; Mercury: 5.017 g/cm^3; Venus: 3.868 g/cm^3. No systematic evaluation of precision has been performed and Stacey [30] commented: "It is difficult to assign uncertainties because of unknown compositional variations and temperatures, especially for Mercury and Venus, without moment of inertia control, but they are clearly large enough to justify neglect of a crust". Accordingly, for this book, we quote uncompressed densities for the Earth (3.96 g/cm^3), Moon (3.27 g/cm^3) and Mars (3.70 g/cm^3) at three significant figures and for Mercury (5.0 g/cm^3) and Venus (3.9 g/cm^3) at two significant figures. These values are slightly lower (up to 0.1 g/cm^3), especially for Mercury and Venus than commonly quoted values.

On the basis of these values, a few simple conclusions can be reached. The uncompressed densities and therefore bulk compositions of the terrestrial planets vary considerably with the Moon having the lowest value (consistent with the presence of only a very small core) and Mercury the largest (consistent with an immense core, due to loss of silicate material during a late-stage giant impact). Mars almost certainly has a significantly lower (by ~6–7%) uncompressed density compared to the Earth, consistent with its volatile-rich and more oxidized state. The uncompressed densities of Venus and Earth are probably indistinguishable within uncertainties and combined with their very similar size, indicate a similar bulk composition and internal mantle/core structure.

1.5 Types of crusts

Once a rocky terrestrial planet has been assembled, what happens next? The release of gravitational energy during the process of collisional accretion from large precursors results in hot, probably mostly molten planets, while radioactivity provides an ongoing heat source. Cooling of the planets drives the tectonic processes that result in the variety of planetary surface features that we observe.

One of the characteristics of planets is that they form crusts, distinct from the bulk composition of the planets; this is the topic of this book. However just as the planets themselves are not identical, not all crusts are equal. Three types of crusts on rocky planets, conveniently divided into "primary", "secondary" and "tertiary" may be distinguished [31]. Icy crusts that occur further away from the Sun may be either primary or secondary.

Primary crusts are formed as a consequence of initial planetary differentiation, caused, for example, by melting during accretion. Primary crusts form on short timescales (approximately 10^8 years) following accretion. One of the distinguishing features of primary crusts is that they can contain high concentrations of incompatible elements. This is because primary crusts are derived from mostly molten planets and so are able to garner trace elements from large volumes of planetary mantles. The ancient, heavily cratered crust of Mars in the southern highlands is a probable example. The putative Hadean crust of the Earth might also have been one, if remnants had been preserved. The highland crust of the Moon that now constitutes 8–9% of the Moon, was produced directly following the formation of the Moon and forms the type example. The icy crusts of Europa and Callisto, produced by early differentiation, probably also belong in this category.

Secondary crusts form on much longer timescales. These products of partial melting of the silicate mantles of our rocky planets are various species of basalts. Typical examples include the oceanic crust of the Earth, the present surface of Venus, the volcanic outpourings that gave rise to the Tharsis plateau and the northern plains on Mars and the lunar maria. The basalts that form the dark maria on the Moon were derived from partial melting from the lunar interior over a period exceeding 10^9 years. However, they constitute only about 0.1% of lunar mass and so require only minimal amounts of mantle melting. Likewise the oceanic crust of the Earth constitutes only 0.1% of the mass of the planet; but as it is continuously formed by partial melting in the mantle, the total volume extruded over geological time has perhaps amounted to 10% of that of the Earth [32]. Because magmas that form secondary crusts are derived from partial melting of a very limited volume of planetary mantles, secondary crusts are enriched in incompatible elements to a much lesser degree than primary crusts.

Tertiary crusts are formed by dehydration or melting of secondary crusts. The continental crust of the Earth remains the only current example in our Solar System. Because such crusts build up over long periods of time, they can contain enrichments of incompatible elements resembling primary crusts, due to the continuous recycling of the secondary crust. It has taken the Earth over four thousand million years to form the continental crust. It seems difficult to produce such tertiary crusts, as no evidence has appeared elsewhere in the Solar System of the existence of such silica-rich crusts analogous to that of the terrestrial continental crust. But that platform constitutes only a trivial fraction (0.4%) of the mass of the Earth, while the useful granitic upper crust produced by intra-crustal melting amounts to perhaps 0.1%. Indeed microclassifiers might wish to form another subdivision to include the production of such zoned crusts.

In the absence of recycling of the oceanic crust that occurs on the Earth, crustal growth on the other rocky planets in our Solar System is essentially irreversible, a process that results mostly in surfaces covered with basalt, the so-called "stagnant-lid" regime. On the Earth, the continental crust remains buoyant, destroyed only by erosion, unlike the oceanic crust that is rapidly recycled back into the mantle.

1.6 Geochemical processes during crust formation

The formation of crusts serves to concentrate incompatible elements toward the surface. These processes accelerate mantle cooling, as the incompatible elements include the radiogenic heat-producing elements potassium, uranium and thorium. On account of this, it is a common error to overestimate the abundances of such elements in calculating bulk planetary compositions, as these are inevitably based on and biased toward near-surface samples. We discuss some examples of this problem in Chapter 12.

What are the reasons for the concentration of certain elements in planetary crusts? The two principal factors influencing planetary compositions in the inner Solar System are volatility on the one hand and their siderophile, chalcophile or lithophile character on the other. Volatility is a major factor in determining the bulk planetary composition and as we have seen, is responsible for the scarcity of gases, ices and

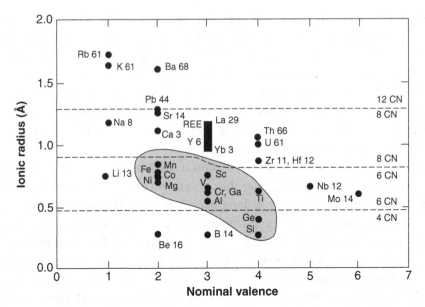

Fig. 1.6 The enrichment of a given element in the continental crust (shown here in percent relative to the primitive mantle) depends on how much its ionic radius and valency differ from those of the abundant elements (shaded area) in the mantle. The dashed lines separate the regions where the cations are in 4-, 6- or 12-fold coordination with O^{2-} in silicates.

volatile elements in the terrestrial planets. After a terrestrial planet has formed, the initial differentiation into a rocky mantle and a metallic core causes element fractionation based on the relative effects of their siderophile, chalcophile and lithophile affinities.

The composition of the silicate mantle is subsequently dominated by the geochemistry of the lithophile elements that form ionic bonds with oxygen. Thus, during crustal formation, the principal factors influencing the distribution of the lithophile elements are those that control their entry into lattice sites in silicate minerals (Fig. 1.6). This figure clearly illustrates that the enrichment of elements in the continental crust depends on the size and valency of the lithophile elements and illustrates the significance of these parameters during crystallization of silicate melts.

Olivine and pyroxene dominate the mineralogy of the terrestrial upper mantle and the mantles of the Moon, Mars and Venus, because iron and magnesium are abundant products of nucleosynthesis in stars. The lithophile elements are often divided into "compatible" and "incompatible" elements. Compatible elements are those that enter the Fe and Mg sites in these mantle minerals. Elements whose size or valency exclude them from the dominant mantle minerals are termed "incompatible". During the formation and crystallization of magmas, such elements (including the heat-producing elements potassium, uranium and thorium) are concentrated and so eventually find their way into the crust. It is because of this combination of cosmochemical and geochemical factors that crusts, despite their volumetric insignificance relative to planets, are so important in planetary evolution. However, estimates of crustal, or even planetary, compositions can be biased towards overestimates of the abundances of the incompatible elements because of this near-surface concentration.

1.6.1 Europium as a universal tracer

In many respects, the rare earth element (REE) europium, an element whose abundance rarely exceeds a few parts per million in crustal rocks of any planet, turns out to be one of the most useful elements in geochemistry and cosmochemistry. By its enrichment or depletion relative to the other REE, one can trace much of the history of processes in the Solar System and the Earth. But the element is also useful to astrophysicists. Europium is formed in stars almost entirely by the r-process of nucleosynthesis. The atomic emission lines of europium at 4129 and 4205 Å are readily measured in stellar spectra, so that its abundance gives a measure of the contribution to the element abundances by the nucleosynthesis of the r-process isotopes, particularly in the earliest stars. By measuring Th/Eu ratios (using the Th line at 4019 Å), an estimate of the age of the star can be established by the depletion in thorium (^{232}Th has a half-life of 14.2 Gyr) relative to the abundance of europium [33].

Next, europium records the existence of a diverse series of events in the early Solar System that are recorded in the earliest samples to which we have access: the calcium–aluminium inclusions (CAIs) in primitive meteorites. These include high temperature evaporation and condensation processes that separated europium relative to the more refractory lanthanides on account of its relative volatility.

However, when processes of planetary differentiation begin, europium with its ability to record oxidizing or reducing environments, is of extraordinary usefulness. Thus in basaltic meteorites that are derived from the asteroid 4 Vesta, depletions and enrichments in europium record the details of crystal–liquid fractionation occurring in basaltic lavas.

The depletion of europium in lunar basalts and its complementary enrichment in the feldspar-rich lunar crust was a major key to our understanding that much of the Moon was melted and that a feldspar-rich crust had developed very early in lunar history. Such events did not occur on the Earth, and are probably unique in the Solar System, with a possible exception on Mercury. To account for the amount of feldspar in the lunar crust, close to 50% of the aluminium and europium in the bulk had to be concentrated in the crust, shortly after the formation of the Moon.

Flotation of feldspar crystals in a dry magma ocean is the only viable hypothesis. All basalts derived from the lunar interior display a complementary depletion in europium, due to an overall depletion of their source regions by the prior crystallization of the plagioclase feldspar now in the highland crust. In retrospect, this behavior of europium was one of the most significant geochemical observations on lunar rocks.

The distribution of REE in the crust of the Earth, that has been a major clue to the evolution of the continental crust, further highlights the importance of europium. The uniform REE patterns observed in continental-crustal sediments on the Earth, with the regular depletion in europium, provide not only a key to the problems of estimating the composition of the crust, but provide a crucial piece of evidence about its evolution. Europium is trivalent under the oxidizing conditions at the surface of the Earth, so that the observed depletion of that element in the upper crust records a previous history of that element as a divalent ion under more reducing conditions. This depletion in europium thus provides evidence that the present upper crust of the Earth was formed by the production of granitic melts deep within the crust under reducing conditions; leading to the retention of europium in plagioclase in the deep crust, with a corresponding depletion at the present surface. The general absence of europium depletion in Archean crustal terrains demonstrates the difference between Archean and Post-Archean crustal development. Overall, an extraordinarily useful element [34].

Synopsis

Planets differ fundamentally from stars in that they form "bottom-up" in the nebula by essentially random processes, from any combination of the original gases, ices and rock. The early formation of Jupiter dominates the evolution of the rest of the Solar System. The solar nebula itself has a limited life span of a few million years. But in the inner nebula, the gases, ices and many volatile elements (e.g. potassium, lead, bismuth, evaluated most readily by K/U or K/Th ratios) were swept out to a snow line around 5 AU by early intense solar activity. This volatile-element depletion, relative to CI, included biologically important elements such as carbon, nitrogen, potassium and phosphorus and influenced the abundance of ^{40}K, an important early heat source. The planetesimals were next assembled from the left-over dry rocky rubble, remnants of which now form the asteroid belt. Many were melted and separated metal from silicate. Our present supply of meteorites provides useful, if inexact analogs. Mars and Mercury formed very rapidly within a few million years, but Earth and Venus took between 10–100 Myr to complete accretion, although a 30 Myr age of uncertain status is often quoted from the Hf–W isotopic systematics.

The accretion was hierarchical, so that the largest objects (Mars-sized) accreted last, one such event on Earth occurring at the right velocity and angle to form the Moon. So the formation of the Earth was essentially accidental. Venus or Earth were equally likely outcomes, or alternatively totally different planets.

Earth and Venus have similar uncompressed densities, implying similar bulk compositions but Mars is 6–7% less dense and has a higher volatile-element content. Both Mars and Mercury provide examples of objects that failed to accrete into larger planets. The energies involved in accretion are sufficient to cause planetary-wide melting with core and mantle formation.

Three major classes of crust, primary, secondary and tertiary, may be distinguished among the variety observed. Primary crusts result from initial planetary melting, secondary crusts are derived by partial melting in mantles and tertiary crusts by reprocessing of secondary crusts. The distribution of elements between cores, mantles and crusts is governed by the invariant laws of physics and chemistry while the concentration of elements into crusts depends mostly on the difference in ionic radius and valency from the common mantle elements (e.g. iron, magnesium).

Finally, the rare earth element europium turns out to be the single most informative member of the periodic table for the study of crusts and planetary evolution on account of its unique combination of distinctive valence behavior, ionic radius and volatility.

Notes and references

1. Laplace, P. S. (1809) *The System of the World* (trans. J. Pond), Richard Phillips, Book 2, p. 293.
2. Definitions of planets have proven as elusive as definitions of life. The debate over the status of Pluto by a committee of the International Astronomical Union (IAU) is an interesting example of a struggle between politics, sentiment and science. Indeed trying to define a planet runs into the philosophical difficulty of attempting to classify any set of randomly assembled products. A bewildering array of objects formed in the early solar nebula, that included dust, asteroids, Trojans, centaurs, comets, TNOs and our eight planets from tiny Mercury to huge Jupiter. All differ from one another in some salient manner, just as do the 160 satellites. The significant question is how did they form and evolve, not what pigeonhole they can be forced into. So the decision of the IAU that there are eight major planets and a host of minor planets seems an appropriate compromise. But each Solar-System body is distinct so that qualifiers such as terrestrial planets, gas giants and extra-solar planets will always be needed.
3. Of the more than 160 satellites surrounding the planets, seven have sizes that approach the size of the smallest planet Mercury (within a factor of two in diameter and approximately a factor of five in volume), and include (in decreasing size) Ganymede, Titan, Callisto, Io, Moon, Europa and Triton. In addition, Eris and Pluto in the Edgeworth–Kuiper Belt are only slightly smaller than these satellites with diameters of 2400 and 2300 km respectively. The largest asteroid Ceres, at 900 km diameter, is a factor of two smaller again. Accordingly, while the number of planets with crusts, at four, is small, the number of large bodies with crusts is more respectable.

4. However in detail, the rocky fraction may not be well mixed. See Ranen, M. C. and Jacobsen, S. (2006) Barium isotopes in chondritic meteorites: Implications for planetary reservoir models. *Science* **314**, 809–12; and Andreasen, R. and Sharma, M. (2006) Solar nebula heterogeneity in p-process samarium and neodymium isotopes. *Science* **314**, 806–809; and the discussion in Chapter 9.

5. A good review by 88 authors covering all aspects from the pre-solar epoch, through disk formation, nebular processing, accretion and differentiation of planetesimals to the formation of the terrestrial planets can be found in *Meteorites and the Early Solar System II* (eds. D. S. Lauretta and H. Y. McSween, 2006), University of Arizona Press. Its A4 size and 3 kg mass do not make for bedtime reading. Several other equally massive volumes cover much of the same territory. *Chondrites and the Protoplanetary Disk* (eds. A. N. Krot *et al.*, 2005), Astronomical Society Pacific Conference Series 341; *Protostars and Planets IV* (eds. V. Mannings *et al.*, 2000), University of Arizona Press; *Meteorites, Comets and Planets*. Treatise on Geochemistry, vol. 1 (ed. A. M. Davis, 2004), Elsevier; *Origin of the Earth and Moon* (eds. R. M. Canup and K. Righter, 2000), University of Arizona Press; *Chondrules and the Protoplanetary Disk* (eds. R. H. Hewins *et al.*, 1996), Cambridge University Press; and *Encyclopedia of the Solar System*, 2nd edn. (eds. P. R. Weissman *et al.*, 2007), Academic Press–Elsevier.

6. Haisch, K. E. *et al.* (2001) Disk frequencies and lifetimes in young clusters. *Astrophys. J.* **553**, L153–6.

7. Guillot, T. *et al.* (2004) The interior of Jupiter, in *Jupiter* (eds. F. Bagenal *et al.*), Cambridge University Press, pp. 35–57.

8. Pickett, M. K. and Durisen, R. H. (2007) Numerical viscosity and the survival of gas giant protoplanets in disk simulations. *Astrophys. J.* **654**, L155–8.

9. Thommes, E. W. *et al.* (2002) The formation of Uranus and Neptune among Jupiter and Saturn. *Astron. J.* **121**, 2862–83.

10. e.g. Stevenson, D. J. and Lunine, J. I. (1988) Rapid formation of Jupiter by diffuse redistribution of water vapor in the solar nebula. *Icarus* **75**, 146–55; see review by Taylor, S. R. (2001) *Solar System Evolution: A New Perspective*, Cambridge University Press, ch. 7 and 8; Lunine, J. I. *et al.* (2004) The origin of Jupiter, in *Jupiter* (eds. F. Bagenal *et al.*), Cambridge University Press, pp. 19–34; Guillot, T. *et al.* (2004) The interior of Jupiter, in *Jupiter* (eds. F. Bagenal *et al.*), Cambridge University Press, pp. 35–57.

11. The original models for a relatively volatile-rich Mars came from trace-element distributions in meteorites derived from Mars. Wänke, H. and Dreibus, G. (1988) Chemical composition and accretion history of the terrestrial planets. *Phil. Trans. Royal Soc. London* **A325**, 545–57. However, there are several reasons, such as their mostly young ages and highly depleted incompatible element abundances, to suspect that most of these meteorites provide a rather biased and perhaps unrepresentative view of Mars. e.g. McLennan, S. M. (2003) Large-ion lithophile element fractionation during the early differentiation of Mars and the composition of the Martian primitive mantle. *MPS* **38**, 895–904. Nevertheless, recent mapping of the martian surface by the 2001 Mars Odyssey gamma-ray spectrometer has confirmed that Mars is moderately volatile-element enriched compared to the Earth. Taylor, G. J. *et al.* (2006) Bulk composition and early differentiation of Mars. *JGR* **111**, doi: 10.1029/2005JE002645. Thus, the mean K/Th ratio of the martian surface is about 5500 with relatively little variation, and for a Th/U ratio of about 3.8, these values indicate a K/U ratio of about 20 000.

12. Rubidium-87 decays to ^{87}Sr with a half-life of 4.88×10^{10} years.

13. The causes of this depletion are much debated. Initial suggestions that the depletion was due to condensation or evaporation in a hot nebula have been ruled out by the absence of

predicted isotopic fractionation (e.g. in K) and the astrophysical evidence that nebular disks are cool (a few hundred Kelvin). So the nebula was cool, not hot and the notion of elements condensing from a hot nebula is no longer tenable. Observations of midplane temperatures in disks around T Tauri stars give values mostly less than 600 K at 1 AU (Valenti, J. A. (1993) T Tauri stars in blue. *Astron. J.* **106**, 2024–50; Hartmann, L. (1998) Accretion and evolution of T Tauri stars. *Astrophys. J.* **495**, 385–400). However close to the Sun, refractory inclusions that constitute a few percent of meteorites were formed in the X-wind environment at high temperatures (Shu, F. *et al.* (1996) Toward an astrophysical theory of chondrites. *Science* **271**, 1545–52). These may be transported out into the far reaches of the nebula. See Cielsa, F. (2007) Outward transport of high-temperature materials around the mid-plane of the solar nebula. *Science* **318**, 613–15. Probably the volatile element depletion was due to early intense solar activity that swept away, along with the gases and ices, those volatile elements that were not present in grains. A likely scenario is that the volatile elements were dispersed along with the ices and gases by early stellar winds. A useful summary is given by Yin, Q.-C. (2005) From dust to planets: The tale told by moderately volatile elements, in *Chondrites and the Protoplanetary Disk*. Astronomical Society Pacific Conference Series 341, pp. 632–44.

14. We note in passing that, according to the Hf–W isotopic data, core formation on asteroids may have occurred less than 1.5 Myr after T_{zero} (4567 Myr ago) and so predate chondrule formation, well constrained to occur more than 2 Myr after T_{zero}. This might imply separate origins for the differentiated asteroids and chondrules. Kleine, T. *et al.* (2005) Early core formation in asteroids and late accretion of chondrite parent bodies: Evidence from ^{182}Hf–^{182}W in CAIs, metal rich chondrites and iron meteorites. *GCA* **69**, 5805–18. Of course, these dates may refer to events occurring in different locations in the nebula, so that there may be some overlap in time with these processes.

15. e.g. Chambers, J. E. (2004) Planetary accretion in the inner solar system. *EPSL* **223**, 241–52; Lunine, J. I. *et al.* (2004) The origin of Jupiter, in *Jupiter* (eds. F. Bagenal *et al.*), Cambridge University Press, pp. 19–34; Raymond, S. N. (2006) High-resolution simulations of the final assembly of Earth-like planets I. Terrestrial accretion and dynamics. *Icarus* **183**, 265–82; Raymond, S. N. *et al.* (2007) High-resolution simulations of the final assembly of Earth-like planets 2: Water delivery and planetary habitability. *Astrobiol.* **7**, 66–84.

16. e.g. Taylor, S. R. and Norman, M. D. (1990) Accretion of differentiated planetesimals to the Earth, in *Origin of the Earth* (eds. H. E. Newsom and J. H. Jones), Oxford University Press, pp. 29–43; Mittlefehldt, D. W. *et al.* (1998) Non-chondritic meteorites from asteroidal bodies, in *Planetary Materials* (ed. J. J. Papike), *Rev. Mineral.* **36**, Mineralology Society of America, pp. 4.103–4.130; Carlson, R. W. and Lugmair, G. W. (2000) Timescales of planetesimal formation and differentiation based on extinct and extant radioisotopes, in *Origin of the Earth and Moon* (eds. R. Canup and K. Righter), University of Arizona Press, pp. 25–44.

17. The date of the formation of the Solar System is given conventionally by the ages of the oldest refractory calcium–aluminium-rich inclusions (CAIs) in meteorites at 4567 ± 2 Myr and referred to as T_0 or T_{zero}. Amelin, Y. *et al.* (2002) Lead isotopic ages of chondrules and calcium-aluminum-rich inclusions. *Science* **297**, 1678–83.

18. Canup, R. and Asphaug, E. (2001) Origin of the Moon in a giant impact near the end of the Earth's formation. *Nature* **412**, 708–12; Canup, R. M. (2004) Dynamics of lunar formation. *Ann. Rev. Astron. Astrophys.* **42**, 441–75.

19. e.g. Halliday, A. N. and Kleine, T. (2006) Meteorites and the timing, mechanisms and conditions of terrestrial planet accretion and early differentiation, in *Meteorites and the*

Early Solar System II (eds. D. S. Lauretta and H. Y. McSween), University of Arizona Press, pp. 775–801; Halliday, A. N. *et al.* (2000) Tungsten isotopes, the timing of metal–silicate fractionation, and the origin of the Earth and Moon, in *Origin of the Earth and Moon* (eds. R. Canup and K. Righter), University of Arizona Press, pp. 45–62; Jacobsen, S. (2005) The Hf–W isotopic system and the origin of the Earth and Moon. *Ann. Rev. Earth Planet. Sci.* **33**, 531–70. Estimates between 30 and 50 Myr from this isotopic system are often supposed to date the Moon-forming event; Kleine, T. *et al.* (2005) Hf–W chronometry and the age and early differentiation of the Moon. *Science* **310**, 1671–4. But later data do not substantiate the conclusions of this paper. See Touboul, M. *et al.* (2007) Late formation and prolonged differentiation of the Moon inferred from W isotopes in lunar metals. *Nature* **450**, 1206–9. See also Kramers, J. D. (2007) Hierarchical Earth accretion and the Hadean Eon. *J. Geol. Soc. London* **164**, p. 11, who comments that the 30 Myr interval from T_{zero} often thought to document the time of core formation on the Earth is actually recording the time when "the silicate matter that is now in the Earth (but was then probably still in different precursors) acquired on average its present Hf/W ratio" and unless the giant impact "involved a rehomogenization of the core and mantle of the whole Earth, which is not shown by the models" the Hf–W system "does not date any event".

20. Grieve, R. A. F. (1998) in *Meteorites: Flux with Time and Impact Effects* (eds. M. M. Grady *et al.*), Geological Society of London Special Publication 140, p. 120.

21. Lissauer, J. J. (1999) Chaotic motion in the solar system. *Rev. Modern Phys.* **71**, 835–45.

22. Canup, R. and Agnor, C. B. (2000) Accretion of the terrestrial planets and the Earth–Moon system, in *Origin of the Earth and Moon* (eds. R. Canup and K. Righter), University of Arizona Press, p. 113; Levison, H. *et al.* (1998) Modeling the diversity of outer planetary systems. *Astron. J.* **116**, 1998–2014.

23. Drake, M. J. and Righter, K. (2002) Determining the composition of the Earth. *Nature* **416**, 39–44; Ikoma, M. and Genda, H. (2006) Constraints on the mass of a habitable planet with water of nebular origin. *Astrophys. J.* **648**, 696–706.

24. e.g. Morbidelli, A. *et al.* (2000) Source regions and time scales for the delivery of water to Earth. *MPS* **35**, 1309–20; Lunine, J. I. *et al.* (2004) The origin of Jupiter, in *Jupiter* (eds. F. Bagenal *et al.*), Cambridge University Press, pp. 19–34; Lunine, J. L. (2006) Origin of water ice in the solar system, in *Meteorites and the Early Solar System II* (eds. D. S. Lauretta and H. Y. McSween), University of Arizona Press, pp. 309–319; Raymond, S. N. *et al.* (2007) High-resolution simulations of the final assembly of Earth-like planets 2: Water delivery and planetary habitability. *Astrobiol.* **7**, 66–84; Kasting, J. F. and Howard, M. T. (2006) Atmospheric composition and climate on the early Earth. *Phil. Trans. Royal Soc.* **B361**, 1731–42; Lunine, J. I. (2006) Physical conditions on the early Earth. *Phil. Trans. Royal Soc.* **B361**, 1721–31.

25. The Galilean satellites likely formed in a gas-starved accretion disk where major collisions and accretional heating was minimal, in great contrast to the accretion of the terrestrial planets; see Canup, R. M. and Ward, W. R. (2002) Formation of the Galilean satellites: Conditions of accretion. *Astron. J.* **124**, 3404–23; and Canup, R. M and Ward, W. R. (2006) A common mass scaling for satellite systems of giant planets. *Nature* **441**, 834–9. Further comments appear in Chapter 13. The capture of Triton was probably responsible for the destruction of any regular system of satellites around Neptune.

26. Taylor, S. R. (1999) On the difficulties of making Earth-like planets. *MPS* **34**, 317–29. There are few grounds to expect that the random sequence of events that resulted in producing our own satisfactory planet is likely to be repeated in detail elsewhere. See also Ward, P. D. and Brownlee, D. (1999) *Rare Earth*, Copernicus; and Taylor, S. R. (1998) *Destiny or Chance: Our Solar System and its Place in the Cosmos*, Cambridge University Press.

27. e.g. Goldschmidt, V. M. (1954) *Geochemistry*, Oxford University Press, ch. 2.
28. Righter, K. *et al.* (2006) Compositional relationships between meteorites and terrestrial planets, in *Meteorites and the Early Solar System II* (eds. D. S. Lauretta and H. Y. McSween), University of Arizona Press, pp. 803–28.
29. e.g. Javoy, M. (1995) The integral enstatite chondrite model of the Earth. *GRL* **22**, 2219–22.
30. Stacey, F. D. (2005) High pressure equations of state and planetary interiors. *Rep. Prog. Phys.* **68**, 341–83.
31. Taylor, S. R. (1989) Growth of planetary crusts. *Tectonophysics* **161**, 147–56.
32. Taylor, S. R. (1994) Large scale basaltic volcanism on the Moon, Mars and Venus, in *Volcanism (Radhakrishna Vol.)* (ed. K. V. Subbarao), Wiley Eastern, pp. 2–20; Head, J. W. and Coffin, M. F. (1997) Large igneous provinces: A planetary perspective, in *Large Igneous Provinces: Continental, Oceanic and Planetary Flood Volcanism*. American Geophysical Union, Geophysical Monograph 100, pp. 411–38.
33. McWilliam, A. (1997) Abundance ratios and galactic chemical evolution. *Ann. Rev. Astron. Astrophys.* **35**, 503–56.
34. Why is europium such a significant element? It is formed by the r-process of nucleosynthesis and occupies a unique position among the rare earths or lanthanides, with an electronic structure [Xe] $4f^7 6s^2$. In the lanthanides, the inner 4f electron subshell, rather than the outer 5d subshell is being filled. Thus as additional electrons are added, they are less effective in shielding the nucleus, so that the ionic radius decreases with increasing elemental mass, leading to the well-known lanthanide contraction from La (La^{3+} $r = 1.16$ Å) through Lu (Lu^{3+} $r = 0.977$ Å). In europium, the 4f subshell is stable, half-filled with seven electrons, so that under reducing conditions, the divalent state is stable, only the $6s^2$ electrons being removed. A third electron is lost only under oxidizing conditions. The ionic radius of Eu^{2+} (1.25 Å) is 17% larger than Eu^{3+} (1.066 Å) and similar to that of Sr^{2+} (1.26 Å) leading to similar geochemical behavior, distinct from the other REE that are trivalent in most geochemical environments.

2

A primary crust: the highland crust of the Moon

> Comparisons with the Earth's geologic style, though inevitable, have
> proved to be treacherous guides to the Moon.
>
> *(Don Wilhelms) [1]*

Every school child is aware that the Moon is not a planet. So why begin this
discussion on planetary crusts with examples from a planetary satellite? The reason
is that the two types of crusts on the Moon, that form the lighter highlands and the
darker maria, are among our best examples of primary and secondary crusts. Their
origin and evolution are better understood than those of any other examples in the
Solar System, including the Earth. In addition, the Moon, in contrast to the Earth,
forms a classic example of a one-plate planet, which is the norm for our Solar System.

2.1 The composition of the Moon

The mean lunar radius is 1737.1 km, which is intermediate between that of the two
jovian satellites of Jupiter, Europa ($r = 1561$ km) and Io ($r = 1818$ km). The Moon is
much smaller than the jovian satellite Ganymede ($r = 2634$ km), which in turn is the
largest satellite in the Solar System and like the saturnian satellite Titan, is larger
than Mercury. Although the jovian satellites and also Titan are comparable in mass,
the Moon/Earth ratio is the largest satellite-to-parent ratio in the planetary system, a
consequence of a distinctive origin. The Charon/Pluto ratio is larger, but Pluto, an
icy planetesimal, is less than 20% of the mass of the Moon and is one of the
largest Trans-Neptunian Objects in the Kuiper Belt, rather than being a planet.
Charon is only relevant here as another example of the collisional origin of a
satellite.

The mass of the Moon is 7.35×10^{25} g, which is 1/81 of the mass of the Earth. The
lunar density is 3.346 g/cm^3, a fact that has always excited interest on account of the
Moon's proximity to the Earth, which has a much higher density of 5.54 g/cm^3 [2].
This major difference in density posed a serious dilemma for workers seeking to

Table 2.1 *The major element composition of the bulk Moon and that of the lunar highland crust (wt% oxides). Data from Note 4*

Oxide	Bulk Moon	Lunar highland crust
SiO_2	47.0	46.0
TiO_2	0.3	0.3
Al_2O_3	6.0	28.0
FeO	13.0*	4.5
MgO	29.0	4.5
CaO	4.6	16.0
Na_2O	0.09	0.45
K_2O	0.01	0.075
Σ	100.0	99.8

* 2.3% Fe or FeS is allotted to a lunar core and 10.7% to a lunar mantle; Mg# [molar (Mg/Mg + Fe)] = 0.80.

understand the origin of these two closely associated rocky bodies. The low density of the Moon tells us that it is depleted in metallic iron relative to the inner planets. The high value for the moment of inertia of our satellite ($I/Mr^2 = 0.3935$) means that there is only a slight increase in density toward the center, so that the Moon has only a tiny core (350 km radius) at best.

Although the Earth contains about 28% metallic iron, the Moon has less than about 2 or 3%. However, the lunar mantle has a high FeO content (present in silicates) so that the bulk Moon iron content, expressed as FeO, is 13%, that is 50% more than for the current estimates of 8% FeO in the terrestrial mantle. Along with its depletion in metallic iron, the Moon also has a low abundance of the other siderophile elements [3].

Compositional data for the bulk Moon and the lunar highland crust are given in Tables 2.1 and 2.2 [4]. Notable differences between the Moon and the Earth are the bulk FeO content and the Al_2O_3 abundance that is reflected in the abundances of the other refractory elements. These are enriched in the Moon by a factor of 1.5 compared to the Earth or of about 3 compared to CI.

Figure 2.1 shows the relative compositions of the Earth and Moon, normalized to the CI abundances. This shows the two extreme features of the lunar composition that make it unique. These are the enrichment in refractory elements and the depletion in volatile elements. The depletion in the Earth relative to CI correlates with volatility. But although the Moon is more highly depleted in volatile elements than the Earth, curiously this depletion is apparently uniform relative to the Earth [5]. Clearly the Moon has a composition that cannot be made by any single-stage process from the material of the primordial solar nebula and calls for a distinctive mode of origin. Thus the cause of the enrichment in refractory elements and the

Table 2.2 *The elemental composition of the bulk Moon and that of the lunar highland crust. Data from Note 4*

Element	Bulk Moon	Lunar highland crust
Li (ppm)	0.83	2
Be (ppm)	0.21	–
B (ppm)	0.1	–
Na (wt%)	0.06	0.33
Mg (wt%)	17.5	2.71
Al (wt%)	3.17	14.8
Si (wt%)	21.9	21.5
K (ppm)	83	600
Ca (wt%)	3.28	11.4
Sc (ppm)	19	5
Ti (ppm)	1800	1800
V (ppm)	150	21
Cr (ppm)	4200	500
Mn (wt%)	0.12	–
Fe (wt%)	10.6	3.5
Co (ppm)	20	–
Ni (ppm)	400	–
Cu (ppm)	15	–
Zn (ppm)	5	2.0
Ga (ppm)	1.0	–
Ge (ppm)	0.14	–
As (ppm)	0.007	–
Se (ppm)	0.09	–
Rb (ppm)	0.28	1.7
Sr (ppm)	34	140
Y (ppm)	6.3	13.4
Zr (ppm)	17	63
Nb (ppm)	1.3	4.5
Mo (ppb)	1.4	–
Ru (ppb)	30	–
Rh (ppb)	10	–
Pd (ppb)	22	–
Ag (ppb)	0.8	–
Cd (ppb)	1.0	–
In (ppb)	0.12	–
Sn (ppb)	5.0	–
Sb (ppb)	0.6	–
Te (ppb)	14	–
Cs (ppb)	12	100
Ba (ppm)	11	70
La (ppm)	1.10	5.3
Ce (ppm)	2.87	13
Pr (ppm)	0.411	1.8
Nd (ppm)	2.13	7.4
Sm (ppm)	0.69	2.0

Table 2.2 (*cont.*)

Element	Bulk Moon	Lunar highland crust
Eu (ppm)	0.26	1.0
Gd (ppm)	0.92	2.3
Tb (ppm)	0.17	0.35
Dy (ppm)	1.14	2.3
Ho (ppm)	0.255	0.53
Er (ppm)	0.75	1.51
Tm (ppm)	0.11	0.22
Yb (ppm)	0.74	1.4
Lu (ppm)	0.11	0.21
Hf (ppm)	0.51	1.4
W (ppm)	0.008	–
Re (ppb)	1.6	–
Os (ppb)	25	–
Ir (ppb)	23	–
Pt (ppb)	40	–
Au (ppb)	7	–
Hg (ppb)	0.7	–
Tl (ppb)	0.2	–
Pb (ppm)	0.004	1.0
Bi (ppb)	0.17	–
Th (ppb)	115	900
U (ppb)	30	240

uniform depletion of the very volatile elements may be a consequence of the condensation of the Moon from vapor following the giant impact [6].

The initial interest in the bulk lunar composition was to test the hypothesis that the Moon was derived from the silicate mantle of the Earth, a notion dating back to George Darwin, resolving the difference in density between the two bodies. Although hotly debated following the Apollo sample return, the siren-like attractions of the model have diminished, despite the similarity in oxygen isotopes between the two bodies [7]. This model has not survived the demonstration of significant differences (e.g. FeO content) and the acceptance of the single-impact hypothesis for lunar origin that derive 85% of the Moon from the mantle of the impactor (Theia), not from the Earth.

The bulk composition of the Moon can thus be understood within the general framework of the standard model for the evolution of the inner Solar System. All inner bodies in the Solar System are depleted in volatile elements relative to the composition of the primordial rocky component (CI) of the solar nebula (Chapter 1). Initial differences have been exacerbated by massive collisions that have produced extremes. The composition of the metal-poor Moon in contrast to metal-rich Mercury represents examples of the range in compositions that result from the

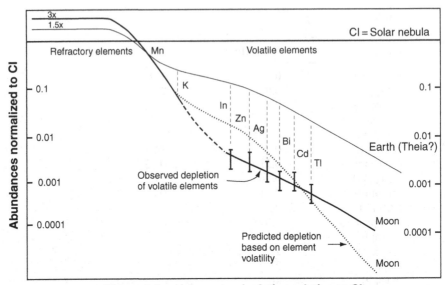

Fig. 2.1 The composition of the Moon compared with that of the Earth, both normalized to CI carbonaceous chondrites. The material in the impactor mantle (Theia), from which the Moon was derived, was inner Solar-System material already depleted in volatile elements at T_{zero}. The abundances in the Earth provide an analogue for its composition. The significant point is that, compared to the Earth (or Theia), the lunar depletion is uniform and not related to volatility, which would produce a much steeper depletion pattern (lower dotted line). Thus, the lunar pattern is interpreted as resulting from a single-stage condensation from vapor (>2500 K) that effectively cut-off around 1000 K [4].

stochastic assembly of planetary-sized bodies from a hierarchy of smaller bodies. It points to the importance of such processes in the inner solar nebula that have led to such random outcomes.

2.2 The lunar surface

Ninety-nine percent of the lunar surface is older than 3 Gyr and more than 80% is older than 4 Gyr. In contrast, 80% of the surface of the Earth is less than 200 million years old. The absence of plate tectonics, water and life, together with the essential absence of an atmosphere, means that the present lunar surface is unaffected by the main agents that shape the surface of the Earth. The major process responsible for modifying the lunar surface is the impact of objects ranging from micron-sized grains to bodies that were tens to hundreds of kilometers in diameter.

As a consequence of this bombardment, the surface of the Moon is covered with a debris blanket, called the regolith. It ranges from fine dust to blocks several meters across. Although there is much local variation, the average regolith thickness on the maria is 4 to 5 m, whereas the highland regolith is about 10 m thick. The regolith is continuously being turned over or "gardened" by meteorite impact. The near-surface structure, revealed by core samples (the deepest was only 3 m at the Apollo 17 site), shows that the regolith is a complex array of overlapping ejecta blankets typically ranging in thickness from a few millimeters up to about 10 cm, derived from the multitude of meteorite impacts at all scales. These sheets have little lateral continuity even on scales of a few meters. The rate of growth of the regolith is very slow, averaging about 1.5 mm per million years or 15 Å per year.

A megaregolith of uncertain thickness covers the heavily cratered lunar high-lands. This term refers to the debris sheets from the craters and particularly those from the large impact basins that have saturated the highland crust. The aggregate volume of ejecta from the presently observable lunar craters and basins amounts to a layer about 2.5 km thick. The postulated earlier bombardment may well have produced megaregolith thicknesses in excess of 10 km, particularly north of the 2500 km diameter South Pole–Aitken Basin. Related to this question is the degree of fracturing and brecciation of the deeper crust due to the large basin collisions. Some workers equate this fracturing with the levelling off of the P-wave seismic velocity (V_P) to a constant 6.8 km/sec at 20–25 km. In contrast to the highlands, bedrock is present at relatively shallow depths (several meters) in the lightly cratered maria.

2.3 Structure of the crust

Re-evaluation of the Apollo seismic data indicates that at the Apollo 12 and 14 sites, the lunar highland crust is 45 ± 5 km thick (rather than earlier estimates of 60 km) [8]. The farside crust averages about 15 km thicker than that of the nearside so that the overall lunar crustal thickness is 52 km [9]. The crust thus constitutes about 8.7% of lunar volume. The maximum relief on the lunar surface is over 16 km while the largest impact basin (South Pole–Aitken) is 12 km deep [10]. In contrast, the relief on the Earth is just over 20 kilometers, ranging from the fossiliferous Ordovician limestones on top of Mt. Everest to the marine muds on the bottom of the Challenger Deep in the Marianas Trench. Mars, again a small body, has a greater range of relief than either the Earth or the Moon, with nearly 30 km from the top of Olympus Mons to the bottom of the Hellas impact basin. Clearly crustal relief is unrelated to planetary size and is independent of processes like plate tectonics.

The mare basalts (Chapter 3) that form the type example of a secondary crust cover 17% of the lunar surface, mostly on the nearside. Although prominent visually, they are usually less than 1 or 2 km thick, except near the centers of the basins so that these basalts constitute only a trivial volume of the crust. Seismic velocities in the crust increase steadily down to 20 km. At that depth, there is a change in velocities within the crust that possibly represents the depth to which extensive fracturing, due to massive impacts, has occurred. At an earlier stage, this velocity change was thought to represent the base of the mare basalts, but these are now known to be much thinner, typically less than a few kilometers in thickness. The lower section of the crust from 20 to 45 km has rather uniform velocities of 6.8 km/sec, corresponding to the velocities expected from the average feldspathic composition of the lunar samples from the highlands.

2.3.1 Tectonics

The dominant features of the lunar crustal surface are the old heavily cratered highlands and the younger basaltic maria, that have flooded into the large impact basins. There is a general scarcity of tectonic features due to internal processes on the Moon, in great contrast to the dynamically active Earth. There are no large-scale tectonic features and the lunar surface acts as a single thick plate that has been subjected to only small internal stresses.

Many overlapping lineaments have been discerned on the surface of the highlands. Although it was once supposed that this "lunar grid" had developed through internal tectonic stresses [11], the lineaments that constitute the grid were formed by the overlap of ejecta blankets from the many multiringed basins and have no tectonic significance [12].

Apart from structures due to cratering, most of the lunar tectonic features are related to stresses associated with subsidence of the mare basins, following their flooding with lava. Wrinkle ridges (or mare ridges) are low-relief, linear to arcuate, broad ridges that commonly form near the edges of the circular maria. They are the result of compressional bending stresses, related to subsidence of the basaltic maria during cooling.

Rilles, which are extensional features similar to terrestrial grabens, are often hundreds of kilometers long and up to 5 km wide. Unlike the wrinkle ridges, they cut only the older maria as well as the highlands and indicate that some extensional stress existed in the outer regions of the Moon prior to about 3.6 Gyr. They should probably be termed "grabens" so as to avoid confusion with the sinuous rilles, such as Hadley Rille, that are formed by flowing lava, presumably through thermal erosion. The set of three rilles, each about 2 km wide, that are concentric to Mare Humorum at about 250 km from the basin center are particularly instructive examples, showing a clear extensional relation to subsidence of the basin.

2.3.2 Lunar stratigraphy

The succession of events on the lunar surface has been determined by establishing a stratigraphic sequence based on the normal geological principle of superposition, a

Table 2.3 *The stratigraphic sequence for the Moon [12]*

System	Age (billion years)	Remarks
Copernican	1.0 to present	The youngest system, which includes fresh ray craters (e.g. Tycho), begins with the formation of Copernicus.
Eratosthenian	1.0–3.1	Youngest mare lavas and craters without visible rays (e.g. Eratosthenes).
Imbrium	3.1–3.85	Extends from the formation of the Imbrium Basin to the youngest dated mare lavas. Includes Imbrium Basin deposits, Orientale and Schrödinger multiring basins, most visible basaltic maria, and many large impact craters, including those filled with mare lavas (e.g. Plato, Archimedes). Sometimes divided into Early Imbrium from the formation of the Imbrium Basin (3.85 Gyr) to the formation of the Orientale Basin (3.82 Gyr), and the Late Imbrium from 3.82 to 3.10 Gyr.
Nectarian	3.85–3.92	Extends from the formation of the Nectaris Basin to that of the Imbrium Basin. Contains 12 large, multiring basins and some buried maria.
Pre-Nectarian	Pre 3.92	Basins and craters formed before the Nectaris Basin. Includes 30 identified multiring basins.

fundamental contribution due to Gene Shoemaker [13]. Geological maps based on this concept have been made for the entire Moon, notably by Don Wilhelms [12]. Relative ages have been established by crater counting, and isotopic dating of returned samples has enabled absolute ages to be assigned to the various units. The formal stratigraphic sequence is given in Table 2.3.

The Nectarian (3850–?3920 Myr) and Pre-Nectarian systems (?3920–?4520 Myr) on the Moon cover the period of time represented on Earth by the Hadean Eon, where the terrestrial rock record is missing. Like meteorites, the Moon has thus been a kind of Rosetta Stone in providing us with insights into the early history of the Solar System not available on the Earth.

The Pre-Nectarian System includes a number of impact basins, the oldest being the doubtful Procellarum Basin that covers much of the lunar nearside. Most of the Nectarian System resides on the farside, coincident with the Feldspathic Highlands Terrane (FHT). Tectonic activity is absent while the only evidence of volcanism is the rare presence of basaltic clasts in highland breccias. The Nectarian System includes all events from the formation of the Nectaris Basin to that of the Imbrium Basin and includes 11 other large impact basins, such as Crisium [14].

The Imbrium System is divided into the Lower Imbrium Series (3.85–3.82 Gyr) that extends to the formation of the Orientale Basin and the Upper Imbrium Series (3.82–3.1 Gyr) dates from that event and includes most of the lava plains down to the age of the youngest dated lavas. The limits of the Eratosthenian System (3.2–0.8 Gyr) are more difficult to define precisely. It includes some basaltic plain units and all the craters without visible rays down to the formation of the crater Copernicus, about 1 Gyr ago. The Copernican System dates from the formation of that crater down to the present.

2.4 Craters and multiring basins

One of the most diagnostic features of the lunar surface, that is in great contrast to the surface of the Earth's crust, is the ubiquitous presence of impact craters at all scales, from micron-sized "zap pits" to multiring basins. Although the volcanic origin of the lunar craters was popular before the Apollo missions, "two landforms so clearly different in gross geometry as terrestrial calderas and lunar craters hardly can be proved to be analogous on the basis of minor morphologic characteristics" as Richard Pike has remarked [15].

Although the correct explanation for the origin of the lunar craters had already been reached by G. K. Gilbert in 1893 and Ralph Baldwin in 1949, this topic was the subject of ongoing controversy until the Apollo sample return. The question still surfaces occasionally in popular articles. As meteorites and other impacting bodies could be expected to strike the Moon at all angles, the circularity of the lunar craters was long used as an argument against impact and in favor of a volcanic origin. It was eventually realized that bodies impacting the Moon at velocities of several km/sec explode on impact and form a circular crater regardless of the angle of impact, except for very oblique impacts. The morphology of the craters resembles that of terrestrial explosion craters and is quite distinct from the landforms of terrestrial volcanic centers [15].

The ultimate form resulting from impact is the multiring basin, which may have six or more rings. The type lunar example is the Orientale Basin (Fig. 2.2). This structure is 920 km in diameter (about the size of France), with several concentric mountain rings that have a typical relief of about 3 km, with steep inward-facing scarps. These formed in a few minutes following the impact of a body perhaps 50 km in diameter. The central portion has been flooded much later with a thin filling (0.6 km) of mare basalt. Thirty such basins have been recognized on the Moon with another 14 probable ones.

The largest confirmed example is the South Pole–Aitken Basin (centered at 180° E, 56° S) that is 2500 km in diameter and 12 km deep. The presence of a larger Procellarum Basin (3200 km diameter, centered at 23° N, 15° W) covering much of the nearside is questionable.

There has been much controversy over the origin of multiring basins. However a consensus has emerged that the mountain rings are fault scarps, formed by collapse

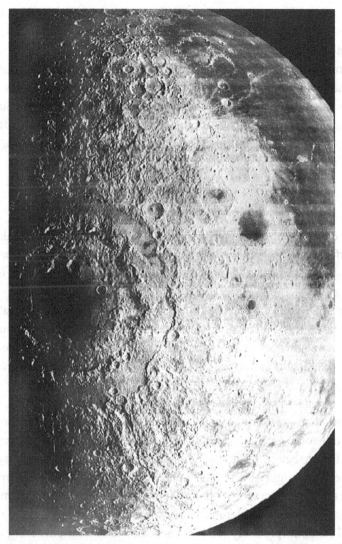

Fig. 2.2 Mare Orientale, the classical example of a multiring basin. The diameter of the outer mountain ring, Montes Cordillera, is 900 km. The structures radial to the basin are well developed. The small area of mare basalt to the northeast is Grimaldi, while the western shore of Oceanus Procellarum fills the northeastern horizon. NASA Orbiter IV 187 M.

into a deep transient crater formed by the initial impact. The depth of excavation of the lunar basins decreases with increasing basin diameter. A transient cavity forms during the initial stage of the impact, but most excavated material comes from shallower depths. Thus no unequivocal lunar mantle material has been recognized in the returned samples from the lunar highland crust and so the transient depth of excavation of the largest basins does not appear to have exceeded 50 km [16].

The scale of impact melting in the development of the highland crust remains problematical. Although impact melt normally constitutes only a few percent of the crater volume, evidence from the Sudbury impact event in Canada reveals that a large initial melt pool resulted that eventually formed the Sudbury igneous complex (SIC). Crystallization of impact melt pools following major basin-forming impacts on the Moon might have resulted in the formation of "igneous" rocks. Such events on the Hadean Earth may also have produced small amounts of felsic differentiates, perhaps accounting for the controversial evidence from zircon grains that range in age from 4.36 to 3.9 Gyr [17].

2.4.1 A lunar cataclysm?

The intense cratering of the lunar highlands and the absence of a similar heavily cratered surface on the Earth were long recognized as due to an early "pre-geological" bombardment although its cause and when it happened have been strongly debated.

In contrast, the lightly cratered basaltic mare surfaces, on which the cratering rate is about 200 times less, had escaped this catastrophe and were clearly younger. The ages of the mare surfaces, dated from the sample return to be between 3.1 and 3.8 Gyr, showed that the cratering flux declined to that observed terrestrially within a factor of two.

Although it was also established that the intense cratering of the highlands occurred more than 3.8 Gyr ago, the radiometric ages of the ejecta blankets from the large collisions tend to cluster around 3.9 Gyr, with the dates for the Imbrium collision being 3.85, and for Nectaris 3.95 Gyr, although ages up to 4.1 Gyr have been suggested for the latter. This is a surprisingly narrow range and indicates a rapid increase in the cratering flux before 3.8 Gyr. Impact melt breccias from lunar highland rocks and impact glasses from lunar meteorites from several random (probably farside) sites show a similar clustering of ages in this time interval [18].

The non-cataclysmic explanation is that the cratering represents the final sweep-up of large objects during the accretion of the Moon. The clustering of ages has been suggested to be an artefact of the sampling by the Apollo and Luna missions of nearside young impact basins [19]. Arguments for the sweep-up of the accretionary tail include the large number of apparently old very degraded basins. This stratigraphic evidence for an extended sequence of collisions has been elegantly described by Wilhelms [12] but such a sequence of impacts may also occur within a brief timeframe (basins form on a timescale of minutes) so the visible lunar-basin sequence might well form within the 200 Myr interval recorded by the radiometric ages.

The major problem with this scenario of an accretionary tail is that extrapolation from the rate at 3.8 Gyr back to 4.5 Gyr results in the accretion of a Moon several orders of magnitude larger than observed. It seems probable that accretion of the Moon was essentially complete and that the Moon was at its present size by about 4450 Myr, around the time of the crystallization of the feldspathic highland crust. Indeed the most likely scenario for lunar origin assembles the Moon in periods ranging from a few hours to 100 yr following the giant impact.

The alternative model, accepted here, suggests that the bombardment is due to a late spike in the cratering flux, the "lunar cataclysm" [20]. Arguments in favor of the cataclysm include the scarcity of impact melts older than 4 Gyr and the lead isotope data, which indicate a major resetting of the lead ages at about 3.9 Gyr. Although it is often argued that the sampling from the Apollo missions is dominated by Imbrium ejecta, lunar meteorites have provided fresh insights. These provide a random sampling of the surface but display few impact melts older than 3.92 Gyr supporting the notion of a "cataclysm".

The cataclysm has wider implications as it appears to have affected the entire inner Solar System [18]. Impact heating events on the asteroid 4 Vesta also appear to be concentrated between 4.1 and 3.4 Gyr, giving support to the concept of a cataclysmic period of cratering throughout the inner Solar System although this example perhaps lasted for a longer period than in the case of the Moon [21].

However this model of a late spike of large impactors raises a serious question. It is a major dynamical problem to store such bodies for several hundred million years after the formation of the Solar System. If the bodies originated in the main asteroid belt, how were they suddenly perturbed into lunar-crossing orbits? One possibility is that a large asteroid broke up [22]. Another is that some objects escaped being swept up into the planets and remained as "loose cannons" until they collided, producing a shower of impactors. These could also have been sent sunwards from the Kuiper Belt by the outward migration of Uranus and Neptune and so the late heavy bombardment, in this scenario, was due mostly to comets. Alternatively these bodies in turn precipitated a shower from the asteroid belt. These questions remain open [23].

A principal reason for discussing this late lunar bombardment that occurred around 4 Gyr in a book dealing with crusts is that it seems inevitable that such a bombardment also affected the Earth, Mars and Mercury (any evidence from Venus has been obscured). But while Mars and Mercury exhibit ancient cratered terrains, no such terrains are known on Earth. For example, there was no decisive petrographic and geochemical evidence of a late heavy bombardment around 3.85 Gyr in the Early Archean rocks at Isua, Greenland [24]. If the cataclysm model is correct, most of the Hadean Eon was much less turbulent than commonly imagined.

2.5 Composition of the lunar highland crust

The lunar highland crust is our best understood example of a primary crust that formed directly following upon accretion. A consequence of the massive bombardment that pulverized the lunar highlands is that the rocks returned by the Apollo missions from the lunar highlands are impact melts and breccias, usually consisting of rock fragments or clasts set in a fine-grained matrix. Lunar breccias are divided into monomict, dimict, and polymict breccias, consisting, respectively, of a single rock type, two distinct components, and a variety of rock types and impact melts. Polymict breccias, involving several generations of breccias, are the most common rock type returned from the lunar highlands. They are further subdivided into fragmental breccias, glassy melt breccias, crystalline melt breccias (or impact melt breccias), clast-poor impact melts, granulitic breccias and regolith breccias.

In addition to the information gained from the returned Apollo and Luna samples, our understanding of the lunar crust has been aided by the discovery of lunar meteorites of which about 30 have been found so far [25]. Many appear to be from the lunar farside; they are distinct from the nearside highland samples returned by Apollo 14, 15, 16 and 17 and Luna 20. However, their major element composition is close to that of estimates of the average highland crust. They confirm, as do the Galileo, Clementine and Prospector missions, the essentially anorthositic nature of the lunar highland crust.

The key to establishing the bulk composition of the crust is to integrate the sample data with the global coverage provided by the orbital data, tying this into the Apollo sample sites [26]. Data for the average composition of the highland crust are given in Tables 2.1 and 2.2.

The lunar highland crust is ancient. The most generally accepted age is 4460 ± 40 Myr that is taken to represent the crystallization of anorthosites from the lunar magma ocean – see Chapter 3 [27]. The upper crust is plagioclase-rich (averaging 30% Al_2O_3). Evidence from the composition of the uplifted central peaks of impact craters indicates that this composition extends to at least 30 km [28].

It is a general article of faith, although without much supporting evidence, that the highland crust becomes more basic with depth [28]. So the crust is often divided, on the basis of geophysical modeling, into an upper anorthositic and a lower noritic layer. One must be cautious with such models, in view of the failure of terrestrial geophysical interpretations to predict the deep crustal structure that was revealed in the Russian, German and Chinese super-deep drill holes. The lunar model is perhaps driven by terrestrial analogy, where the lower continental crust is much more basic than the upper crust. However this difference is due to intra-crustal melting on the Earth that produced a granitic upper crust and a more basic residual lower crust. Such processes of intra-crustal differentiation seem unlikely to have operated on the Moon [29].

Table 2.4 *Major element composition of typical samples (lunar sample numbers in parentheses) from the lunar highlands (wt%). Data from Note 4*

Oxide	Anorthosite (15415)	Gabbroic anorthosite (68415)	Anorthositic gabbro (15455)	Norite (78235)	Impact breccia (14310)
SiO_2	44.1	45.5	44.5	49.5	47.2
TiO_2	0.02	0.32	0.39	0.16	1.24
Al_2O_3	35.5	28.6	26.0	20.9	20.1
FeO	0.23	4.25	5.77	5.05	8.38
MnO	–	0.06	–	0.08	0.11
MgO	0.09	4.38	8.05	11.8	7.87
CaO	19.7	16.4	14.9	11.7	12.3
Na_2O	0.34	0.41	0.25	0.35	0.63
K_2O	–	0.06	0.10	0.061	0.49
Cr_2O_3	–	0.10	0.06	0.23	0.18
P_2O_5	0.01	0.07	–	0.04	0.34
Σ	100.0	100.1	100.0	99.8	98.8

Although the bombardment of the lunar highlands has destroyed much original textural evidence, it has had little effect on their bulk chemistry, so enabling geochemists to unravel the evolution of the Moon. Moreover, some monomict breccias have low siderophile-element contents indicating little contamination from the meteoritic bombardment. These are considered to be "pristine" rocks that represent the original igneous components making up the highland crust [30, 31].

Three of these pristine constituents make up the lunar highland crust, namely, ferroan anorthosites that are the dominant component, KREEP and the Mg-suite. Analyses of typical crustal samples are given in Tables 2.4–2.6.

2.5.1 Anorthosites

The most abundant rock type is ferroan anorthosite that constitutes at least 80% of the highland crust. The pristine clasts in lunar meteorites are mostly ferroan anorthosites (Fig. 2.3).

Anorthosite is mainly (95%) composed of calcium-rich plagioclase (anorthite, An_{95} to An_{97}) and has a strong enrichment in the rare earth element, europium, typical values of Eu/Eu* being about 50 (Fig. 2.3) [32]. The lunar crust thus contains at least half of the europium in the Moon, resulting in a mantle depleted in that element. This was one of the keys that unlocked our understanding of the evolution of the Moon.

Low-Mg pyroxene is the next most abundant mineral, but mafic (Mg- and Fe-rich) minerals are merely minor constituents in this nearly monomineralic

Table 2.5 *Trace element composition of typical samples (lunar sample numbers in parentheses) from the lunar highlands (ppm). Data from Note 4*

Element	Anorthosite (15415)	Gabbroic anorthosite (68415)	Anorthositic gabbro (15455)	Norite (78235)	Impact breccia (14310)
I. Large cations					
Cs	0.02	–	0.12	0.064	0.54
Rb	0.22	1.7	1.1	0.92	12.8
K	151	650	830	510	4080
Ba	6.3	76	42	80	630
Sr	173	182	220	–	250
Pb	0.27	0.78	1.0	–	6.2
K/Rb	690	380	750	550	320
Rb/Sr	0.001	0.009	0.005	–	0.05
K/Ba	24	8.5	20	6.4	6.5
II. Rare earth elements					
La	0.12	7.3	3.0	–	56
Ce	0.28	18.3	6.7	9.2	144
Pr	–	–	–	–	17
Nd	0.18	11.3	3.73	–	87
Sm	0.046	3.09	0.88	5.4	24
Eu	0.81	1.11	1.67	1.49	2.15
Gd	0.05	3.78	0.90	1.03	28
Tb	–	–	0.14	–	5.1
Dy	0.044	4.18	0.84	–	33
Ho	–	–	0.17	2.26	6.5
Er	0.019	2.57	0.46	–	20
Yb	–	2.29	0.36	1.47	18
Lu	–	0.34	0.06	1.64	2.5
Y	–	22	4.8	0.24	175
La/Yb	3.4	3.0	3.0	–	3.1
Eu/Eu*	51.6	0.99	5.58	–	0.25
III. Large high valence cations					
U	0.0017	0.32	0.05	1.97	3.10
Th	0.0036	1.26	0.23	0.19	10.4
Zr	–	100	11	0.59	840
Hf	–	2.4	0.17	–	21
Nb	–	5.6	0.95	–	52
Th/U	2.1	3.94	4.6	–	3.4
Zr/Hf	–	42	65	3.1	40
Zr/Nb	–	18	11.5	–	16
IV. Ferromagnesian elements					
Cr	–	700	440	–	1250
V	–	20	16	–	36
Sc	–	8.2	–	–	20
Ni	1.0	180	12	–	270
Co	–	11	10	12	17

Table 2.5 (cont.)

Element	Anorthosite (15415)	Gabbroic anorthosite (68415)	Anorthositic gabbro (15455)	Norite (78235)	Impact breccia (14310)
Cu	–	12	1.3	–	5.0
Fe (%)	0.18	3.30	4.49	–	6.51
Mn	–	470	–	3.93	850
Zn	0.26	4.8	1.9	600	1.8
Mg (%)	0.05	2.54	4.86	1.5	4.75
Ga	–	2.0	2.6	7.12	3.2
Li	–	5	–	–	22

Table 2.6 *Volatile- and siderophile-element composition of typical samples (lunar sample numbers in parentheses) from the lunar highlands (ppb except where given as ppm). Data from Note 4*

Element	Anorthosite (15415)	Gabbroic anorthosite (68415)	Anorthositic gabbro (15455)	Norite (78235)	Impact breccia (14310)
Os	–	–	–	–	10.5
Ir	–	4.6	0.024	0.14	10.5
Re	0.0008	0.43	0.006	0.012	1.0
Ni (ppm)	1.0	180	12	12	270
Au	–	2.7	0.042	0.42	4.3
Sb	0.70	530	0.22	0.08	4
Ge	1.2	73	9.4	19	110
Se	0.23	98	8.3	7.5	120
Te	–	–	2.6	–	–
Ag	1.7	–	1.7	0.40	–
Zn (ppm)	0.26	4.8	1.9	1.5	1.8
In	0.18	–	0.05	–	30
Cd	0.57	2.8	1.0	2.9	8.4
Bi	0.1	–	0.14	0.05	–
Tl	0.09	–	0.054	0.02	–
Br	2.3	–	35	6	–

feldspathic rock. The anorthosites are typically coarsely crystalline with cumulate textures. Their age of 4460 Myr is taken to represent their crystallization age from the lunar magma ocean and the flotation of the feldspathic highland crust as "rockbergs". Alternatively, this date may represent an "isotopic closure age" during cooling of the crust. The volume of feldspar in the crust (that is nearly 9% of the Moon) means that at least half of the Eu in the Moon resides in the crust.

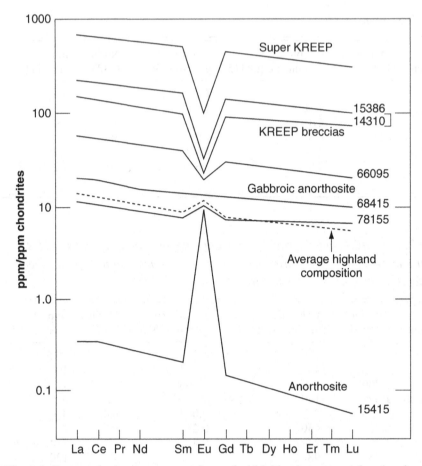

Fig. 2.3 Rare earth element patterns for typical highland crust samples showing enrichment in europium due to feldspar and depletion due to the presence of KREEP. Data from Ref. 4.

2.5.2 KREEP

In addition to europium, there is a strong enrichment in the highland crust of incompatible elements. This component has been termed KREEP and it dominates the trace-element chemistry of the highland crust. It originated as the residual melt, two percent or so, that contained the incompatible trace elements that were excluded from the major mineral phases (olivine, orthopyroxene, clinopyroxene, plagioclase, ilmenite) during crystallization of the magma ocean (Chapter 3). The KREEP component is enriched in potassium, rare earth elements, and phosphorus, hence the name, as well as a host of other incompatible elements. It is commonly applied as an adjective to refer to highland rocks that contain this characteristic trace-element signature [33]. This residual phase was the last to crystallize, at about 4.36 Gyr and apparently pervaded the crust, with which it was intimately mixed by cratering [34].

Extreme enrichment of the REE up to 1000 times CI (or solar nebular abundances) is known. This concentration of trace elements amounts to a significant part of the total lunar budget and must have been derived from a major portion of the Moon on geologically short timescales. Thus it provides strong evidence for the magma ocean hypothesis that involved large-scale lunar melting.

The KREEP component is not uniformly distributed across the lunar crust. It appears to be scarce on the lunar farside and, most significantly, in the deeply excavated South Pole–Aitken Basin. Although it was the residual melt that was sandwiched between the top of the cumulate pile and the bottom of the anorthositic crust during the crystallization of the magma ocean, KREEP does not seem to have formed a globe-encircling layer. Instead the residual melt appears to have been concentrated on the nearside as a lens or pod underlying Oceanus Procellarum.

The REE patterns in the lunar highland crust thus reflect the presence of two dominant components, anorthosite and KREEP. Except for the depletions in Eu due to KREEP or enrichments in Eu caused by feldspar, these REE patterns are closely parallel, while the other element ratios are quite uniform (Fig. 2.3). If multiple igneous events occurred during the formation of the highland crust, one might expect to see a wide diversity of REE patterns such as are observed in terrestrial igneous rocks. Thus the evidence for the frequently invoked "serial magmatism" is weak.

An enigmatic rock type, KREEP basalt, with only a few undisputed examples, is highly enriched in incompatible elements (K, REE, P) but has a more primitive Mg/(Mg + Fe) ratio. This combination of primitive and evolved components suggests that these samples were derived, like the members of the Mg-suite (see below), from different sources and they may ultimately be derived from melt pools, even if they are not direct impact melts. Probably the Apennine Bench formation is composed of KREEP basalt. This formation appears to have formed close in time to the excavation of the Imbrium Basin and may be related to that event.

However, the continuing bombardment of the crust may have produced large volumes of impact melt. As noted earlier, a terrestrial analog is the Sudbury impact basin in Canada, where the SIC formed from the differentiation of a pool of impact melt, entirely of crustal rocks, that originally was about 90 km in diameter and 3 km deep. Differentiation of this crustal melt resulted in a differentiated sequence resembling that of layered intrusions, except that the compositional layering is simple, as might be expected from the monotonic crystallization of a single melt pool. Most layered intrusions on Earth have received pulses of new magma that have upset such an orderly crystallization sequence. Also in contrast, the SIC contains only traces of olivine but is about 60% granophyre, reflecting its crustal origin [35]. Impacts on the Moon produce less melt than on the Earth, but the early lunar crust was hotter, perhaps contributing to melt production as on Venus.

So the possibility exists that much of the complexity read into the lunar crust by terrestrially oriented petrologists is the result of the crystallization of such impact melts, rather than reflecting serial magmatism or multiple intrusions, with such activity continuing for some hundreds of millions of years. Impact-produced melts offer a simpler solution to the dilemma of an energy source for these events. When mare basalts eventually erupted (Chapter 3) they involved only a trivial percentage (1%) of the volume of the Moon and so require minimal energy sources.

2.5.3 The Mg-suite

A minor constituent of the lunar crust, perhaps amounting to 10% in the areas sampled by the Apollo missions, is referred to as the Mg-suite. It appears to be concentrated on the nearside rather than distributed globally within the crust, so probably it is of local significance. Among its members are dunites, gabbroic anorthosites, norites, spinel troctolites and troctolites, none of which seem to be genetically related. The Mg-suite rocks range in age from about 4.44 Gyr down to about 4.2 Gyr; typical ages are 100–200 Myr younger than those of the ferroan anorthosites. They have higher Mg/(Mg + Fe) ratios compared to the ferroan anorthosites, but it is the Mg# that is high rather than the Mg content. If the Mg-suite rocks are indigenous igneous rocks, their parent magmas must have formed in regions of the lunar mantle much higher in Mg# than the source regions of the mare basalts [36].

The Mg-suite is petrographically distinct from the older ferroan anorthosites and does not appear to be related directly to the crystallization from the magma ocean. It is particularly enigmatic, as in contrast to normal igneous rocks, it contains two contradictory components. The high Mg/(Mg + Fe) ratios indicate a primitive source but the Mg-suite is also distinguished by high concentrations of incompatible elements, that are typical of highly differentiated igneous rocks. The source of this highly evolved component is KREEP and indeed the Mg-suite seems to be associated with KREEP-rich regions. It seems apparent that some kind of mixing of KREEP with some more primitive magma has occurred; but where this occurred and the source of the "primitive" Mg-rich component are both unclear.

Conventional theories based on terrestrial petrology propose that the Mg-suite originated as plutons that intruded the crust in separate igneous events. However, in common with other crustal samples, all Mg-suite rocks have parallel REE (less europium) patterns, a characteristic compatible with mixing, but not a pattern that is predicted to occur in separate igneous intrusions. Furthermore, it is of interest that the Mg-suite contains Mg-rich orthopyroxene, a mineral that is lacking in mare basalt, so that the primitive component of the Mg-suite originates in a location distinct from the source region of the mare basalts.

During crystallization of the magma ocean (Chapter 3), Mg-rich minerals (e.g. olivine and orthopyroxene) are among the first to crystallize and accumulate on the bottom of the magma chamber, at depths exceeding 400 km. It is sometimes suggested that massive overturning has occurred to bring these deep-seated cumulates within reach of the surface. However, the magma ocean had completed crystallization by 4400 Myr with only the KREEP component remaining liquid until about 4360 Myr. There is no obvious source of energy for remelting early refractory Mg-rich cumulates nor is the low Ni content of the Mg-suite compatible with this notion. So the idea that the primitive component of the Mg-suite comes from deep within the early cumulate zones in the mantle appears unfounded and the high concentrations of Ni in olivine predicted by that model are missing from the Mg-suite.

An alternative suggestion to overturning is that the Mg-suite is produced from impact-generated melts. The impacting bodies would have a high Mg# and acquire the KREEP component during impact melting and mixing. What seems clear is that normal petrological processes of magma generation and intrusion as understood from terrestrial experience did not produce the Mg-suite. No satisfactory explanation having appeared, the problem remains open.

2.5.4 Lunar highland terranes

Geochemical mapping carried out by the Clementine and Lunar Prospector missions has resulted in a significant advance in our understanding of the detailed structure of the lunar highland crust. Based on the FeO and Th abundances measured by these missions, the crust can be divided into three major terranes [37] (Fig. 2.4).

These are respectively,

(i) Feldspathic Highlands Terrane (FHT). This is the primordial feldspathic lunar crust that formed about 4460 ± 40 Myr ago by flotation from the magma ocean. It has been subdivided into FHT-A (anorthositic) and FHT-O, which is somewhat richer in FeO and poorer in Al_2O_3.

(ii) Procellarum KREEP Terrane (PKT). This was formed when the residual incompatible-element-rich liquid from the last stages of crystallization of the magma ocean was intruded into the primitive feldspathic crust [33]. Apparently the KREEP liquid collected mostly under the nearside during solidification of the magma ocean, thus accounting for the localized occurrence of KREEP. Little evidence of a significant KREEP component appears on the lunar farside crust or in the South Pole–Aitken Basin.

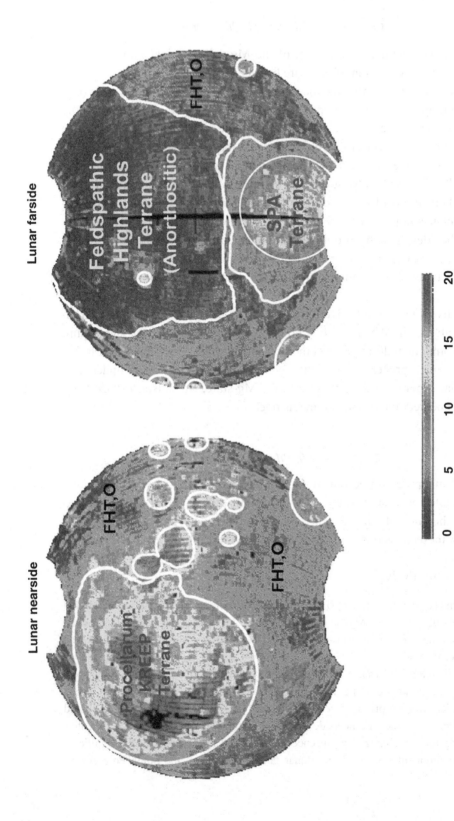

Fig. 2.4 Major lunar crustal terranes. The Procellarum KREEP Terrane (PKT) is on the nearside, with thorium concentrations greater than 3.5 ppm. The South Pole–Aitken Terrane (SPAT) has an outer region corresponding to basin ejecta. The Feldspathic Highlands Terrane (FHT) corresponds to the thickest part of the crust, concentrated on the lunar farside. FHT,O consists of those regions where basin ejecta or cryptomare obscure the feldspathic surface (courtesy Brad Jolliff, Washington University).

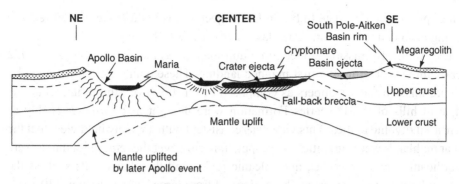

Fig. 2.5 The South Pole–Aitken Basin, 2500 km in diameter and 12 km deep on which are superimposed two later impact basins, all showing uplifted mantle.

(iii) South Pole–Aitken Terrane (SPAT). This constitutes the South Pole–Aitken Basin whose formation stripped off most of the upper crust over that region, while the ejecta blanket increased the thickness of the farside anorthositic crust. Its importance lies in the opportunity that it provides to examine the deep crust and possibly the uppermost mantle of the Moon. The interior of the South Pole–Aitken Basin, the deepest basin on the Moon, has a more mafic composition relative to the more feldspathic lunar highlands, but it is not clear that it has excavated into the lunar mantle. It would be of great interest to study this area in detail, as no excavated mantle samples have ever been identified in the returned Apollo samples. The South Pole–Aitken Basin, where most of the upper crust is missing, has been preserved for over 4.1 Gyr without significant isostatic compensation occurring. As this is the oldest and largest recognized lunar basin, the lack of compensation indicates that the crust and interior have been strong enough to support this structure for most of lunar history. There is little sign of the residual KREEP component in this location, despite the depth of excavation. This reinforces the model that the residual KREEP melt was not uniformly distributed.

Figure 2.5 shows the mantle uplift beneath the South Pole–Aitken Basin as well as that partially superimposed later uplift resulting from the excavation of the Apollo Basin. Significantly this mantle uplift has been sustained for over four billion years.

2.5.5 The Cayley Plains: a cautionary tale

The dark maria contrast strongly with the lighter lunar highlands. However, many light plains occur in the highlands that are similar in albedo to the rest of the highlands terrain. While the origin of the heavily cratered regions was clear, the occurrence of these widespread plains units was particularly puzzling to geologists mapping the Moon before the Apollo missions. These smooth areas sometimes occurred as crater fill (a particularly good example is the 80 km diameter crater Wargentin) but often as widely separated pools. These were of differing ages and some appeared to post date Imbrium Basin ejecta. Mostly these were interpreted as having been deposited as siliceous lavas or volcanic tuffs [38].

The Apollo 16 mission landed, on a unit mapped as the Cayley Plains, in 1972 nearly three years after the first manned landing. These plains were interpreted by the pre-mission lunar mappers as composed of ash flows or ignimbrite, while the adjacent hills of the Descartes Formation were thought to be a lunar analog of terrestrial rhyolite domes. This view was consistent with a widely held view that the lunar highlands were "granitic" in composition. However the returned samples were anorthositic impact breccias, not volcanic rocks, and the plains units such as the Cayley Formation were debris sheets derived from large basin collisions [39].

These sheets were probably fluidized and included significant impact melt as well as much locally derived material that was incorporated as the impact sheet ploughed across the lunar surface [40]. Thus although the volcanic explanation of the Cayley Plains was "eminently logical" [41], it was nevertheless erroneous, forming a cautionary tale to interpreters of planetary surfaces based on remote sensing.

2.5.6 *Tektites and the Moon*

Although it is now decisively established from isotopic and chemical evidence that tektites are derived from the surface of the Earth by meteoritic or asteroidal impact, the notion that tektites were derived from the Moon enjoyed considerable, even widespread support before the Apollo missions. Many workers supposed that tektites represented samples of the lunar highland crust. The notion that the lunar highlands were "granitic" derives partly from this controversy and so it is worthy of a brief comment here [42].

However, the debates that had raged, particularly in the 1960s, over a lunar versus a terrestrial origin, were settled in favor of the Earth by the first sample return from the Moon in 1969. Tektites are mostly derived on the Earth from melting of silica-rich sedimentary rocks during meteoritic, asteroidal or cometary impact. Their derivation from sedimentary rather than igneous rocks is clearly shown by their composition. Thus potassium shows an inverse correlation with SiO_2, the reverse of the positive correlation displayed in igneous rocks. This indeed was a key observation, as few supposed that there were sedimentary rocks on the Moon. Because the debate still surfaces occasionally, readers interested in these glassy objects will find a useful review of the evidence for a terrestrial origin in Koeberl (1986) [43].

Synopsis

The Moon is enriched in refractory elements, is "bone-dry" and depleted in volatile elements relative both to the Earth and to CI. It has a low density and is depleted in metallic Fe (2–3%), but has an overall FeO content of 13% compared to 8% for the terrestrial mantle. Both the origin of the low-density Moon and the distinctive character of high-density Mercury have resulted from collisions. The Moon is a one-plate body. The surface is sculpted by impacts of meteorites, asteroids and comets. Tectonic features are rare and are due mostly to subsidence of mare basalts.

The highland crust, unique in the Solar System, is about 50 km thick with 16 km relief. The relative stratigraphy based on craters and basins has been tied to absolute ages from the Apollo sample return. The impact-dominated topography is due mostly to a cataclysmic bombardment that lasted from 4 to 3.85 Gyr. The source of the projectiles remains conjectural but is possibly connected with the outward migration of the outer planets.

The highland crust forms the type example of a primary crust. Samples from the highland crust are mostly multiple breccias as a result of the bombardment. Impact melts are common. The highland crust has an average Al_2O_3 content around 30% and is dominated by anorthosite. It was pervaded by KREEP that was the residual liquid from the crystallization of the magma ocean and so is strongly enriched in incompatible elements.

Despite the petrological complexity, the chemical compositions of the rocks are little affected and reflect the dominance of feldspar and KREEP. Thus the REE patterns of highland samples are sub-parallel, except for enrichment or depletion in Eu due to mixtures of feldspar or KREEP. A younger Mg-suite of uncertain origin is a mixture of primitive and highly evolved components. Although often ascribed from terrestrial experience to late intrusions, the similarity of REE patterns to the other highland rocks and absence of energy sources except from impacts, makes this terrestrially based petrological analog less appealing. But the lack of stratigraphic controls makes further evaluation of the Mg-suite difficult.

Three crustal terranes have been distinguished, respectively FHT, PKT and SPAT. The KREEP component appears to have been ponded under the lunar nearside as the magma ocean crystallized (Chapter 3). Many smooth plains (notably the Cayley Plains), initially thought by many to be volcanic ash flows, are debris sheets from basin-forming impacts. Finally the lunar sample return convincingly settled a long-standing controversy by demonstrating that the tektites were not derived from the Moon.

Notes and references

1. Wilhelms, D. E. (1993) *To a Rocky Moon*, University of Arizona Press, p. 75.
2. Konopliv, A. S. *et al.* (1998) Improved gravity field of the Moon from Lunar Prospector. *Science* **281**, 1476–80. The density of the Moon is also intermediate between that of Europa ($d = 3.014$ g/cm^3) and Io, the innermost of the Galilean satellites of Jupiter, with a density of 3.529 g/cm^3. The other satellites in the Solar System are mostly ice–rock

mixtures and so are much less dense. But such comparisons of objects formed by different processes have little fundamental significance. See Chapter 14.

3. The very low concentrations of siderophile elements in mare basalts are consistent with the removal of these trace elements into a small core during melting but where this process occurred is a matter of debate. Iron meteorites form close to T_{zero} in asteroidal-size bodies and the planets were formed later from the accretion of such planetesimals. In the current model for the formation of the Moon by the impact of a Mars-sized body with the Earth, both bodies have previously differentiated into metallic cores and silicate mantles. Thus the pattern of siderophile-element depletion in the Moon may have a long and complex history, occurring not only in the impacting body (named Theia) from which most of the Moon is derived, but also in precursor planetesimals and finally from a possible "late veneer" following accretion. This late veneer appears to have produced the uniform but low ($0.008 \times CI$) concentrations of highly siderophile elements (Os, Ir, Ru, Pt, Pd, Re) in the terrestrial mantle. The abundances of these elements have been estimated to be present in the lunar mantle at 0.0002 times CI, 40 times lower than in the terrestrial mantle. As the Moon was undoubtedly present during the arrival of the late veneer, this is interpreted to indicate that the thick lunar crust must have shielded the mantle from the influx. Day, J. M. D. *et al.* (2007) Highly siderophile element constraints on accretion and differentiation of the Earth–Moon system. *Science* **315**, 217–19.

4. The primary data for lunar compositions are contained in the 45 volumes of Proceedings of the Lunar Science Conference published in yearly sets from 1970 to 1992. Thereafter the data have appeared mostly in *Geochimica et Cosmochimica Acta*, *Earth and Planetary Sciences* or in the *Journal of Geophysical Research*. Secondary sources include: Taylor, S. R. (1975) *Lunar Science: A Post-Apollo View*, Pergamon Press; Taylor, S. R. (1982) *Planetary Science: A Lunar Perspective*, LSI, that includes data for representative lunar samples; *Lunar Source Book* (eds. G. H. Heiken *et al.*, 1991), Cambridge University Press, that contains only statistical summaries rather than analyses of individual samples; Papike, J. J. *et al.* (1998) Lunar samples, in *Planetary Materials*, Reviews in Mineralogy and Geochemistry 36, ch. 5, pp. 5.1–5.234; and *New Views of the Moon* (eds. B. Jolliff *et al.*, 2006), Reviews in Mineralogy and Geochemistry 60. The values in the tables are adapted from Taylor, S. R. (1982) with later modifications. The principal changes for the major elements are higher Si and lower Mg values rather than the CI ratios that have been commonly assumed. These are based on geophysical evaluation of the lunar interior from seismic data by Kuskov, O. L. *et al.* (2002) Geochemical constraints on seismic properties of the lunar mantle. *PEPI* **134**, 175–89 and others. The lunar Mg# [molar (Mg/Mg + Fe)] is estimated to be about 0.80, significantly lower than that of the terrestrial mantle value of 0.89. Khan, A. *et al.* (2006) Are the Earth and the Moon compositionally alike? Inferences on lunar composition and implications for lunar origin and evolution from geophysical modeling. *JGR* **111**, doi: 10.1029/2005JE002608. The lunar samples are highly reduced so that ferric iron is absent. Trace element data have also been re-evaluated following the assessment of Taylor, S. R. *et al.* (2006) The Moon: A Taylor perspective. *GCA* **70**, 5904–18. The Moon is bone-dry and no indigenous H_2O has been detected at ppb levels. Polar ice, presumably derived from comets, was reported by the Clementine mission and supported by the discovery of hydrogen at the lunar poles by the Prospector mission. Nozette, S. *et al.* (2001) Integration of lunar polar remote-sensing data sets: Evidence for ice at the lunar south pole. *JGR* **106**, 23,253–66. However a reassessment of the data indicates that ice sheets in the regolith, if present, are only a few centimeters thick while the hydrogen may in fact be mostly elemental, derived from the solar wind. Campbell, B. A. *et al.* (2003) Radar mapping of the lunar poles. *Nature* **426**, 137–8; Campbell, B. A. *et al.* (2006) No

evidence for thick deposits of ice at the lunar poles. *Nature* **443**, 835–7. The detailed volatile-element budget of the Moon is difficult to ascertain. The best study was by Wolf and Anders (1980) who compared the distribution of volatile elements in terrestrial oceanic basalts with those in the low-Ti lunar basalts. Wolf, R. and Anders, E. (1980) Moon and Earth: Compositional differences inferred from siderophiles, volatiles and alkalis in basalts. *GCA* **44**, 2111–24. The moderately volatile elements are also depleted relative to the Earth. This depletion is an order of magnitude less than that observed for the more volatile elements discussed above. Manganese is another moderately volatile element that has the same abundance in the Earth and Moon (Fig. 2.1) although both bodies are depleted relative to CI. The low initial strontium isotopic values ($^{87}Sr/^{86}Sr$) in the Moon indicate that much loss of volatile elements occurred in early events close to T_{zero} (4567 Myr) when a general depletion of the gases and volatile elements occurred throughout the inner solar nebula. Further loss occurred during the formation of the Moon so that the Moon has suffered a double depletion of the volatile elements.

5. See the section on the origin of the Moon in Chapter 3. The Moon is enriched in refractory elements such as Ti, U, Al, and Ca relative both to the Earth and to CI. This conclusion is consistent with geophysical studies of the lunar interior, and with the early appearance of plagioclase (An_{95-97}) during crystallization of the magma ocean. The Galileo, Clementine and Prospector missions in the 1990s indicated that the thick highland crust is dominated by anorthosite. This means that the bulk lunar composition requires 6% Al_2O_3, compared with a value of about 3.6% for the terrestrial mantle. As the refractory elements, such as Al, Ca, Ti and the REE have not been fractionated from one another during the formation of the Earth and Moon and most meteorites, the high abundance of Al implies a similar enrichment in the Moon in the other refractory elements. The only example of relative fractionation among refractory elements occurs in the calcium–aluminium inclusions (CAI) that form a minor component in some meteorites and that originated in extreme conditions near the Sun (X-wind environment) in the early solar nebula. See Shu, F. H. *et al.* (1996) Toward an astrophysical theory of chondrites. *Science* **271**, 1545–52.

6. See Chapter 3 and Taylor, S. R. *et al.* (2006) The Moon: A Taylor perspective. *GCA* **70**, 5904–18.

7. However this similarity may be a consequence of the formation process itself rather than implying a genetic connection. See Pahlevan, K. and Stevenson, D. J. (2007) Equilibration in the aftermath of the lunar-forming giant impact. *EPSL* **262**, 438–49.

8. Khan, A. *et al.* (2000) A new seismic velocity model for the Moon from a Monte Carlo inversion of the Apollo lunar seismic data. *GRL* **27**, 1591–8.

9. Wieczorek, M. A. and Zuber, M. T. (2001) The composition and origin of the lunar crust: Constraints from central peaks and crustal thickness modeling. *GRL* **28**, 4023–6; Chenet, H. *et al.* (2006) Lateral variations of crustal thickness from the Apollo seismic data set. *EPSL* **243**, 1–14; Wieczorek, M. A. *et al.* (2006) in *New Views of the Moon* (eds. B. Jolliff *et al.*), Reviews in Minerology and Geochemistry 60, p. 273.

10. Smith, D. E. *et al.* (1997) Topography of the Moon from the Clementine LIDAR. *JGR* **102**, 1591–612.

11. e. g. Fielder, G. H. (1961) *Structure of the Moon's Surface*, Pergamon Press.

12. Wilhelms, D. E. (1987) *The Geologic History of the Moon*, US Geological Survey Professional Paper 1348.

13. Shoemaker, E. M. and Hackman, R. J. (1962) Stratigraphic basis for a lunar time scale, in *The Moon* (eds. Z. Kopal and Z. K. Mikhailov), Academic Press, pp. 289–300. For a general survey, see Stöffler, D. *et al.* (2006) Cratering history and lunar chronology, in *New Views of the Moon* (eds. B. Jolliff *et al.*), Reviews in Minerology and Geochemistry 60, pp. 519–96.

14. Two ages, 3850 Myr and 3770 Myr have been proposed for the age of the formation of the Imbrium Basin; the former, following Wilhelms (1987) is adopted here: Wilhelms, D. E. (1987) *The Geologic History of the Moon*, US Geological Survey Professional Paper 1348, but see Stöffler, D. *et al.* (2005) *New Views of the Moon* (eds. B. Jolliff *et al.*), Reviews in Mineralogy and Geochemistry 60, ch. 5, for the detailed arguments about the age of the Imbrium Basin.

15. Pike, R. J. (1980) *Geometric Interpretation of Lunar Craters*, US Geological Survey Professional Paper 1046-c, p. 17; Baldwin, R. (1949) *The Face of the Moon*, University of Chicago Press, Fig. 12.

16. Ejecta blankets incorporate much local material as they travel across the surface in a manner analogous to a base surge. Apart from the ejecta blankets, numerous blocks from large impacts travel with sufficient velocity to produce secondary craters. These must be carefully distinguished from primary craters, not a simple task, to avoid confusion in the dating of lunar surfaces by crater counting. Shock pressures up to 1000 kbar (100 GPa) cause a variety of effects from the development of planar features in minerals (>100 kbar) to whole-rock melting (500–1000 kbar). Above about 1500 kbar, the rocks are vaporized. Vapor masses of a few times projectile mass and melt masses about 100 times the projectile mass may be formed. Impact melts compose 30–50% of all samples returned from the lunar highlands.

17. Nemchin, A. A. *et al.* (2006) Oxygen isotopic signature of 4.4-3.9 Ga zircons as a monitor of differentiation processes on the Moon. *GCA* **70**, 1864–72.

18. Kring, D. A. and Cohen, B. A. (2002) Cataclysmic bombardment throughout the inner solar system. *JGR* **107**, doi: 1029/2001JE001529.

19. Hartmann, W. K. (2003) Megaregolith evolution and cratering cataclysm models – Lunar cataclysm as a misconception (28 years later). *MPS* **38**, 579–93.

20. Ryder, G. (1990) Lunar samples, lunar accretion and the early bombardment of the Moon. *EOS* **71**, 322–3; Ryder, G. *et al.* (2000) Heavy bombardment of the Earth at 3.85 Ga: The search for petrographic and chemical evidence, in *Origin of the Earth and Moon* (eds. R. Canup and K. Righter), University of Arizona Press, pp. 475–92; Norman, M. D. (2006) Identifying impact events within the lunar cataclysm from $^{40}Ar–^{39}Ar$ ages and compositions of Apollo 16 impact melt rocks. *GCA* **70**, 6032–49; Kring, D. A. and Cohen, B. A. (2002) Cataclysmic bombardment throughout the inner solar system. *JGR* **107**, doi: 1029/2001JE001529.

21. Bogard, D. D. and Garrison, D. H. (2003) $^{39}Ar–^{40}Ar$ ages of eucrites and the thermal history of asteroid 4 Vesta. *MPS* **38**, 669–710. This conclusion is reinforced by comparing the size distribution of craters on Mercury, the Moon, Mars and Venus. In turn, these are related to the size distribution of the impacting projectiles. Two populations appear from this study. The craters on younger (post 3.8 Gyr) surfaces (lunar maria, Mars northern plains, Mercury post-Caloris and Venus) match the size distribution of the near-Earth asteroids (NEAs) while the craters and basins formed during the cataclysm around 3.8–4.0 Gyr match the size distribution of the main-belt asteroids. These surfaces include the lunar highlands, the Mercury highlands and the heavily cratered southern highlands on Mars. These two populations have been labeled 1 (older) and 2 (younger). This is suggestive evidence that the Population 1 impactors come from the asteroid belt rather than from comets. Strom, R. G. *et al.* (2005) The origin of planetary impactors in the inner Solar System. *Science* **309**, 1847–50.

22. Gladman, B. J. *et al.* (1997) Dynamical lifetimes of objects injected into asteroid belt resonances. *Science* **277**, 197–201; Zappalà, V. *et al.* (1998) Asteroid showers on Earth after Family breakup events. *Icarus* **134**, 176–9.

23. Gomes, R. *et al.* (2005) Origin of the cataclysmic Late Heavy Bombardment period of the terrestrial planets. *Nature* **435**, 466–9.

24. Ryder, G. *et al.* (2000) Heavy bombardment of the Earth at 3.85 Ga: The search for petrographic and chemical evidence, in *Origin of the Earth and Moon* (eds. R. Canup and K. Righter), University of Arizona Press, pp. 475–92; Koeberl, C. (2006) The record of impact processes on the early Earth: A review of the first 2.5 billion years. *Processes on the Early Earth* (eds. W. U. Reimold and R. L. Gibson), GSA Special Paper 405, pp. 1–22.

25. Korotev, R. L. *et al.* (2003) Feldspathic lunar meteorites and their implications for compositional remote sensing of the lunar surface and the composition of the lunar crust. *GCA* **67**, 4895–923.

26. Taylor, S. R. (1973) Geochemistry of the lunar highlands. *The Moon* **7**, 181–95.

27. This age is based on the Sm–Nd isotopic array for mafic phases from the noritic ferroan anorthosites, including sample numbers 60025, 67215, 67016, and 62236. Norman, M. D. *et al.* (2003) Chronology, geochemistry, and petrology of a ferroan noritic anorthosite clast from Descartes breccia 67215: Clues to the age, origin, structure, and impact history of the lunar crust. *MPS* **38**, 645–61.

28. Wieczorek, M. A. and Zuber, M. T. (2001) The composition and origin of the lunar crust: Constraints from central peaks and crustal thickness modeling. *GRL* **28**, 4023–6; Wieczorek, M. A. and Phillips, R. J. (1997) The structure and compensation of the lunar highland crust. *JGR* **102**, 10,933–43.

29. One possible model relates the increase in thorium in the deep South Pole Basin to the crystallization of the magma ocean that could account for compositional changes with depth in the crust, although it more likely represents a random distribution of KREEP in which much of the lunar budget of thorium is concentrated. Wieczorek, M. A. and Zuber, M. T. (2001) The composition and origin of the lunar crust: Constraints from central peaks and crustal thickness modeling. *GRL* **28**, 4023–6.

30. Heiken, G. H. *et al.* (eds., 1991) *Lunar Source Book*, Cambridge University Press, ch. 6.

31. This concept is due to Warren, P. H. (1977) In quest of primary highlands rock. *LPSC* **VIII**, 988 (abs).

32. In Eu/Eu*, Eu is the measured concentration of the element; Eu* is the theoretical concentration when that element shows no enrichment or depletion relative to neighboring REE.

33. Taylor, S. R. and Jakes, P. (1974) The geochemical evolution of the Moon. *Proc. Lunar Sci. Conf.* **5**, 1287–305. For a general survey, see Shearer, C. K. *et al.* (2006) Thermal and magmatic evolution of the Moon, in *New Views of the Moon* (eds. B. Jolliff *et al.*), Reviews in Mineralogy and Geochemistry 60, pp. 365–518. We apologize for the use of the acronym KREEP, but the term is thoroughly entrenched.

34. An evaluation of the radiometric ages is given in Chapter 3 in the section on the magma ocean, 3.4.

35. Grieve, R. A. F. *et al.* (1991) The Sudbury Structure: Controversial or misunderstood? *JGR* **95**, 22,753–64; Stöffler, D. *et al.* (1994) The formation of the Sudbury Structure, in *Large Meteorite Impacts and Planetary Evolution* (eds. B. O. Dressler *et al.*), GSA Special Paper 293, pp. 303–18; Norman, M. D. (1994) Sudbury Igneous Complex: Impact melt or endogenous magma? Implications for lunar crustal evolution, in *Large Meteorite Impacts and Planetary Evolution* (eds. B. O. Dressler *et al.*), GSA Special Paper 293, pp. 331–41; Therriault, A. M. *et al.* (2002) The Sudbury Igneous Complex: A differentiated impact melt sheet. *Econ. Geol.* **97**, 1521–40. However, high water content in target rocks, not expected on the Moon, may inhibit production of impact melt. See Artemieva, N. (2007) Possible reasons of shock melt deficiency in the Bosumtwi drill cores. *MPS* **42**, 883–94.

36. Warren, P. H. (1986) Anorthosite assimilation and the origin of the Mg/Fe related bimodality of pristine lunar rocks. *JGR* **91**, D331–43.

37. Haskin, L A. (1998) The Imbrium impact event and the thorium distribution at the lunar highlands surface. *JGR* **103**, 1679–89; Haskin, L. A. *et al.* (2000) The materials of the lunar Procellarum KREEP Terrane; a synthesis of data from geomorphological mapping, remote sensing and sample analyses. *JGR* **105**, 20,403–15; Jolliff, B. L. *et al.* (2000) Major lunar crustal terranes, surface expressions and crust-mantle origins. *JGR* **105**, 4197–216.

38. Wilhelms, D. E. (1970) *Summary of Lunar Stratigraphy – Telescopic Observations*, US Geological Society Professional Paper 599F, p. 27.

39. Taylor, S. R. (1982) *Planetary Science: A Lunar Perspective*. LPI, p. 36.

40. A clear terrestrial example is provided by the ejecta blanket from the Ries Crater in Germany that crosses a geological boundary south of the crater. Much of this material, not present at the crater site, is torn up by the base surge and incorporated into the ejecta blanket: Hörz, F. and Banzholzer, G. S. (1980) *Lunar Highlands Crust*, LPI, p. 211.

41. Wilhelms, D. E. (1976) Mercurian volcanism questioned. *Icarus* **28**, 551. See also Chapter 4 on Mercury for a similar problem.

42. The largest "lunar granites" are tiny fragments (the largest is 1.8 g) that are derived by differentiation of mare basalts.

43. Koeberl, C. (1986) Geochemistry of tektites and impact glasses. *Ann. Rev. Earth Planet. Sci.* **14**, 323–50; Taylor, S. R and Koeberl, C. (1994) The origin of tektites. *Meteoritics* **29**, 739–44.

3

A secondary crust: the lunar maria

One knew that the Moon had a lower specific gravity than the Earth; one knew, too, that it was sister planet to the Earth and that it was unaccountable that it should be different in composition. The inference that it was hollowed out was clear as day

(H. G. Wells) [1]

3.1 The maria

The dark lunar maria form a type example of a secondary crust, derived by partial melting from the mantle during ongoing planetary evolution. These enormous plains cover 17% ($6.4 \times 10^6 \, \text{km}^2$) of the surface of the Moon and constitute the familiar dark areas that form the features of the "Man in the Moon" and various other imaginary figures. But despite their prominent visual appearance, the maria form only a thin veneer on the highland crust (Fig. 3.1).

The lavas are mostly less than 500 meters thick, reaching thicknesses of up to 4 km only in the centers of the circular maria such as Imbrium. They cover an area that is only a little more extensive than that of the submerged Ontong Java basaltic plateau, which lies northeast of the Solomon Islands in the Pacific Ocean. The total volume of the maria, about $1.8 \times 10^7 \, \text{km}^3$, is trivial compared to the anorthositic crust or the whole Moon. This compares with the total volume of the current terrestrial oceanic crust of $1.7 \times 10^9 \, \text{km}^3$, two orders of magnitude larger [2].

While we observe only the surface outcrop, some magma might not have reached the lunar surface but became trapped within the thick feldspathic crust. However, very few clasts of basaltic composition have been found in the samples from the highland crust, either from the Apollo missions or from the lunar meteorites. Similarly, basaltic clasts that might represent intrusions into the highland crust have not yet been identified in the returned samples. Some areas of mare basalts (so-called cryptomaria) are covered by ejecta blankets from multiring basins; their

Fig. 3.1 The relationship between the anorthositic lunar highlands and the basaltic lunar maria is well shown in this view of Mare Ingenii on the lunar farside. The large circular crater, filled with mare basalt, is Thomson Crater, 112 km diameter, in the northeast sector of Mare Ingenii, 370 km diameter, that is centered at 34° S, 165° E. The crater in the right foreground is Zelinsky Crater, 54 km diameter, excavated in the old highland crust. The stratigraphic sequence, from oldest to youngest, is (1) formation of white feldspathic highland crust, (2) excavation of Ingenii Basin, (3) formation of Thomson Crater, (4) formation of Zelinsky Crater, (5) flooding of Ingenii Basin and Thomson Crater with basaltic lava and (6) production of small impact craters on the smooth mare surface, including a chain of secondary craters. NASA AS15-87-11724.

presence is revealed by the haloes of dark basalt ejected from impact craters that have punched through the light-colored highland plains units of anorthositic composition into the underlying basalts. They amount to about 10% of mare basalt volume.

The individual lava flows may be up to 1200 km long but are thin, typically only 10–40 m thick. Dark deposits formed by "fire fountaining" during pyroclastic eruptions are found particularly on the southern border of Mare Serenitatis. Another interesting feature of the maria is the lack of shield volcanoes, a fact that stands in great contrast to the large basaltic domes on the Earth, Venus and particularly on Mars. This lack of volcanic constructional forms is reminiscent of plateau basalts on Earth. The maria are remarkably flat. Differences in elevation are less than 150 m over distances as much as 500 km. The slopes that lie between 1:500 to 1:200 are imperceptible [3].

The smoothness of the lunar plains made it very difficult to identify the nature of the maria in pre-Apollo times. This led, among the more sober models, to such interesting and ingenious suggestions that they consisted of sediments, dried-up lake beds or even asphalt [4]. The nature of the maria was finally confirmed as being due to basaltic lavas from the samples returned by the Apollo 11 mission.

The formation of the extensive smooth plains is probably due to several factors. These include a combination of high eruption rates and the low viscosity of the lunar lavas, which is about an order of magnitude less than that of terrestrial basalts, and is close to that of engine oil at room temperature [5].

Precise elevation data have revealed that the maria are not all at the same level [3]. This fact indicates that the mare basalts are derived from independent eruptions from diverse sources at differing depths in the interior that flooded into the nearest available depressions. The maria are mostly subcircular in form, because they often fill the giant multiring basins that were originally excavated by impact (Mare Imbrium is the size of Texas). Other maria are less regular in form. Oceanus Procellarum is the type example of an irregular mare, where the lavas have flooded widely over the highland crust. However, this mare may be filling parts of the controversial Procellarum Basin, 3200 km in diameter, that may have occupied most of the lunar nearside (see Chapter 2).

The scarcity of maria on the farside of the Moon is due to the greater crustal thickness. An exception is part of the area of the deep depression of the South Pole–Aitken Basin (2500 km in diameter, Fig. 2.5) on which is superimposed the Ingenii impact basin (650 km in diameter), now occupied in part by the lavas of Mare Ingenii (Fig. 3.1). However, most of the South Pole–Aitken Basin, that is much deeper than the nearside maria, is not flooded with lava so that this giant impact did not initiate widespread volcanism [6]. This is all clear evidence for localized mantle sources for the mare basalts unrelated to surface structures, rather than some Moon-wide melting of the lunar interior, that might have resulted in flooding of lava to a uniform level across the face of the Moon [7].

Another significant observation is that the mare basalts are heterogeneous in composition. Over 25 distinct varieties of mare basalts are known, even from our limited sampling. Even though some of this diversity is due to near-surface fractionation, the lunar mantle is heterogeneous and the basalts come from sources that vary significantly in mineralogy. This observation contrasts strongly with the much more uniform character of basalts erupted from the terrestrial mantle, that are dominated by mid-ocean ridge basalts (MORB) with lesser additions from oceanic island basalts (OIB, Chapter 8).

3.1.1 Mare basalt ages

The mare basalts are very old by terrestrial standards. The oldest ages for returned lunar mare basalts come from Apollo 14 breccias. These aluminous low-Ti basaltic clasts found in the breccias range in age from 3.9 to 4.3 Gyr. The oldest basalt sample returned from a maria is Apollo sample number 10003, one of the very first samples collected by the Apollo 11 crew. It is a low-K basalt from Mare Tranquilitatis with an age of 3.86 ± 0.03 Gyr. This gives a younger limit for the age of the Imbrium collision, as the lavas of Mare Tranquilitatis overlie the ejecta blanket from the Imbrium collision. In turn, the lavas on Mare Imbrium bury the ejecta blankets from the Serenitatis and Tranquilitatis basins, along with those from all the other basins, including those from Orientale, the youngest large basin. These stratigraphic truths show that the production of the lavas was unrelated to the formation of the basins.

An ilmenite basalt, sample number 12022, with an age of 3.08 ± 0.05 Gyr is the youngest dated mare basalt. Ilmenite basalts are examples of low-Ti basalts, are generally younger than high-Ti basalts, clearly sampling a different mantle source. Stratigraphically younger flows, some of which appear to embay obviously more recent craters, may be as young as 1 Gyr but are of very limited extent. The most voluminous period of eruption of lavas appears to have been between about 3.8 and 3.1 Gyr ago. This restricted period of lunar igneous activity is in great contrast to the continuing production of the analogous oceanic crust on the Earth.

Isotopic measurements show that the mantle source regions of the mare basalts formed at about 4400 Myr and this age has been taken to represent the solidification of much of the magma ocean.

3.2 Composition of the mare basalts

Mare basalts are much more heterogeneous in composition compared with the relative uniformity of the various species of MORB that form most of the oceanic crust on the Earth (Chapter 8). In comparison with terrestrial basalts, the silica contents of mare basalts are low (37–45%) and the lavas are more iron-rich (18–22% FeO).

Data for typical mare basalt samples are given in Tables 3.1–3.4. The lunar basalts are notably high in Ti, Cr, and Fe/Mg ratios and low in Ni, Al, Ca, Na, and K compared with terrestrial counterparts while they have an extraordinary range in TiO_2.

They mostly contain less than 12% Al_2O_3 and are depleted both in volatile and siderophile elements compared to terrestrial analogues. The ratio of volatile (e.g. K) to refractory elements (e.g. U) is low. Thus lunar K/U ratios average about 2500 compared to terrestrial values of about 10 000–12 500. Their REE patterns all display a characteristic depletion in europium (Fig. 3.2). This is one of the several pieces of evidence that the mare basalts come from a previously differentiated interior, rather than being melted from a primitive undifferentiated lunar composition.

Table 3.1 *Major element compositions of lunar mare basalts (wt%; lunar sample numbers in parentheses). Data from Chapter 2, Note 4*

Oxide	Green glass (15426)	Low-Ti Apollo 15 olivine (15016)	Low-Ti Apollo 12 ilmenite (12051)	High-Ti Apollo 17 (75055)	High-K Apollo 11 (10049)	High-Al Apollo 14 (14053)
SiO_2	45.2	44.1	45.3	40.6	41.0	46.4
TiO_2	0.38	2.28	4.68	10.8	11.3	2.64
Al_2O_3	7.5	8.38	9.95	9.67	9.5	13.6
FeO	20.0	22.7	20.2	18.0	18.7	16.8
MnO	0.26	0.32	0.28	0.29	0.25	0.26
MgO	17.5	11.3	7.01	7.05	7.03	8.48
CaO	8.5	9.27	11.4	12.4	11.0	11.2
Na_2O	0.13	0.27	0.29	0.43	0.51	–
K_2O	0.03	0.04	0.06	0.08	0.36	0.10
Cr_2O_3	0.53	0.85	0.31	0.27	0.32	–
Σ	100.0	99.5	99.5	99.5	100.0	99.5
Mg/(Mg+Fe)	0.59	0.47	0.38	0.41	0.40	0.48

The lunar basalts are highly reduced, with oxygen fugacities of 10^{-14} at 1100 °C or about a factor of 10^6 lower than those of terrestrial basalts at any given temperature. This is a characteristic of their source region. Ferric iron is effectively absent.

The differences in composition of the mare basalts are mostly due to source region heterogeneity, with only minor evidence for near-surface fractionation. Variations in the amount of partial melting from a uniform source, subsequent fractional crystallization or assimilation cannot account for the observed diversity. Some mare basalts are vesicular, evidence for a now vanished gas phase, usually thought to be carbon monoxide.

The basic classification of the lunar basalts is chemical, with finer subdivisions based on mineral composition. The basalts are divided into low-Ti, high-Ti, and high-Al basalts. The low-Ti basalts include VLT (very-low-Ti), as well as olivine, pigeonite and ilmenite basalts. The high-Ti basalts include high-K, low-K, and VHT (very-high-Ti) basalts. Probably there is a continuous variation in titanium contents by over a factor of 30 from less than 0.4% to a maximum of over 16%.

The major minerals in the mare basalts are pyroxene, olivine (Mg-rich), plagioclase (Ca-rich) and opaques, mainly ilmenite. All lunar minerals are anhydrous and the Moon is completely dry. Only a trivial amount of mare basalt has been extracted from the mantle and none has been returned on this one-plate body. Thus the bulk composition of the lunar mantle, unlike that of the Earth, has been little affected by the eruption or recycling of basalts.

Table 3.2 *Trace-element abundances in low-Ti lunar basalts (lunar sample numbers in parentheses; data in ppm except where indicated in %). Data from Chapter 2, Note 4*

Element	Green glass (15426)	Apollo 15 olivine (15016)	Apollo 12 ilmenite (12051)
I. Large cations			
Cs	–	0.034	0.04
Rb	0.58	0.81	0.91
K	170	330	500
Ba	18	61	74
Sr	41	91	148
K/Rb	290	410	550
Rb/Sr	0.01	0.01	0.01
K/Ba	11	5.4	6.8
II. Rare earth elements			
La	1.20	5.58	6.53
Ce	–	15.6	19.2
Nd	2.2	11.4	16.3
Sm	0.76	4.05	5.68
Eu	0.21	0.97	1.23
Gd	0.91	5.4	7.89
Tb	0.21	0.9	–
Dy	1.1	5.74	9.05
Ho	–	1.1	–
Er	0.8	3.1	5.57
Yb	0.93	2.62	5.46
Lu	0.14	0.32	–
Y	7.2	21	48
La/Yb	1.29	2.13	1.20
Eu/Eu*	0.76	0.63	0.56
III. Large valence cations			
U	0.02	0.12	0.26
Th	0.08	0.50	1.0
Zr	22	86	130
Hf	0.57	2.6	–
Nb	1.5	10	7
Th/U	4.0	4.2	3.85
Zr/Hf	38	33	–
Zr/Nb	15	8.6	18.3
IV. Ferromagnesian elements			
Cr	–	6400	–
V	165	200	100
Sc	38	39	58
Ni	153	–	6
Co	75	–	–
Cu	–	11	6
Fe (%)	16.4	17.7	15.8
Mn	2000	2450	2150
Zn	–	4	2
Mg (%)	10.0	6.78	4.21
V/Ni	1.08	–	17
Cr/V	–	32	–

Table 3.3 *Trace-element abundances in high-Ti and high-Al lunar basalts (lunar sample numbers in parentheses; data in ppm except where indicated in %). Data from Chapter 2, Note 4*

Element	High-Ti Apollo 17 (75055)	High-K Apollo 11 (10049)	High-Al Apollo 14 (14053)
I. Large cations			
Cs	0.019	–	0.09
Rb	0.58	6.2	2.20
K	660	3000	830
Ba	76	330	145
Sr	190	161	100
K/Rb	1140	484	380
Rb/Sr	0.003	0.04	0.02
K/Ba	8.7	9.1	5.7
II. Rare earth elements			
La	6.27	28.8	13.0
Ce	21.5	83	34.5
Nd	23.9	63	21.9
Sm	10.1	22.3	6.56
Eu	2.09	2.29	1.21
Gd	15.7	29.3	8.59
Tb	–	–	–
Dy	18.1	33.4	10.5
Ho	–	–	–
Er	10.7	30.9	6.51
Yb	9.79	30.2	6.0
Lu	–	–	0.90
Y	112	–	55
La/Yb	0.64	1.43	2.2
Eu/Eu*	0.51	0.27	0.49
III. Large valence cations			
U	0.14	0.81	0.59
Th	0.45	3.1	2.1
Zr	270	–	215
Hf	7.2	17	9.8
Nb	25	–	15.7
Th/U	3.3	3.8	3.55
Zr/Hf	38	–	22
Zr/Nb	11	–	13.7
IV. Ferromagnesian elements			
Cr	–	–	2860
V	–	–	135
Sc	83	81	55
Ni	2	–	14
Co	–	–	25
Cu	–	–	–
Fe (wt%)	14.1	14.6	13.1
Mn	2230	1930	2000
Zn	7	–	3.4
Mg (wt%)	4.23	4.22	5.09

Table 3.4 *Abundances of siderophile, volatile and chalcophile elements in lunar mare basalts (lunar sample numbers in parentheses; data in ppb except for Ni and Zn in ppm). Data from Chapter 2, Note 4*

Element	Low-Ti Apollo 15 olivine (15016)	Apollo 12 ilmenite (12051)	High-Ti Apollo 17 (75055)	High-Al Apollo 14 (14053)
Ir	0.018	0.09	0.035	0.017
Re	0.003	–	0.0031	0.007
Ni (ppm)	–	6	2	14
Au	0.025	0.0075	0.007	0.11
Sb	3.8	–	0.99	0.64
Ge	4.4	–	2.5	–
Se	114	200	120	140
Te	2.4	–	–	–
Ag	0.84	0.81	0.76	0.60
Zn (ppm)	1.1	0.53	1.5	2.1
In	0.34	1.2	0.57	–
Cd	2.1	–	1.9	–
Bi	0.36	0.53	–	0.29
Tl	0.32	0.37	0.37	1.4

3.2.1 The interior of the Moon

The crustal thickness at the Apollo 12 and 14 sites is 45 ± 5 km while the overall lunar crustal thickness is estimated to be 52 km (8.7% of lunar volume) [8]. The high value of the lunar moment of inertia (Chapter 2) requires a slight density increase toward the center, in addition to the presence of a low-density crust [9].

The lunar mantle, that resembles a gigantic layered intrusion, contrasts strongly with the terrestrial mantle (See Chapter 8). The structure of the lunar mantle has been difficult to evaluate on account of the problems encountered in interpreting the lunar seismograms in which the identification of arrival times is a matter of debate. Down to about 1100 km, the average P-wave velocity is 7.7 km/sec and the average S-wave velocity is 4.45 km/sec. Most models postulate a pyroxene-rich upper mantle that is distinct from an olivine-rich lower mantle beneath about a depth of 500–600 km. The main foci for moonquakes lie deep within the lower mantle at about 800–1 000 km. The outer 800 km has a very low seismic attenuation, indicative of a volatile-free rigid lithosphere. Solid-state convection is thus extremely unlikely in the upper mantle.

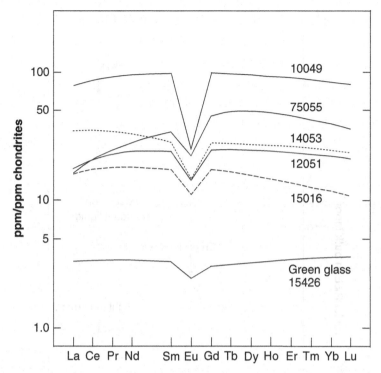

Fig. 3.2 Rare earth element patterns in lunar mare basalts showing the depletion in europium and LREE. Data from Chapter 2, Ref. 4.

Below about 800 km, both P- and S-waves become attenuated. P-waves are transmitted through the center of the Moon, but S-waves are missing, possibly suggesting the presence of a melt phase. It is unclear, however, whether the S-waves were not transmitted or were so highly attenuated that they were not recorded [10].

The lunar seismic data, mass and moment of inertia have been inverted to constrain the composition and mineralogy of the lunar mantle [11]. Both the lunar bulk composition and the Mg# of 0.80 are distinct from that of the terrestrial mantle (Mg# = 0.89). The seismic data are also consistent with whole-Moon melting that produces Mg-rich olivine cumulates as the first products of the magma ocean. The lower mantle has high Mg and lower Fe contents, compared to the lower Mg values and higher Fe for the upper mantle, again consistent with the mineralogy and with a break in composition between the upper and lower mantles. This evidence for a mineralogically zoned mantle thus agrees with magma ocean models and the petrological requirements for producing mare basalts from a mineralogically zoned interior [12].

Does the Moon have a core? Seismic and magnetic data are consistent with the presence of a small metallic core, but a sulfide core and even a metal-free core are

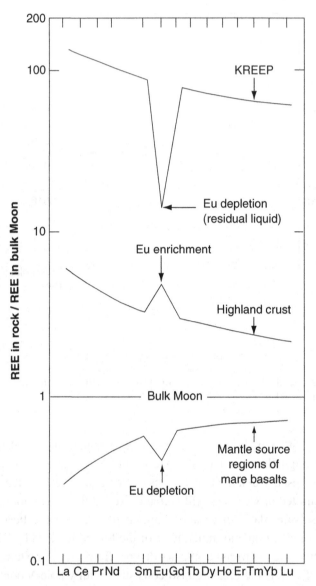

Fig 3.3 The abundances of the rare earth elements in the source regions of the mare basalts, the highland crust and KREEP, relative to the bulk Moon concentrations. These patterns result from the preferential entry of divalent europium (similar in radius to strontium) into plagioclase feldspar. This mineral floats to form the highland crust, and so depletes the interior in Eu. Mare basalts, subsequently erupted from this region deep within the Moon, bear the signature of this earlier depletion in Eu. The KREEP component is the final residue of the crystallization of the magma ocean. It is strongly depleted in Eu due to prior crystallization of plagioclase. It is enriched in the other rare earth elements and incompatible trace elements (e.g. K, U, Th, Ba, Rb, Cs, Zr, P) that are excluded from olivine, pyroxene and ilmenite during the crystallization of these major mineral phases of the magma ocean.

not ruled out and the geophysical data could just as well indicate a silicate core for the Moon. Data from the Lunar Prospector magnetometer have refined the estimate for the size of a metallic core to one with a radius of 340 km [13]. The revised moment of inertia value also places narrow limits, 220–370 km diameter for an Fe core and 350–590 km diameter for an FeS core [14]. An assessment of the lunar laser ranging data, coupled with Doppler tracking of the Lunar Prospector craft showed that the core was likely molten, about 350 km in diameter, with a density of 7.2 g/cm^3 and so consistent with a partly or wholly molten iron core [15].

3.3 Origin of the mare basalts

Mare basalts appear to have originated by partial melting, at temperatures of about 1200 °C, deep in the lunar interior, probably at depths between 200 and 400 km. Like its relatives elsewhere in the Solar System, the Moon has produced melts from its mantle long after its formation.

No basalts that could be construed as melts from a primitive lunar interior have ever been identified. In contrast, the lunar basalts were derived from the zones and piles of cumulate minerals developed, at various depths, during crystallization of the magma ocean. The isotopic systematics of the mare basalts indicate that the source region had crystallized by 4400 Myr. Partial melting occurred in these regions of diverse mineralogy some hundreds of millions of years later producing tiny amounts of various species of basalt. As melting in the interior to produce the mare basalts involved only a trivial volume of the Moon, this magmatism was probably due to the slow build-up of heat in isolated pockets within distinct mineralogical zones in the mantle due to the local presence of the heat-producing elements potassium, uranium and thorium. Lunar thermal history models have appeared frequently but none have gained general acceptance [16].

As the Moon is bone-dry and because the mantle had crystallized by 4400 Myr, convection in the solid interior seems unlikely to be very significant. Moreover, the dramatic overturns of the cumulate pile of Ti-rich minerals, that are often postulated, seem constrained by the very low abundance of titanium (0.3% TiO_2) in the bulk Moon. Hence the presence of a dense ilmenite phase that constituted a tiny fraction of the bulk Moon seems likely to have had only a trivial influence on a mostly solid mantle; therefore such overturning seems to require special conditions [17].

As noted previously, two other observations support a rigid mantle. The mountain arcs, such as the Apennines, have been supported for around 4000 Myr. The mascons, caused by mantle uplift under the basins, have likewise remained in place for similar periods, as beneath the South Pole–Aitken Basin. These structures place serious constraints on models that demand a less than rigid mantle.

The heterogeneity of the mantle is reinforced by the great chemical and isotopic variability of the mare basalts. This is unlike our experience on Earth where the principal magma being erupted at mid-ocean ridges, MORB, is much more uniform than the many species of mare basalts that have appeared from our obviously incomplete sampling of our satellite. The heterogeneity of lunar basalts, in contrast, argues strongly for a lack of convective mixing in the lunar mantle. Probably this was in great contrast to the crystallization of the terrestrial mantle that followed the melting induced by the giant lunar-forming impact.

3.3.1 An impact origin?

It seems necessary once again to refute the notion that the basaltic lavas that flood the lunar maria originated as a direct consequence of the massive impacts that formed the giant basins and their formation was "triggered by the impacts themselves" [18]. Such models typically envisage that the excavation of the crater depressurizes mantle material so that it melts in place. An alternative is that the mantle rebound under the crater (that contributes to the gravity high or mascon) might warp "isotherms at the lithosphere–asthenophere boundary, which may initiate convection in which adiabatic melting may occur" [19]. But it seems unlikely that the dry Moon possesses an asthenosphere in the terrestrial sense.

We have also noted that the mantle uplifts under the basins, like the mountain ranges that surround them, have remained supported by a cold mantle for most of geological time. It has also been pointed out for a long time from the stratigraphic evidence that the mare basalt, which fills the impact basins, is unrelated to the formation of the basin, and merely floods into the low-lying depressions much later [20]. This is clear both from the stratigraphy of the lunar surface, the ages of the basins and the ages of the lavas. Most of the basalts were erupted up to hundreds of millions of years after the impacts that excavated the basins.

During the impact, substantial volumes of the highland crust may be melted, but this impact melt would be anorthositic and should not be confused with the mantle-derived basaltic mare lavas, that differ in composition, age and source. The Imbrium Basin forms a particularly clear example. Several large craters, including Plato, Iridum and Archimedes, all visible through field glasses, were formed after the basin-forming collision but before the mare flooding. There are many large basins on the Moon, of which the huge South Pole–Aitken Basin is the prime example, in which there is little or no mare basalt fill.

However, in addition to these geological realities, geophysical calculations also demonstrate that the energies involved in producing the spectacular basins, although prodigious on human scales, are much too small to produce melting in the mantles of either the Moon or the Earth [21].

Nevertheless, the idea that the impacts were responsible for the eruption of the basalts that flood some of the basins seems to have achieved the status of a myth, analogous with the notion of an early world-encircling crust of granite on the Earth.

3.4 The magma ocean

The geochemical evidence is decisive that at least half and probably the whole Moon was molten at or shortly following its accretion. This stupendous mass of molten rock is referred to as the "magma ocean" and a very energetic mode of origin of the Moon, such as delivered by the giant impact hypothesis, is required to account for it.

The concept of the magma ocean has proven robust. Several decisive pieces of evidence require that the Moon was mostly melted at or shortly following accretion. The first is the presence of the 50–60 km thick anorthositic crust. This volume of feldspar with 36% Al_2O_3 requires the concentration in the crust of around 50% of the Al (and Eu) in the Moon.

The second is the high concentration of many incompatible elements at the lunar surface. The near-surface concentrations of aluminium, europium and of the many incompatible elements in KREEP constitute a significant proportion of the entire lunar budget for these elements. Their abundances are not related to element volatility but are due to crystal–liquid fractionation and so indicative of an internal origin. This near-surface concentration occurred shortly after the formation of the Moon so requiring a catastrophic event rather than the slow uniformitarian pro-cesses that resulted in the similar concentrations of incompatible elements in the continental crust of the Earth [22]. (Fig. 3.3)

The third point is that the mare basalts that are derived from the deep lunar interior are all depleted in europium. As discussed later, early crystallization of the plagio-clase that floated to form the highland crust, depleted the mantle, the later source of the mare basalts, in europium [23]. Finally all these events happened within a short timeframe so that the Moon formed a thick crust closely following its formation. The Earth in contrast has taken four Gyr to build a much smaller crust relative to planetary volume.

The crystallization of such a large body is difficult to constrain from our limited terrestrial experience. A reasonable scenario is that olivine and orthopyroxene were the first minerals to crystallize. These dense phases sank to form a zone of deep cumulate minerals. The aluminium and calcium content of the magma increased as a result of the removal of the phases dominated by iron and magnesium so that plagioclase crystallized. This phase was less dense than the bone-dry melt and floated, forming "rockbergs" that eventually coalesced to form the lunar highland crust, apparently around 4460 Myr [24]. (Fig. 3.4)

The first-order variation in thickness from nearside to farside is probably a relic of primordial convection currents in the magma ocean. Excavation by large basin impacts has subsequently imposed additional substantial variations in crustal thickness. The magma ocean was enriched in calcium and aluminium over typical terrestrial values, a conclusion from the Apollo samples that was reinforced by the Galileo, Clementine and Lunar Prospector data. The implication is that the Moon was enriched in these and other refractory elements compared to our estimates of the terrestrial mantle (Chapters 2, 8).

As crystallization of the magma ocean proceeded, those elements unable to enter the major mineral phases olivine, ortho- or clinopyroxene, plagioclase, and ilmenite were concentrated in a near-surface residual melt zone (KREEP), that eventually constituted about 2% of the volume of the magma ocean. Most of the incompatible trace elements such as K, REE (less Eu), P, Th, U, Ba, Zr, Hf, and Nb, were concentrated in the residual melt. Many of these elements are concentrated in near-surface rocks by factors of a thousand relative to bulk-Moon or primitive nebular values. This again constitutes evidence for large-scale lunar melting as these elements have to be sequestered from a large volume of the Moon on a short timescale [12]. The uniformity of elemental ratios in KREEP and the concordant model ages around 4360 Myr also argue for a single-stage process and so for a magma ocean rather than a disconnected series of igneous events.

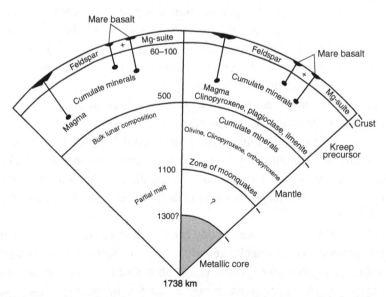

Fig. 3.4 Two alternatives for the internal structure of the Moon. On the left, only half the Moon was melted and differentiated and the deep interior has primitive lunar composition. Some partial melting has occurred due to the presence of K, U and Th. On the right, the Moon was totally melted and differentiated, forming a small metallic core.

Possibly a small iron core that incorporated the siderophile elements formed in the center of the Moon. The crystallization of the magma ocean was probably asymmetric, as shown by the variations in crustal thickness and the apparent concentration of the residual KREEP melt under the lunar nearside. Crystallization of the main mineral phases (olivine, pyroxenes, ilmenite) in the lunar mantle was complete, in most interpretations of the isotopic data, by about 4400 Myr and the final KREEP residuum was solid by about 4360 Myr [25].

The crystallization sequence portrayed here was far from peaceful. During all this time, the outer portions of the Moon were subjected to a continuing bombardment, which broke up and mixed the various components of the highland crust. Thus although the lunar crust is often thought to be formed simply from the accumulation of feldspar crystals floating on the magma ocean, the crust is much more complex as discussed in Chapter 2. During its formation, a massive bombardment continued, producing impact melts, mixing and pulverizing the crust and mixing in the KREEP component.

The solidification of the magma ocean was followed by the formation of the enigmatic Mg-suite. As discussed in Chapter 2, the origin of this suite is unclear as its unrelated members combine extremely primitive $Mg/(Mg + Fe)$ ratios with high concentrations of incompatible trace elements (KREEP) typical of highly evolved magmas. An origin by mixing (from impacts?) is indicated by the REE patterns, that parallel those of the other highland rocks. This observation raises serious problems for the popular notion of serial magmatism.

Probably some local overturning of the deeper cumulate pile may have occurred, but such events did not homogenize the interior. As we have seen, this region eventually produced a wide variety of mare basalt compositions.

3.4.1 The depletion in europium

The lunar basalts have one characteristic in common in that they all display a depletion in Eu that tells of the previous extraction of that element into the feldspar now residing in the highland crust. Many models had been proposed to explain the depletion of Eu in mare basalts and it is worth noting some of these here. These have included many exotic explanations that form a tribute to the human imagination [26]. The wide variety of suggestions include:

i. The whole Moon was depleted in europium.
ii. Europium was volatilized from lava lakes.
iii. Near-surface fractionation removed plagioclase (with Eu) from the lavas.
iv. Europium was selectively retained in the lunar interior during partial melting.
v. Europium was selectively retained in accessory minerals in the lunar interior.

vi. The magma ocean hypothesis is incorrect [27].

vii. Crystal-chemical oxidation–reduction effects are the cause [28].

viii. The lunar crust is not enriched in Eu [29].

ix. Discrepancies in anorthosite ages invalidate the magma ocean [30].

x. The only model that has survived is that the lunar mantle was depleted in divalent Eu by the early crystallization of feldspar that floated, in the dry magma, and formed the lunar crust. This depleted the lunar interior from which the mare basalts were subsequently erupted [12, 26].

The depletion in the source regions of the mare basalts, coupled with the presence of the thick feldspathic crust with its complementary enrichment in europium is the best evidence for a magma ocean that encompassed at least half and probably the whole Moon shortly following the accretion of the satellite. The recognition of the complementary patterns of chondrite-normalized REE contents of mare basalts and crustal materials was the key to understanding the basic relationship between the Moon's crust and mantle.

The lunar orange and green glasses that come from the deepest mantle sources (probably around 400 km) are among the most primitive lunar samples available. They are often claimed to be samples of a primitive lunar interior on the basis of their low Pb/U ratios. Nevertheless, they are all depleted in europium, and this diagnostic signature indicates derivation from a previously melted source. So in summary, the magma ocean hypothesis and the concept of the derivation of mare basalts from cumulate zones that crystallized from the ocean, remain robust.

The study of the lunar surface has also given us particularly clear evidence of the nature and formation of both primary and secondary planetary crusts. The Moon presents a special case of a primary crust composed essentially of anorthosite, which floated on an anhydrous magma ocean as a consequence of whole-Moon melting during accretion. There is no evidence of a similar feldspathic crust forming on the Earth. Nor did the Moon form a granitic crust, despite pre-Apollo expectations based on erroneous interpretations of the composition of tektites.

3.4.2 Depth of melting

There is some disagreement over how much of the Moon was melted. Most models invoke either melting half the Moon, to a depth of 500 km, or melting the entire Moon. The lack of surface features that might have resulted from the freezing of the magma ocean has sometimes been used to limit the depth of melting to a few hundred kilometers. But the crystallization of the magma ocean took place very quickly after accretion [31]. Surficial structures would not have survived the many cratering episodes, culminating in the late heavy bombardment 500 Myr later (The lobate scarps on Mercury formed at the same time as the heavy cratering episodes).

Basic constraints are the near-surface concentrations of Al in the thick lunar crust and of many incompatible elements in KREEP. Their abundances require derivation of Al, Eu and KREEP from at least half the Moon.

Isotopic balance calculations indicate that at least 85% of the Moon was melted while the giant impact model for lunar origin envisages complete melting. It is difficult enough to produce the highland crust and reasonable geochemical estimates for the bulk Moon if it is only half melted. Lesser degrees of melting exacerbate the problem. Our assessment is that in order to explain both the lunar enrichment in refractory elements and the curious depletion of the volatile elements, the entire Moon was fully melted during accretion.

3.5 Large-impact model for lunar origin

How did the Moon form? This has been one of the enduring questions, engaging some of the best intellectual resources of *Homo sapiens*. It is relevant here because the problem bears directly on lunar crustal development. However, none of the models proposed prior to the Apollo landings survived the confrontation with the data from the returned samples. This has provided an interesting demonstration of the power of a small amount of data over some centuries of theoretical reasoning [32].

One of the crucial features of the Earth–Moon system is that its angular momentum is anomalously high when compared to that of the other planets. This is a rock on which many hypotheses for the origin of the Moon have foundered. It was first pointed out by Al Cameron that this high angular momentum can only arise through the impact with the Earth of a Mars-sized body [33] and this model has survived [34]. Other notions that still occasionally surface form the Moon as a normal outcome of the accretion of the planet. Such models fail the crucial test that they should also produce a comparable satellite for our "twin planet", Venus.

In the single-impact model, both the Earth and the impactor (named Theia) are assumed to be differentiated, having already formed a metallic core and silicate mantle. As the event is perceived to be the last major impact on the Earth during its hierarchical accretion from a multitude of planetesimals, this is not an unreasonable assumption. Following the collision, the core of Theia fell into the Earth, leaving material from its rocky mantle spun out in a disk, so accounting for the small metal content and low density of the Moon. The collision was sufficiently energetic to form a molten Moon that is required by the geochemical evidence. Only a small fraction of material from the Earth's mantle is incorporated in the Moon. Instead, the simulations show that the Moon was mostly (85%) derived from material from the mantle of the impactor, Theia [34]. The high lunar FeO abundance (13%) relative to the terrestrial mantle (8%) is attributed to inheritance from the mantle of Theia.

The unusual enrichment of refractory elements and interesting depletion in volatile elements in the Moon was noted in Chapter 2. Despite this enrichment, the bulk composition of the Moon displays no fractionation among the REE or the other refractory lithophile elements (RLE), unlike that observed in the refractory inclusions (CAI) in meteorites. Evaporation of volatile elements can only enrich the refractory element levels by a few percent, not by a factor of three over the CI abundances, whereas condensation is open-ended. The uniform enrichment of the Moon in RLE is consistent with a one-step condensation from a vapor cloud below 2500 K (Fig. 3.5). So is this curious and probably unique chemistry a consequence of condensation rather than evaporation during the giant impact?

The impactor, Theia, is assumed to have had an initial depletion of volatile elements, imposed at T_{zero}, similar to that of the Earth. Additional loss appears to have occurred following the impact, as the Moon is much more depleted than the Earth (or Theia?). But curiously the depletion of the volatile elements in the Moon relative to the Earth (Theia?) appears to be uniform (Fig. 2.1). Volatilization during the impact should have resulted in a steeper decline directly related to volatility. Instead the volatile-element abundances appear to parallel those in the Earth. Recall that the initial depletion of inner Solar-System bodies (including the Earth and Theia?) in these elements relative to CI was due to a nebular-wide process that occurred at T_{zero} and was a direct function of the volatility of the elements.

Condensation from a high-temperature vapor between 2500 K and 1000 K would enrich the Moon in refractories but if accretion of the proto-lunar material cut off around 1000 K, material below that temperature would be mostly dispersed as a gas phase. What little was accreted to the Moon would then parallel its initial abundance in Theia (= Earth?). Such a loss of volatile elements both at T_{zero} and during the impact can explain their extreme depletion in the Moon and their apparently uniform concentration parallel to that in the Earth [35]. Clearly a fertile field for research.

The similarity in the oxygen isotopes between the Earth and Moon has often been used to argue for derivation of the Moon from the Earth. However, there are other possible explanations that might homogenize the oxygen isotopes, consistent again with high-temperature processes [36].

Just as the conditions of formation of the Moon were unique, so the composition of the Moon seems to be unique. Like that of Mercury, the composition of the Moon more likely reflects the extraordinary conditions of its formation. If this indeed is subsequently shown to be so, attention must then focus on the conditions surrounding the giant impact. But during the giant impact, the Earth was also melted with consequences for the geochemical evolution of the terrestrial mantle that have not yet been fully evaluated [37]. So much work remains to be done on the chemical and isotopic consequences of the giant impact.

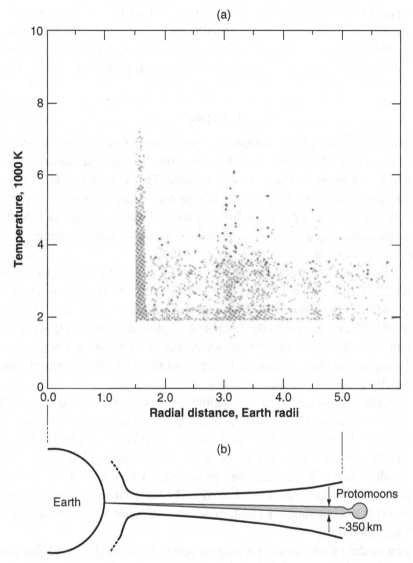

Fig. 3.5 (a) Temperature distribution from 1.5 to 6 Earth radii, 40 hours after the impact, of particles originally in the impactor. Note that the temperatures exceed 2500 K at 5 Earth radii [34]. (b) Schematic diagram of the Earth–Moon at the same scale as (a), showing that the Moon formed in a high-temperature environment in this scenario [35].

The age of this event and of the crystallization of the magma ocean remain debatable. There is a significant gap between reported Hf–W ages of 30 Myr after T_{zero} [38] and those Sm–Nd ages of 4460 Myr for the crystallization of the feldspathic highland crust, 4400 Myr age for the crystallization of the mare basalt source regions and 4360 Myr age for the crystallization of KREEP. These latter dates are conventionally taken to represent the crystallization of the crust, the solidification of the bulk of the magma ocean and of the last dregs [25, 39]. As the interpretation of the Hf–W isotopic data remains equivocal [38], we accept the Sm–Nd ages here.

Synopsis

The lunar maria form a type example of a secondary crust. The maria, although prominent visually, form only a thin veneer on the thick highland crust and constitute less than about 1% of crustal volume. The lavas are of low viscosity and form remarkably flat plains rather than central volcanic constructs. They fill various impact basins and craters to differing levels. Over 25 distinct varieties have been sampled, pointing to derivation from heterogeneous mantle sources, in contrast to the uniform composition of terrestrial MORB.

The lavas are ancient, mostly erupted between 3 and 4 Gyr. In comparison with terrestrial MORB, they have higher FeO (20%), lower Al_2O_3 and cover a much wider range in TiO_2 from 0.4 to 16% consistent with derivation from a heterogeneous mantle. They are strongly depleted in siderophile and volatile elements. Their REE patterns, with trivial exceptions, all display a depletion in europium. The limited geophysical data indicate a layered mantle. The Moon probably has a small metallic core.

Mare basalts originated by partial melting at depths of 200–400 km but their production involved only a tiny volume of the mantle of the Moon. No satisfactory thermal models have appeared and heat sources must have been highly localized (K, U and Th ?) to account for the spread in age, composition and the small volume of the basalts. Their origin is disconnected in time (by hundreds of Myr) from the excavation of the impact basins that they fill. The lunar interior has been cold and rigid for 4 Gyr and has supported both mascons underneath and the mountain arcs above the giant basins.

Melting of the Moon, forming a magma ocean, occurred during its formation. Crystallization of this vast mass resembled that of an enormous layered intrusion. A plagioclase-rich crust floated and a mineralogically zoned interior crystallized. The incompatible elements that could not enter olivine, pyroxene or ilmenite concentrated in the residual melt (KREEP) that ponded under the nearside and pervaded the feldspathic crust, mixed in by the bombardment. The europium depletion in the mare basalts was due to the prior removal of the element into the plagioclase now in the highland crust.

The origin of the Moon itself was due to a giant impact with a Mars-sized body (Theia) whose metallic core accreted to the Earth. Theia supplied most (85%) of the lunar material, despite a persistent myth of derivation from the terrestrial mantle.

The enrichment in refractory elements and depletion in volatile elements relative both to the Earth and CI, is attributed to condensation of the Moon between 2500 K and 1000 K from a volatile depleted impactor. Although this particular outcome was unique, major collisions were common during the hierarchical accretion of the terrestrial planets and might have led to other outcomes. However, the Moon remains unique in the Solar System and perhaps elsewhere.

Notes and references

1. Wells, H. G. (1901) *The First Men on the Moon*, George Newnes, p. 87.
2. Head, J. W. (1976) Lunar volcanism in space and time. *Rev. Geophys. Space Sci.* **14**, 265–300. A matter of terminology needs to be noted here. The dark basaltic lavas that fill the basins are the maria as in Mare Imbrium or Mare Ingenii (see Fig. 3.1). The circular basins, as in the Imbrium Basin, were formed much earlier by impact. Although sometimes thought to be the cause of the eruption of the mare basalts, the impacts have nothing to do with the later generation of the mare basalts.
3. Smith, D. E. *et al.* (1997) Topography of the Moon from the Clementine Lidar. *JGR* **102**, 1591–612.
4. This latter suggestion was indeed a possibility if the Moon were a large carbonaceous chondrite, as suggested by the low density of the Moon. Urey, H. C. (1952) *The Planets*, Yale University Press. This book directed scientific attention to the chemistry of the Solar System in contrast to the previous emphasis on its physical parameters. Urey emphasized the importance of the Moon, that he regarded as a primitive body. His views had a significant impact on the scientific community while his advocacy was a major factor in persuading NASA to explore the Moon.
5. Head, J. W. and Coffin, M. F. (1997) Large igneous provinces: A planetary perspective. *AGU Geophys. Monograph* **100**, 411–38; Hiesenger, H. and Head, J. W. (2006) New views of lunar geoscience, in *New Views of the Moon* (eds. B. Jolliff *et al.*), Reviews in Mineralogy and Geochemistry 60, p. 41. The presence of the thick anorthositic crust may also have acted as a density trap. If rising lavas were trapped below this zone of neutral buoyancy, overpressurization might have resulted in very large volumes of magma being rapidly erupted.
6. See discussion in Sections 3.3.1 and 10.6. See also Ivanov, B. A. and Melosh, H. J. (2003) Impacts do not initiate volcanic eruptions. *Geology* **31**, 869–72.
7. It has often been suggested that small dome-like features such as the Gruithuisen Domes that are about 20 km in diameter, might be lunar equivalents of terrestrial rhyolite domes. Early measurements suggested that they were too low in thorium to match the tiny mg mass fragments of lunar "granites" that formed as differentiates from mare basalts. Refined processing of remote sensing data suggests higher Th concentrations so that these small features, that include two other locations, might indeed be the long-sought felsic differentiates. The usual caveats apply to such conclusions based on remote sensing data and terrestrial analogs. Hagerty, J. J. *et al.* (2006) Refined thorium abundances for lunar red spots: Implications for evolved nonmare volcanism on the Moon. *JGR* **111**, doi: 10.1029/2005JE002592.
8. Wieczorek, M. A. *et al.* (2006) The constitution and structure of the lunar interior, in *New Views of the Moon* (eds. B. Jolliff *et al.*), Reviews in Mineralogy and Geochemistry 60, p. 273.
9. Konopliv, A. S. *et al.* (1998) Improved gravity field of the Moon from Lunar Prospector. *Science* **281**, 1476–80. A homogeneous sphere has a moment of inertia of 0.400. The

value for the Earth, with its dense metallic core that constitutes 32.5% of the mass of the Earth, is 0.3315. For a general survey, see Wieczorek, M. A. *et al.* (2006) The constitution and structure of the lunar interior, in *New Views of the Moon* (eds. B. Jolliff *et al.*), Reviews in Mineralogy and Geochemistry 60, pp. 221–364; Khan, A. and Mosegaard, K. (2001) New information on the deep lunar interior from an inversion of lunar free oscillation periods. *GRL* **28**, 1791–4.

10. Nakamura, Y. (2005) Farside deep moonquakes and the deep interior of the Moon. *JGR* **110**, doi: 10.1029/2004JE002332.

11. Kuskov, O. L. *et al.* (2002) Geochemical constraints on the seismic properties of the lunar mantle. *PEPI* **134**, 175–89; Khan, A. and Mosegaard, K. (2001) New information on the deep lunar interior from an inversion of lunar free oscillation periods. *GRL* **28**, 1791–4; Khan, A. *et al.* (2006) Are the Earth and the Moon compositionally alike? Inferences on lunar composition and implications for lunar origin and evolution from geophysical modeling. *JGR* **111**, doi: 10.1029/2005JE002608.

12. e.g. Taylor, S. R. and Jakes, P. (1974) The geochemical evolution of the Moon. *Proc. Lunar Sci. Conf.* 5, 1287–305.

13. Hood, L. L. *et al.* (1999) Initial measurement of the lunar induced magnetic dipole moment using Lunar Prospector magnetometer data. *GRL* **26**, 2327–30.

14. Konopliv, A. S. *et al.* (1998) Improved gravity field of the Moon from Lunar Prospector. *Science* **281**, 1476–80; Wieczorek, M. A. *et al.* (2006) The constitution and structure of the lunar interior, in *New Views of the Moon* (eds. B. Jolliff *et al.*), Reviews in Mineralogy and Geochemistry 60, pp. 303–26.

15. Khan, A. *et al.* (2004) Does the Moon possess a molten core? Probing the deep lunar interior using results from LLR and Lunar Prospector. *JGR* **109**, doi: 10.1029/2004JE002294.

16. New thermal models have appeared yearly for the past 35 years; this suggests some lack of basic understanding. Stevenson, D. J. and Bittker, S. S. (1990) Why existing terrestrial planet thermal history calculations should not be believed (and what to do about it). *LPSC* **XXI**, 1200–1.

17. e.g. Elkins-Tanton, L. T. *et al.* (2002) Re-examination of the lunar magma ocean cumulate overturn hypothesis: Melting or mixing is required. *EPSL* **196**, 239–49.

18. Elkins-Tanton, L. T. *et al.* (2004) Magmatic effects of the lunar late heavy bombardment. *EPSL* **222**, 17. This notion has even been extended to explain the origin of large igneous provinces on Earth. See discussion in Chapter 10. Green, D. H. (1972) Archean greenstone belts may include terrestrial equivalents of lunar maria? *EPSL* **15**, 263–70; Jones, A. P. *et al.* (2002) Impact induced melting and large igneous provinces. *EPSL* **202**, 551–61; Glikson, A. Y. (1999) Oceanic mega-impacts and crustal evolution. *Geology* **27**, 387–90; Abbott, D. H. and Isley, A. E. (2002) Extraterrestrial influences on mantle plume activity. *EPSL* **205**, 53–62; and Alt, D. *et al.* (1988) Terrestrial maria: The origins of large basaltic plateaux, hotspot tracks and spreading ridges. *Jr. Geol.* **96**, 647–62.

19. Elkins-Tanton, L. T. *et al.* (2004) Magmatic effects of the lunar late heavy bombardment. *EPSL* **222**, 17.

20. Urey, H. C. (1952) *The Planets*, Yale University Press.

21. Ivanov, B. A. and Melosh, H. J. (2003) Impacts did not initiate volcanic eruptions. *LPSC* **XXXIV**, Abs. 1338; (2003) Large scale impacts and triggered volcanism. Large Meteorite Impact Conference, Nördlingen, Germany, Abs. 4062; (2003) Impacts do not initiate volcanic eruptions. *Geology* **31**, 869–72.

22. Taylor, S. R. (1973) Chemical evidence for lunar melting and differentiation. *Nature* **245**, 203–5. For a general survey, see Shearer, C. K. *et al.* (2006) Thermal and magmatic

evolution of the Moon, in *New Views of the Moon* (eds. B. Jolliff *et al.*), Reviews in Mineralogy and Geochemistry 60, pp. 365–518.

23. A sole exception is some USSR Luna 24 VLT basalt samples where only milligram-size amounts were available, raising serious questions about how representative were the samples that were analysed. Neal, C. R. and Taylor, L. A. (1992) Petrogenesis of mare basalts: A record of lunar volcanism. *GCA* **56**, 2177–211.

24. Plagioclase crystals sink in wet terrestrial magmas. Walker, D. and Hays, J. F. (1977) Plagioclase flotation and lunar crust formation. *Geology* **5**, 425–8.

25. We are indebted to Marc Norman for the following comment: "The 4.36 Gyr age is a ^{143}Nd CHUR-model age on pristine KREEP basalt 15386 by Lugmair G. W. and Carlson R. W. (1978) The Sm–Nd history of KREEP. *Proc. Lunar Planet. Sci. Conf.* **9**, 689–704. It is supported by the ^{142}Nd work of Nyquist, L. *et al.* (1995) (^{146}Sm–^{142}Nd formation time for the lunar mantle. *GCA* **59**, 2817–2837) who determined a closure age of 4.32 Gyr for the lunar mantle. In contrast, the U–Pb and Rb–Sr systems give older model ages. Nyquist, L. and Shih, C.-Y. (1992) (The isotopic record of lunar volcanism. *GCA* **56**, 2213–34) calculated a model age of 4.42 Gyr. They note that the U–Pb model age assumes no change between T_{zero} (4.56) and 4.42, and that a younger model age can be calculated if some Pb isotopic evolution is allowed in that 140 million years. According to Nyquist and Shih (1995) 'The lunar mantle formation age of ~4.32 Ga obtained from the coupled $^{146/147}$Sm–$^{142/143}$Nd system is consistent with previous radiometric estimates of the time of lunar differentiation, agreeing most closely with Sm–Nd CHUR model ages of 4.36 ± 0.06 Gyr found by Lugmair and Carlson (1978) for KREEP basalts'. One possibility might be that the various model ages are dating different events, e.g. the Rb–Sr model age might be dating Rb–Sr fractionation in the magma ocean due to plagioclase crystallization. In this case the correspondence between the Sr model age (4.42 Gyr) and the timing of plagioclase crystallization (4.44 Gyr) might be no coincidence".

26. See Taylor, S. R. (1975) *Lunar Science: A Post-Apollo View*, Pergamon, pp. 154–9, for an extended discussion of some of the suggestions about the causes of the Eu depletion in mare basalts.

27. Walker, D. (1983) Lunar and terrestrial crust formation. *JGR* **88**, B17–25; Longhi, J. and Ashwal, L. D. (1984) A two-stage model for lunar anorthosites: An alternative to the magma ocean hypothesis. *LPSC* **XV**, 491–2.

28. Shearer, C. K. and Papike, J. J. (1989) Is plagioclase removal responsible for the negative Eu anomaly in the source regions of mare basalts? *GCA* **53**, 3331–6. But for a contrary evaluation of this model, see Brophy, J. G. and Basu, A. (1989) Europium anomalies in mare basalts as a consequence of mafic cumulate fractionation from an initial lunar magma. *Proc. Lunar Planet. Sci. Conf.* **20**, 25–30.

29. O'Hara, M. J. (2000) Flood basalts, basalt floods or topless Bushvelds. *Jr. Petrol.* **41**, 1545–651. O'Hara (2000) disputes the magma ocean hypothesis in which the crust formed by plagioclase flotation. The crucial point in his argument is that "there is no positive europium anomaly in the average lunar highland crust" (p. 1545, see also p. 1551). Thus he concludes that "there is no negative europium anomaly in the average mantle to be inherited by later mare basalts" (p. 1545, see also p. 1555). This notion is based on the data from the Apollo 16 site. Korotev, R. L. and Haskin, L. (1988) Europium mass balance in polymict samples and implications for plutonic rocks of the lunar crust. *GCA* **55**, 1795–813. However, later data from the farside highlands by the Clementine and Lunar Prospector missions have demonstrated the overwhelming abundance of anorthosite in the lunar highlands crust that demands a magma ocean, Eu enrichment and a high-alumina Moon. Lucey, P. G. *et al.* (1995) Abundance and

distribution of iron on the Moon. *Science* **268**, 1150–3; Pieters, C. M. and Tompkins, S. (1999) Tsiolkovsky crater: A window into crustal processes on the lunar farside. *JGR* **104**, 16,481–863.

30. Further controversies over the viability of the magma ocean hypothesis have arisen over the interpretation of the Nd isotopic data. But disturbance of the Nd isotopic system has occurred probably due to metamorphism resulting from impacts. Norman, M. D. *et al.* (2003) Chronology, geochemistry, and petrology of a ferroan noritic anorthosite clast from Descartes breccia 67215: Clues to the age, origin, structure, and impact history of the lunar crust. *MPS* **38**, 645–61. It must be recalled that the lunar highland crust grew in a turbulent environment in which many massive impacts caused subsequent melting and recrystallization.

31. Earlier attempts to impose geophysical constraints were based on the lack of features on the lunar surface that indicate either expansion or compression. This was used to limit the initial depth of melting to about 200 km. Solomon, S. C. and Chaiken, J. (1976) Thermal expansion and thermal stress in the terrestrial planets. *Proc. Lunar. Sci. Conf.* **7**, 3229–43. Later studies have negated these conclusions. Tonks, W. B. and Melosh, H. J. (1990) The physics of crystal settling and suspension in a turbulent magma ocean, in *The Origin of the Earth* (eds. H. Newsom, and J. H. Jones), Oxford University Press, pp. 151–74. Kirk and Stevenson comment that the conclusions of Solomon and Chaiken (1976) on the depth of the magma ocean are "insufficiently precise to infer a useful constraint" and that "the Moon could have been initially >50% molten (with the remainder relatively close to the solidus) and yet have experienced little volume change over the last 3.8 Gyr." Kirk, R. L. and Stevenson, D. J. (1989) The competition between thermal contraction and differentiation in the stress history of the Moon. *JGR* **94**, 12,133–44. See also Pritchard, M. E. and Stevenson, D. J. (2000) Thermal aspects of a lunar origin by giant impact, in *Origin of the Earth and Moon* (eds. R. Canup and K. Righter), University of Arizona Press, pp. 179–96; and Ranen, M. C. and Jacobsen, S. B. (2004) A deep lunar magma ocean based on neodymium, strontium and hafnium isotopic mass balance. *LPSC* **XXXV**, Abst. 1802. Don Wilhelms' (1998) personal communication comments that the limits of ± 1 km change to lunar radius used to limit melting to a depth of 200 km "apply only after the end of the massive bombardment at about 3.8 Gyr that is responsible for the major surface features of the lunar surface". The broken and fractured nature of the outer portions of the crust, coupled with the effects of the production of impact basins hundreds of kilometers in diameter around 4 Gyr are likely to erase earlier evidence.

32. See Brush, S. G. (1996) *Fruitful Encounters: The Origin of the Solar System and of the Moon from Chamberlin to Apollo*, Cambridge University Press, pp. 177–258.

33. Cameron, A. G. W. and Ward, W. R. (1976) The origin of the Moon (abs). *LPSC* **VII**, 120.

34. Stevenson, D. J. (1987) Origin of the Moon – The collision hypothesis. *Ann. Rev. Earth Planet Sci.* **15**, 271–315; Cameron, A. G. W. (2000) Higher-resolution simulations of the giant impact, in *Origin of the Earth and Moon* (eds. R. Canup and K. Righter), University of Arizona Press, pp. 133–44; Canup, R. M. (2004) Dynamics of lunar formation. *Ann. Rev. Astron. Astrophys.* **42**, 441–75.

35. See Taylor, S. R., G. J. and L. A. (2006) The Moon: A Taylor perspective. *GCA* **70**, 5904–18.

36. Both the chromium and oxygen isotopic compositions are identical in the Earth and Moon. However this may only mean that high-temperature conditions during the formation of the Moon from the impact with the Earth of a Mars-sized body may have homogenized the O and Cr isotopic compositions. See Pahlevan, K. and Stevenson, D. J. (2007) Equilibration in the aftermath of the lunar-forming giant impact. *EPSL* **262**, 438–49.

37. Stevenson, D. J. (1988) Greenhouses and magma oceans. *Nature* **336**, 587–8.
38. Jacobsen, S. B. (2005) The isotopic system and the origin of the Earth and Moon. *Ann. Rev. Earth Planet. Sci.* **33**, 531–70; Halliday, A. N. and Kleine, T. (2006) Meteorites and the timing, mechanisms and conditions of terrestrial planet accretion and early differentiation, in *Meteorites and the Early Solar System II* (eds. D. S. Lauretta and H. Y. McSween), University of Arizona Press, pp. 775–801; Kleine, T. *et al.* (2005) Hf–W chronometry and the age and early differentiation of the Moon. *Science* **310**, 1671–4 . But later data do not confirm the conclusions of this latter paper. See Touboul, M. *et al.* (2007) Late formation and prolonged differentiation of the Moon inferred from W isotopes in lunar metals. *Nature* **450**, 1206–9. See also Kramers, J. D. (2007) Hierarchical Earth accretion and the Hadean Eon. *J. Geol. Soc. London* **164**, p. 11, who comments that although a date of 30 Myr after T_{zero} (from the tungsten isotopic data) is often thought to represent the time of the giant impact, it may not be dating a unique event.
39. Norman, M. D. *et al.* (2003) Chronology, geochemistry, and petrology of a ferroan noritic anorthosite clast from Descartes breccia 67215: Clues to the age, origin, structure, and impact history of the lunar crust. *MPS* **38**, 645–61; Stöffler, D. *et al.* (2006) Cratering history and lunar chronology, in *New Views of the Moon* (eds. B. Jolliff *et al.*), Reviews in Mineralogy and Geochemistry 60, pp. 519–96.

4

Mercury

The laws of motion of Mercury are extremely complicated; they do not
take place exactly in the plane of the ecliptic

(Pierre-Simon Laplace) [1]

4.1 The planet

Mercury is a unique planet even by the standards of the Solar System. Like Mars, it
is a survivor of many similar bodies, possibly a dozen or more, that were formerly
present in the inner Solar System before the hierarchical assembly of the Earth and
Venus. Thus the investigation of this planet may yield important insights into the
early stages of the accretion of planets between the assembly of kilometer-size
objects and of the Earth-sized bodies (Chapter 1).

Because so little is known about this smallest planet, one might question the
wisdom of including here a separate chapter on Mercury. Although it was tempting
to include this discussion in a section under minor bodies (Chapter 13), we decided
on separate treatment. This conclusion was driven by the similarities between the
mercurian crust and that of the lunar highlands, so that this chapter follows on
naturally from those dealing with the Moon. It also provides some interesting
problems about primary or secondary crusts. Further, the Messenger mission is
already en route to this innermost planet, so that it is useful at this stage to
summarize more thoroughly our current understanding [2].

The geology of Mercury, a one-plate planet like Mars and Venus, shows some
similarities with that of the Moon. Accordingly, the current interpretation of the
geology and stratigraphy of Mercury depends strongly on the well-known lunar
analogs. This has important implications for extrapolating from the well-dated
sequence of events on the Moon to other bodies; the surface of Mercury is therefore
highly significant in the interpretation of the early history of the Solar System.
However, there are many differences as well, contrary to frequently expressed
opinions that Mercury is "just like the Moon".

However, we understand Mercury less well than any other planet in the Solar System, including the ice giants, Uranus and Neptune. The orbit and rotation of Mercury are exotic. Mercury slowly rotates with a period of 59 days while orbiting the Sun every 88 days so that the planet rotates three times during each two orbits of the Sun [2]. The current inclination of the orbit is, at 7°, the highest among the planets while the eccentricity is 0.2056, a value only a little lower than that of Pluto; the value has exceeded 0.325 during the past 4 Gyr [3].

The surface temperature on the sunlit side of this airless planet reaches 725 K, close to that of Venus (743 K). The night-time temperature falls to 90 K. The temperature in some deep craters at the poles may be as low as 60 K so that if water ice were trapped in such locations, it might be stable. Indeed Earth-based radar reflections may indicate the presence of some trapped ice [4].

4.1.1 The composition and internal structure of Mercury

There are few constraints on the bulk composition of Mercury, except that the high density (5.43 g/cm^3; uncompressed density = 5.0 g/cm^3) indicates the presence of a substantial metallic core that must account for 75–80% of the mass of the planet. This compares with a value of 32.5% for mass of the core of the Earth and perhaps 2 or 3 percent for a lunar core. The mercurian core is large so that the mantle constitutes a much thinner silicate shell than on the other terrestrial planets [5].

The combination of a silicate surface whose spectral reflectance is similar to the highland crust of the Moon, a high bulk density and a magnetic field all indicate that the planet is differentiated and contains either a large Fe or Fe–FeS core [6]. The planet possesses a dipolar magnetic field that is apparently driven by an internal dynamo. The field intensity is about 300 nT, about 100 times weaker than that of the Earth's field, with an axial tilt of 10° [7].

The presence of this field, if due to a dipole, requires the core of Mercury to be partially molten. As a core of pure iron is likely to have frozen long ago, cores containing FeS are the most likely alternative. Sulfur is a volatile element and so not expected to be present in some models for the origin of Mercury close to the Sun. However, the demonstration that the core is indeed partially molten [8] makes FeS the most viable candidate, providing yet another problem for the older condensation hypotheses.

Although several model compositions for Mercury have been advanced, none are better than guesses and have error limits so large that they provide no effective geochemical constraints nor provide any predictions that might be seriously tested [9].

4.2 Origin of Mercury

The origin of Mercury has strongly influenced the development of the crust, so that it is relevant to consider the problem at this stage. Various models have been proposed to account for the high density of Mercury. Early explanations involved a physical fractionation of iron from silicate based on magnetic properties or density. These essentially rely on Mercury accumulating from small particles in contrast to the planetesimal hypothesis in which hierarchical accretion of such bodies occurs from an assortment of planetesimals of varying sizes. The success of the planetesimal model has led to the abandonment of these earlier ideas.

The concept of a condensation sequence extending outwards from this dense iron-rich planet through the less dense terrestrial planets out to the gas giants has been a seductive trap for modelers. But the discovery of a partially liquid core, implying the presence of volatile sulfur as FeS makes such hypotheses less viable. One popular model has relied on a chemical separation in the nebula of iron from silicates due to volatility differences [10]. This model builds Mercury from small particles but relies on the small differences in volatility between iron and magnesium–iron silicates that were involved. It suffers from the major defect that the condensation temperature of iron (1360 K) is very close to that of other refractory elements, notably magnesium (1340 K) and silicon (1300 K) that form the other major mantle components of the terrestrial planets. Other models have invoked vaporization of the silicate shell, but have been largely abandoned [11].

The current explanation for the high iron/silicate ratio of Mercury lies within the framework of the planetesimal hypothesis. Bodies the size of Mercury formed within the first few million years after T_{zero} and slowly accreted into the Earth and Venus. Many collisions occurred. One such is postulated to be with a body of 0.20 Mercury masses, with an impact velocity of 20 km/sec that disrupted a proto-Mercury of about twice the present mass of the planet. This giant collision removed much of an earlier more massive silicate mantle, reducing it to centimeter-size fragments [12]. The tougher iron core survived in this scenario and re-accreted a thin coating of the dispersed silicate, the remainder either falling into the Sun or possibly being accreted onto proto-Venus or proto-Earth. Although this model also reassembles the mantle (but not the core) of Mercury from small particles, this is a different scenario from accreting the entire planet from dust ab initio, with fundamentally different consequences for the composition of the planet.

Thus the high density of Mercury is an accidental consequence of its origin, not part of a grand design of the Solar System, that at a first glance extends from the dense inner planets outwards to the low-density giant planets. Although collisions were ubiquitous during the accretion of the inner planets, this large-impact hypothesis that disrupted proto-Mercury might affect obliquities and rotation rates, or form a moon if the collisional parameters are right. The impact energy of the Moon-forming event on the Earth was only about 20% of that needed to disrupt the planet but smaller analogues of Mercury and the Moon probably suffered break-ups.

Among many questions waiting to be answered are whether there was a preferential recondensation of refractory elements following the collision, or whether the reassembly of the planet was isochemical? Our judgement is that a more refractory mantle resulted, perhaps condensing from a vapor phase, as seems to have occurred with the Moon [13]. The K/Th and/or K/U ratio, hopefully to be established by the Messenger mission, will inform us of the relative amounts of volatile and refractory elements present in the planet and so test this idea [14].

Presumably some of the volatile elements survived the traumatic collision as shown by the presence of the alkali ions sodium and potassium in the tenuous mercurian atmosphere, but we require new data before we can even begin to have an adequate understanding of the composition of the planet.

Once the planet was finally assembled, differentiation must have occurred rapidly. Several principal observations support this conclusion: (a) the high density of Mercury, (b) the presence of silicate material and a lunar-like composition and topography at the surface and (c) the absence of younger geological activity. These observations lead to the conclusion that the planet has a high iron content, which must be segregated into a core about 0.75–0.80 of mercurian radius, overlain by a thin silicate mantle and crust. The heavy cratering of the crust must have been early, by analogy with the Moon, and the crust must have been thick enough and cold enough to preserve the record of this bombardment from before 4.0 Gyr.

4.3 Surface structure

There are several major landscape forms on Mercury. These are the heavily cratered terrain, the intercrater plains, the Caloris basin and the smooth plains as well as a unique set of lobate fault scarps. These are curving fault scarps formed by compressional forces [15]. While the origin of the cratered terrain is not in doubt, controversy swirls around the origin of the plains, recalling similar disputes over the formation of the Cayley Plains in the lunar highlands. The plains units on Mercury are of two types. There are both older intercrater plains of Pre-Tolstojan age and younger "smooth" plains of Calorian age. Apart from the production of younger impact craters, there is no sign of more recent geological activity. Table 4.1 gives the stratigraphic sequence for Mercury.

Table 4.1 *Geological sequence and ages for Mercury**

System	Major units	Approx. age
Kuiperian	Crater materials	1.0 Gyr
Mansurian	Crater materials	3.0–3.5 Gyr
Calorian	Caloris group – mountain material, inter-montane plains, hummocky plains, lineated plains, secondary craters	3.9 Gyr
	Calorian plains – smooth plains	
	Crater materials	
	Small-basin material	
Tolstojan	Goya formation – Tolstoj Basin deposits	3.9–4.0 Gyr
	Plains material – smooth, lineated	
	Crater materials	
Pre-Tolstojan	Intercrater plains	>4.0 Gyr
	Multiring basins	
	Crater materials	

* Adapted from Neukum, G. *et al.* (2001) Geologic evolution and cratering history of Mercury. *Planet. Space Sci.* **49**, Table 3.

4.3.1 The heavily cratered terrain

The oldest portions of the surface of Mercury are heavily cratered and bear some resemblance to the lunar highlands in the numbers of craters and impact basins, over 50 of the latter being recognized [16].

However, although the planet, like the Moon, has also undergone an early heavy bombardment, there are few craters with diameters less than 50 km, the smaller ones having been covered during the formation of the intercrater plains [17]. An age around 4000 Myr is usually assigned to the mercurian surface on the basis of the lunar highlands analogue and the absence of similar terrains on the Earth. Although the similarity of the cratering record in the inner Solar System suggests a common population of impactors, attempts to establish a chronology based on cratering that would extend throughout the Solar System have not been successful.

4.3.2 The intercrater plains

The intercrater plains are of Pre-Tolstojan age. They occupy about 45% of the mercurian surface that was visible to Mariner 10 and so form a major geological unit (Fig. 4.1) [18].

About 15 old degraded impact basins, such as Eitoku-Milton, appear dimly. They predate the intercrater plains, suggesting that the deposition of the plains obliterated an older heavily cratered surface. The plains contain a large number of craters in the range of 5–16 km diameter. These appear to be mostly secondary craters resulting from the massive bombardment that formed the heavily cratered terrain. If the

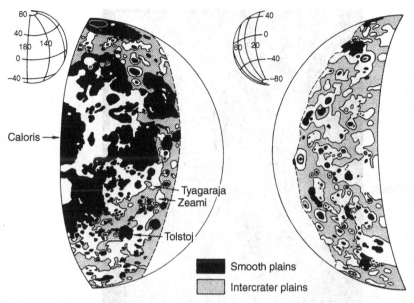

Fig. 4.1 The distribution of the intercrater and smooth plains units on Mercury (adapted from *Basaltic Volcanism on the Terrestrial Planets*, Pergamon Press, 1981, p. 766, Fig 5.5.3).

intercrater plains are volcanic in origin, Mercury must have been producing lavas of low viscosity, representative of a secondary crust. If so, they do not appear to have been iron-rich, as their albedo is twice as high as that of the lunar maria, while the infrared and microwave reflectance spectral data (discussed later) suggest very low surficial abundances of TiO_2 and FeO.

4.3.3 The Caloris Basin: a mercurian cataclysm?

In addition to the old degraded basins and craters that are partly covered by the intercrater plains, a number of fresher and apparently younger impact basins are recognized. The type example is the Caloris Basin, 1550 km in diameter, one of the latest and largest basins on Mercury (Fig. 4.2). This great feature recalls Mare Orientale on the Moon, but is somewhat larger (Orientale is 900 km in diameter) [19]. Antipodal to the Caloris Basin is a hilly terrain. This appears to have formed from seismic waves focused from the Caloris impact [20].

Crater-counting techniques give an estimate of 3.77–3.85 Gyr for the Caloris Basin [21] (Table 4.1). These ages, if valid, provide evidence that the massive cratering episodes recorded on the mercurian surface persisted for several hundred million years following accretion of the planet. However definitive answers to these questions require radiometric dating of returned samples, a woefully distant prospect. Other evidence suggests that the mercurian cratering record is consistent with

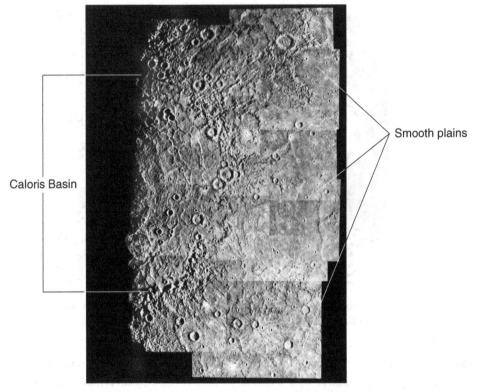

Caloris Basin

Smooth plains

Fig. 4.2 The Caloris Basin (1550 km diameter) showing smooth plains units. NASA PIA 03102.

the concept of a cataclysm. This conclusion is reached by comparing the distribution of crater sizes on Mercury and the Moon that is related to the relative sizes of the impactors [22]. The post-Caloris craters on Mercury match the distribution of sizes on the lunar maria. The sizes of the older basins and craters on Mercury match those on the lunar highlands. If correct, this would indicate that most of the observed cratering record is due to the cataclysm, and that older ages (> 4.0 Gyr) inferred on the surface of Mercury are incorrect [23].

4.3.4 The smooth plains

The smooth plains (Fig. 4.1) are younger than the heavily cratered terrain and mostly occur around the Caloris Basin. They appear to be about the same age as Caloris or perhaps a little younger. Craters are sparse and they closely resemble the Cayley Plains on the Moon. There is no difference in the albedo of the intercrater plains and the smooth plains, the albedo values being 0.15 ± 0.02 for both. In general, there seems to be little variation in albedo across the surface of Mercury.

This lack of strong contrast between the albedos of the heavily cratered terrain and both the plains units is in strong contrast to the differences obvious on the Moon between the rugged light feldspathic highlands and the smooth dark plains of basalt that form the maria. Very large wrinkle ridges occur on the floor of the Caloris Basin that could argue for the presence of lava [24]. However the planet is exposed to an intense solar particle flux, so that caution is warranted in making comparisons with the lunar surface.

4.4 The origin of the plains: a Cayley Plains analog?

The debate over the origin of both sets of the mercurian plains parallels that surrounding the origin of the lunar Cayley Plains and bears on the question of whether there are primary and secondary crusts on Mercury. Some authors regard the smooth plains units as analogues of the lunar maria, formed by fluid volcanic lava flows [25]. Others note that both types of the mercurian plains resemble the Cayley Plains that are common in the lunar highlands [26]. These lunar plains were identified, following the Apollo 16 mission, as debris sheets, impact melt sheets or fluidized ejecta flows from major basin-forming collisions, although they were often previously interpreted as ash flows or ignimbrites derived from the eruption of siliceous volcanics [27].

The large area of the plains units comprises the best evidence for a volcanic origin, as well as the apparent lack of source basins for an origin of basin ejecta. Thus the intercrater plains are extensive and there appears to be an apparent paucity of multiring basins, which could have supplied ejecta. Such ejecta on Mercury have a more restricted ballistic range than on the Moon, due to the higher gravity of Mercury.

Few visible morphological indicators of volcanism can be recognized on Mercury. Perhaps the most persuasive evidence for volcanism on Mercury is the presence of some darker albedo areas within craters (e.g. Tyagaraja) [28]. In addition, recalibration of the Mariner 10 photos indicates that distinctive geological units are present on Mercury [29]. These are interpreted to be consistent with volcanic deposits, thus suggesting that Mercury has had a complex geological history. There is also a doubtful spectral interpretation of basalt [30].

Although wrinkle ridges are present on the smooth plains, they show differences from those on the lunar maria and "in any case, ridges are not necessarily diagnostic of volcanic origin – they may merely indicate deformation of any coherent material" [31].

Likewise, in marked contrast to the lunar maria, there is apparently little differ-ence in age between that of the Caloris Basin and the smooth plains that surround it. There are no post-collision/pre-plains craters in Caloris that are analogous to

Archimedes, Iridum or Plato on the Moon that formed on the Imbrium Basin before it was flooded with mare basalt.

The lunar terrain which most closely resembles that of the mercurian intercrater plains is the so-called Pre-Imbrian "pitted terrain", southwest of the Nectaris Basin (35–65° S; 10–30° E). Their distinction from Cayley Plains is mainly in a higher density of craters. These lunar plains have been suggested to represent an early phase of volcanism as it is difficult to assign them to particular multiringed basins, if they are basin ejecta [32]. However, there seems to be no compelling evidence to interpret this lunar pitted terrain as other than the result of impact-produced debris from basin formation [31]. The lack of identifiable sources may be simply due to the destruction of old basins by new ones, a view consistent with an extended period of basin formation. This is the same problem that is encountered on Mercury. The absence of contrast in albedo between all these units is certainly not consistent at first sight with a basaltic composition for the mercurian plains.

Finally, there is the problem that all analogies with the Moon require caution, as that body has a distinctly lower bulk density and hence a different composition, as well as a different interior structure. Thus the mantle of Mercury may be very different than that of the Moon. Lavas erupted from it may not resemble lunar basalts in albedo. There is accordingly a complex situation with respect to inter-pretations of the mercurian photographs. The Moon provides the only viable analogy, but due to planetary density differences and probable mantle compositional differences, mercurian volcanism, if present, may be sufficiently different to make photogeological interpretations and comparisons difficult [33].

Further caution is needed in interpreting the mercurian surface. Space weathering processes that affect planetary surfaces (discussed in more detail in Chapter 13) are likely to be severe on Mercury. Solar radiation and the flux of solar wind particles and solar energetic particles are an order of magnitude more intense on Mercury than on the Moon, although subject to some shielding by the magnetic field of the planet. Their effects may render conclusions based on our current understanding of reflectance spectroscopy invalid for Mercury. So although we judge that the evidence for extensive basaltic style volcanism on Mercury is slender and the inter-crater plains seem more likely to be debris sheets from basin impacts, this whole debate may be yet another example of arguments at the limits of resolution, analogous to the martian canal problem.

4.4.1 Lobate scarps

Fault scarps characterized by lobate outlines (Fig. 4.3) are unique to Mercury. There are hundreds of them and they vary in length from 20 to 500 km and in height from a few hundred meters to about three kilometers. They appear to have a rather uniform

Fig. 4.3 The northern limb of Mercury, with a prominent east-facing lobate scarp extending for hundreds of km. The linear dimension along the base of the photo is 580 km. NASA P75-61-JPL-654-5-75.

distribution over the photographed portion of the planet. They are reverse thrust faults, formed due to compressive stresses, are demonstrably of tectonic origin and none require a volcanic explanation (e.g. as flow fronts) [34]. They occur mainly on the intercrater plains and on the older parts of the heavily cratered terrain. The scarps cut across the older craters, but younger craters are superimposed on the scarps and so they appear to be Pre-Tolstojan in age [35].

Thus the lobate scarps appear to have formed relatively early in mercurian history. From the analogy of the heavily cratered terrain with that of the lunar highland surface, the scarps must have formed at or before 4.0 Gyr. The decrease in the surface area associated with these scarps ranges from about 6×10^4 to $13 \times 10^4 \, km^2$ that corresponds to a decrease in mercurian radius by about 1 or 2 km [36]. The contraction was probably caused by the cooling and solidification of the mantle and crust. A suggested origin by tidal despinning has not been supported by other workers [37].

4.5 The crust of Mercury

Crustal thickness estimates are uncertain and are too dependent on estimates of composition, radioactive-element content and water content (likely in our opinion to be zero) to be useful [38].

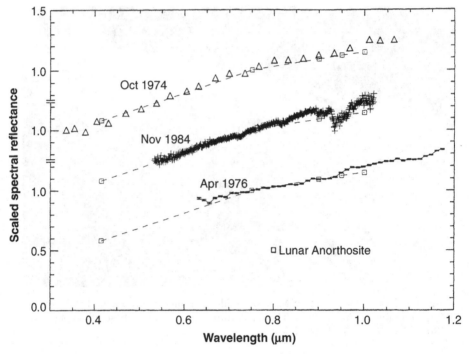

Fig. 4.4 Three reflectance spectra of the surface of Mercury compared with that of lunar anorthosite (> 90% plagioclase). Adapted from Blewett, D. T. *et al* (2002) Lunar pure anorthosite as a spectral analog for Mercury. *MPS* **37**, p. 1250, Fig. 7.

The composition of the enigmatic crust of Mercury is particularly intriguing. One must recall as a cautionary tale the pre-Apollo attempts to decipher the nature of the lunar crust. Information from the Ranger and Orbiter photographs of the Moon's surface did not lead to much understanding, leading to the comment that "a surface cannot be characterized by its portrait … the heightened resolution of the pictures did not resolve the arguments. The Moon remained inscrutable at all scales" [39].

The main evidence for the crustal composition of the planet comes from reflectance spectroscopy that extends from near infrared wavelengths around one micrometer (μm) up to microwave wavelengths of 20 cm [40]. The infrared spectrum for the mercurian surface is closely similar to that of the Apollo 16 highland soil laboratory spectrum (Fig. 4.4).

This similarity indicates that Mercury has a regolith and also that the surface is feldspathic, resembling that of the lunar highlands. However in contrast to the high-calcium feldspars (typically An_{90}) of the Moon, feldspar compositions on Mercury appear more sodic on average and more variable, ranging from An_{50} up to anorthite [41].

Further information is available from the spectra obtained at longer wavelengths. The mid-infrared and microwave data show that the mercurian regolith is 40% more transparent to microwave radiation than that of the lunar highlands and up to three times more transparent than the regolith on the lunar maria [42]. This indicates that the FeO and TiO_2 abundance on the surface of Mercury is significantly lower than that of the lunar highlands; the low abundance of FeO seems to be a generally agreed conclusion while the data suggest that the surface of Mercury "may be even lower in FeO than pure lunar anorthosite" [43].

Thus volcanic activity, expected to be enriched in Fe and Ti, may be minimal on Mercury. If the plains units are not lavas, it is possible that "volcanic heat piping has not played a major role in Mercury" [44]. So it is possible to entertain the hypothesis that Mercury formed a primary feldspathic crust in a similar manner to the Moon, either by flotation of plagioclase or because the entire mantle is refractory and secondary crusts did not develop.

So we are faced with the same dilemmas as with the pre-Apollo Moon and with the problems of interpretation of distant planetary surfaces, particularly one exposed to extreme solar particle fluxes [45]. Although low-albedo deposits might be produced by silicic volcanics, these seem an unlikely possibility on Mercury as their production on the Earth is the end product of much crustal recycling. If Mercury has indeed produced lavas subsequent to crust formation, they seem to be distinct from terrestrial or lunar basalts. So volcanic rocks, if they occur on Mercury, may well be unique. However, we note the cautions needed in interpreting reflectance spectra from this unique body.

4.5.1 Primary and secondary crusts on Mercury?

Can we distinguish between primary or secondary crusts on Mercury? There are two options from the currently limited amount of data. The apparent resemblance between the composition of the mercurian and the lunar highlands crust suggests that the entire crust might be primary. This interprets the various plains units to have originated as debris sheets from basin-forming impacts as we concluded above and that no subsequent melting occurred in the mercurian mantle.

The alternative view is that the heavily cratered terrain represents a primary crust, while the intercrater plains, that cover 45% of the surface photographed by Mariner 10, are lavas forming a conventional secondary crust, similar to the lunar example. Coincident with, or a little younger than the formation of the Caloris Basin, further volcanism produced the smooth plains. The various plains units apparently cover a larger percentage of the mercurian surface than the 17% of the lunar surface covered by maria. Thus if the plains are due to volcanism, extensive partial melting of the mercurian mantle occurred on a shorter timescale than on the smaller Moon. The formation of the plains units ceased around 4 Gyr about the time of the Caloris Basin-forming impact, in contrast to the lunar example where mare volcanism persisted for another billion years. These speculations will hopefully be resolved by Messenger. Because of the inferred violent history of this planet, compositional data from this mission may also test whether refractory lithophile elements have survived in chondritic proportions.

4.5.2 Atmosphere

The point about including this section in a book on planetary crusts is that the presence of sodium, potassium and calcium ions in the atmosphere of Mercury enables some inferences to be drawn about the nature of the crust.

Mercury indeed has a tenuous atmosphere, H, He, O, Ca, Na and K ions having been detected [46]. This is similar to the observations on the Moon of Na and K ions at elevations up to 1200 km above the lunar surface. As these ions are derived by sputtering from the lunar surface, the presence of Ca, Na and K ions is consistent with derivation from a feldspathic surface on Mercury.

Synopsis

Mercury remains an enigma. This one-plate planet is a survivor from an early stage of accretion in the solar nebula. It shows similarities and extreme differences from the Moon, both bodies being products of massive collisions. Its high density implies a large Fe and/or FeS core, partially molten, that constitutes 85% of planetary volume, but its mantle composition remains conjectural. The current model for its formation is that a Mars-sized proto-planet was disrupted by a head-on collision and that most of the silicate mantle failed to re-accrete. A critical question follows. Was this process isochemical or is the mantle refractory?

The oldest visible surface is heavily cratered, covered in part by intercrater plains and followed by a later cratering event of which the Caloris Basin is the major example. Smooth plains, possibly coincident with Caloris complete the stratigraphic sequence. Perhaps the whole sequence is due to the late heavy bombardment; crater-counting ages are equivocal. But perhaps we are simply looking at a primary crust modified by the late bombardment.

The origin of the plains, volcanic lavas or debris sheets from basin-forming collisions remains as controversial as the pre-Apollo 16 debates over the lunar Cayley Plains. Albedo differences across the planet are slight.

A series of scarps coincident with cratering are attributed to a one km contraction of the planet, coincident with the cratering. Reflectance spectra for the crust resemble low-Fe lunar anorthosites, albeit more Na-rich than the Ca-rich lunar feldspars. Both Fe and Ti contents are apparently very low or zero. If the plains are formed by lavas, they do not show the expected lower albedo of basalts. The surface is perhaps a primary crust, sculptured entirely by the late heavy bombardment. All is debatable because of the Cayley Plains analogue while interpretation of the spectra is clouded by possible intense space weathering. So the crustal composition, thickness and origin remain uncertain and many surprises doubtless await the Messenger mission.

Notes and references

1. Laplace, P. S. (1809) *The System of the World* (trans. J. Pond), Richard Phillips, Book 1, p. 69.
2. The standard reference remains *Mercury* (eds. F. Vilas *et al.*, 1988), University of Arizona Press. Detailed descriptions of the planet may be found in the report on the Mariner 10 mission given in The Planet Mercury: Mariner 10 mission. *JGR* **80**, 2341–514 (1975) and in the proceedings of conferences on Mercury given in *Icarus* **28**, 429–609 (1976); *PEPI* **15**, 113–314 (1977) and The Mercury 2001 Workshop in *MPS* **37**, 1165–283 (2002). See also *Planet. Space Sci.* **45**, 1–167 (1997) for information mainly on the magnetic field of Mercury. A special issue of *Planet. Space Sci.* **49**, 1395–632 (2001) provides some valuable summaries of the understanding of Mercury prior to the Messenger (NASA) and the projected Bepi Colombo (ESA) missions as well as details of the missions and spacecraft. A detailed summary of the scientific rationale for the Messenger mission is given by Solomon, S. C. *et al.* (2001) The Messenger mission to Mercury: scientific objectives and implementation. *Planet. Space Sci.* **49**, 1445–65. See also Balogh, A. and Giampieri, G. (2002) Mercury: The planet and its orbit. *Rep. Prog. Phys.* **65**, 529–60 and Taylor, G. J. and Scott, E. R. D. (2004) Mercury, in *Treatise on Geochemistry* (eds. H. D. Holland and K. K. Turekian), Elsevier, vol. 1, pp. 477–85. A more popular account is Strom, R. G. and Sprague, A. L. (2004) *Exploring Mercury: The Iron Planet*, Springer Proxis Books. The photographic coverage of the planet from the Mariner 10 mission was restricted to about 45%. Reliable image resolution over this area is about 1–2 km, comparable to Earth-based telescopic views of the Moon. About 25% of the surface was imaged at Sun angles that were low enough to enable terrain analysis. A very small number of high-resolution (100 m) views are available.
3. The planet was apparently captured into a 3/2 resonance as the most favored outcome of an orbital history that has included high eccentricities in the past. Correla, A. C. M. and Laskar, J. (2004) Mercury's capture into a 3/2 spin-orbit resonance as a result of its chaotic dynamics. *Nature* **429**, 848–50. Variations in the orbit of Mercury have often suggested that another planet, Vulcan, might lurk sunward of Mercury. This notion was prompted by the successful discovery of Neptune in the nineteenth century due to its effects on the orbit of Uranus. This led to a long search for Vulcan. Nothing was found and eventually the problems of the motion of Mercury were resolved by Einstein's general theory of relativity, rather than by the classical approach using Newtonian gravitational astronomy. Baum, R. and Sheehan, W. (1997) *In Search of Planet Vulcan: The Ghost in Newton's Clockwork Universe*, Plenum. For a detailed discussion, see Balogh, A. and Giampieri, G. (2002) Mercury: The planet and its orbit. *Rep. Prog. Phys.* **65**, 529–60.
4. Comets are a debatable source on account of their high impact velocities so close to the Sun. Paige, R. (1992) The thermal stability of water ice at the poles of Mercury. *Science* **258**, 643–6. This observation seems to be more securely based than the debatable presence of ice on the Moon but may only be recording trapped solar wind hydrogen.

Feldman, W. C. *et al.* (2001) Evidence for water ice near the lunar poles. *JGR* **106**, 23,231–52, but see Campbell, B. A. *et al.* (2003) Radar mapping of the lunar poles. *Nature* **426**, 137–8.

5. Spohn, T. *et al.* (2001) The interior structure of Mercury. *Planet. Space Sci.* **49**, 1561–70.

6. Hauck, S. A. *et al.* (2004) Internal and tectonic evolution of Mercury. *EPSL* **222**, 713–28. The identification of iron, rather than nickel, ruthenium, platinum or even lead as the dominant core-forming element in this and other planets is because, as a consequence of nucleosynthesis, iron is by far the most abundant metallic element. It is for similar reasons that magnesium–iron rather than titanium or barium silicates dominate the rocky mantles of the terrestrial planets.

7. Schubert, G. (1988) Mercury's thermal history and the generation of its magnetic field, in *Mercury* (eds. F. Vilas *et al.*), University of Arizona Press, pp. 429–60. See also the review in *Planet. Space Sci.* **45**, 1–167 (1997).

8. Older conjectures about the origin of the magnetism have been supplanted by evidence that the outer core is indeed partially molten, so allowing for dynamo models to generate the field. Margot, J. L. *et al.* (2007) Large longitude libration of Mercury reveals a molten core. *Science* **316**, 710–14. One explanation for the weakness of the field is that the dynamo is deep. Christensen, U. R. (2006) A deep dynamo generating Mercury's magnetic field. *Nature* **444**, 1056–8.

9. See Lodders, K. and Fegley, B. (1998) *The Planetary Scientist's Companion*, Oxford University Press, Table 4.4, p. 106.

10. e.g. Lewis, J. S. (1972) Metal-silicate fractionation in the Solar System. *EPSL* **15**, 286–90; (1988) Origin and composition of Mercury, in *Mercury* (eds. F. Vilas *et al.*), University of Arizona Press, pp. 651–69.

11. Fegley, B. and Cameron, A. G. W. (1987) A vaporization model for iron/silicate fractionation in the Mercury protoplanet. *EPSL* **82**, 207–22.

12. Benz, W. *et al.* (1988) Collisional stripping of Mercury's mantle. *Icarus* **74**, 516–28.

13. Taylor, S. R. *et al.* (2006) The Moon: A Taylor perspective. *GCA* **70**, 5904–18.

14. Planetary Th values are in general more readily established than the more mobile and analytically more challenging U, as experience on Mars shows. Boynton, W. V. *et al.* (2004) The Mars Odyssey gamma-ray spectrometer instrument suite. *Space Sci. Rev.* **110**, 37–83.

15. Neukum, G. *et al.* (2001) Geologic evolution and cratering history of Mercury. *Planet. Space Sci.* **49**, 1507–21.

16. Strom, R. G. and Neukum, G. (1988) The cratering record on Mercury and the origin of the impacting objects in *Mercury* (eds. F. Vilas *et al.*), University of Arizona Press, pp. 336–73; Neukum, G. *et al.* (2001) Geologic evolution and cratering history of Mercury. *Planet. Space Sci.* **49**, 1507–21.

17. Chapman, C. R. and McKinnon, W. B. (1986) Cratering on planetary satellites, in *Satellites* (eds. J. A. Burns and M. S. Matthews), University of Arizona Press, pp. 492–580. The craters are similar in general morphology to those on the Moon. All the features of a surface dominated by impacts such as ejecta blankets, terraces, central peaks and peak rings and secondary crater fields are present. Differences due to gravity, possible target strength, average impact velocity, and the source of the impacting objects lead to minor variations.

18. Trask, N. J. and Guest, J. E. (1975) Preliminary geologic terrain map of Mercury. *JGR* **80**, 2461–77. The oldest impact basin superimposed on the intercrater plains is Dostoevskij, followed by Tolstoj and Beethoven, that is about 625 km in diameter and so comparable in size to Mare Serenitatis on the Moon.

19. Although Caloris is often regarded as the largest basin on Mercury, the Borealis Basin in the north polar region is 1500 km in diameter, while other giants may lurk on the unseen side. McCauley, J. F. (1979) Orientale and Caloris. *PEPI* **15**, 220–50.

20. Hughes, G. H. *et al.* (1977) Global seismic effects of basin-forming impacts. *PEPI* **15**, 251–63; Wieczorek, M. A. and Zuber, M. T. (2001) A Serenitatis origin for the Imbrian grooves and South Pole–Aitken thorium anomaly. *JGR* **106**, 27,853–64; Spudis, P. D. and Guest, J. E. (1988) Stratigraphy and geologic history of Mercury, in *Mercury* (eds. F. Vilas *et al.*), University of Arizona Press, p. 139.

21. Neukum, G. *et al.* (2001) Geologic evolution and cratering history of Mercury. *Planet. Space Sci.* **49**, 1507–21.

22. Strom, R. G. *et al.* (2005) The origin of planetary impactors in the inner Solar System. *Science* **309**, 1847–50.

23. Strom, R. G. and Neukum, G. (1988) The cratering record on Mercury and the origin of the impacting objects, in *Mercury* (eds. F. Vilas *et al.*), University of Arizona Press, pp. 336–73.

24. Milkovich, S. M. *et al.* (2002) Identification of mercurian volcanism: Resolution effects and implications for Messenger. *MPS* **37**, 1209–22.

25. Spudis, P. D. and Guest, J. E. (1988) Stratigraphy and geologic history of Mercury, in *Mercury* (eds. F. Vilas *et al.*), University of Arizona Press, pp. 118–64.

26. Wilhelms, D. E. (1976) Mercurian volcanism questioned. *Icarus*, **28**, 551–8.

27. Neukum, G. *et al.* (2001) Geologic evolution and cratering history of Mercury. *Planet. Space Sci.* **49**, 1507–21; and see Ref. 26 above.

28. Hapke, B. *et al.* (1976) Photometric observations of Mercury from Mariner 10. *JGR* **80**, 2431–43.

29. Robinson, M. S. and Lucey, P. G. (1997) Recalibrated Mariner 10 color mosaics: Implications for mercurian volcanism. *Science* **275**, 197–200.

30. Sprague, A. L. *et al.* (1994) Mercury: Evidence for anorthosite and basalt from mid-infrared (7.3–13.5 μm) spectroscopy. *Icarus* **109**, 156–67.

31. Wilhelms, D. E. (1976) Mercurian volcanism questioned. *Icarus* **28**, 556.

32. Strom, R. G. (1977) Origin and relative age of lunar and Mercurian intercrater plains. *PEPI* **15**, 156–72.

33. Milkovich, S. M. *et al.* (2002) Identification of mercurian volcanism: Resolution effects and implications for Messenger. *MPS* **37**, 1209–22.

34. Watters, T. R. *et al.* (2001) Large-scale lobate scarps in the southern hemisphere of Mercury. *Planet. Space Sci.* **49**, 1523–30.

35. Melosh, H. J. and McKinnon, W. B. (1988) The tectonics of Mercury, in *Mercury* (eds. F. Vilas *et al.*), University of Arizona Press, pp. 374–400.

36. Watters, T. R. *et al.* (1998) Topography of lobate scarps on Mercury: New constraints of the planet's contraction. *Geology* **26**, 991–4.

37. Dzurisin, D. (1978) The tectonic and volcanic history of Mercury as inferred from studies of scarps, ridges, troughs, and other lineaments. *JGR* **83**, 4902.

38. Nimmo, F. (2002) Constraining the crustal thickness of Mercury from viscous topographic relaxation. *GRL* **29**, doi: 10.1029/2001Gl013883; Nimmo, F. and Watters, T. R. (2004) Depth of faulting on Mercury: Implications for heat flux and crustal and effective elastic thickness. *GRL* **31**, doi: 10.1029/2003GL018847.

39. Scott, R. F. (1977) Review of "*Lunar Soil Science*" (I. I. Cherkasov and V. V. Shvarev). *Earth Sci. Rev.* **13**, 379.

40. Vilas, F. (1988) Surface composition of Mercury from reflectance spectrophotometry in *Mercury* (eds. F. Vilas *et al.*), University of Arizona Press, pp. 59–76.

41. Sprague, A. L. and Roush, T. L. (1998) Comparison of laboratory emission spectra with Mercury telescopic data. *Icarus* **133**, 174–83; Sprague, A. L. *et al.* (1994) Mercury:

Evidence for anorthosite and basalt from mid-infrared (7.3–13.5 μm) spectroscopy. *Icarus* **109**, 156–67; Sprague, A. L. *et al.* (1997) Mercury's feldspar connection: Mid-IR measurements suggest plagioclase. *Adv. Space Res.* **19**, 1507–10; Sprague, A. L. *et al.* (2000) Mid-infrared (8.1–12.5 μm) imaging of Mercury. *Icarus* **147**, 421–32.

42. Mitchell, D. L. and De Pater, I. (1994) Microwave imaging of Mercury's thermal emission at wavelengths from 0.3 to 20.5 cm. *Icarus* **110**, 2–32. Jeanloz. R. *et al.* (1995) Evidence for a basalt-free surface on Mercury and implications for internal heat. *Science* **268**, 1455–7.

43. Blewett, D. T. *et al.* (2002) Lunar pure anorthosite as a spectral analog for Mercury. *MPS* **37**, 1255–68. The reflectance spectra are best matched by a "3:1 labradorite (An_{60}) to enstatite regolith" with about 1% FeO and no TiO_2 and "the abundances of both these oxides must be near zero for Mercury". Warell, J. and Blewett, D. T. (2004) Properties of the Hermean regolith: V. New optical reflectance spectra, comparison with lunar anorthosites, and mineralogical modeling. *Icarus* **168**, 257–76.

44. Jeanloz, R. *et al.* (1995) Evidence for a basalt-free surface on Mercury and implications for internal heat. *Science* **268**, 1456. It is sometimes suggested that the angrite class of differentiated basaltic meteorites might be derived from Mercury. e.g. Irving, A. J. *et al.* (2005) Unique angrite NWA 2999: The case for samples from Mercury. AGU Fall Meeting, San Francisco, CA, December 5–9, 2005, Abst. P51A-0898. Angrites have ages close to T_{zero} but are basic–ultrabasic rocks with FeO contents around 10–20% (e.g. Mittlefehldt, D. *et al.* (1998) in *Planetary Materials* (ed. J. J. Papike), Reviews of Mineralogy and Geochemistry 36, Table 40, p. 4-138), making it exceedingly unlikely that they are derived from the low-FeO mercurian surface.

45. Hapke, B. (1993) *Theory of Reflectance and Emittance Spectroscopy*, Cambridge University Press; Hapke, B. (2001) Space weathering from Mercury to the asteroid belt. *JGR* **106**, 10,039–73. However the magnetic field of Mercury, although weak, may shield the surface.

46. Sprague, A. L. *et al.* (1997) Distribution and abundance of sodium in Mercury's atmosphere, 1985–1988. *Icarus* **129**, 506–27; Killen, R. M. *et al.* (2005) The calcium exosphere of Mercury. *Icarus* **173**, 300–11.

5

Mars: early differentiation and planetary composition

> These lines of evidence indicate that Mars and the earth were formed of a mixture of iron and silicate phases which was nearly uniform, and that the earth has formed a core during geologic time and Mars has not.
>
> *(Harold Urey) [1]*

Mars is the only body in the Solar System, apart from the Earth and Moon, to which we devote more than one chapter in this enquiry of planetary crusts. Information now available for Mars, from telescopic observations, orbiters, landed missions and martian meteorites, is enormous and accordingly details now known about the martian crust are considerable [2]. An important finding is that Mars has been geologically active throughout its history and yet still retains a rock record dating back to about 4.5 Gyr, the age of the oldest martian meteorite. Sedimentary deposits are recognized both in some of the oldest and youngest exposed terrains. Accordingly, Mars may well have the most completely preserved geological record of any terrestrial planet.

For both the Moon and Earth, chapters are broken out according to crustal types (primary, secondary, tertiary) and age (Hadean, Archean, Post-Archean). For Mars, we take a different approach. Mars differentiated into core, mantle and crust very early in its history, likely due to magma ocean processes. Unlike Earth, there is unambiguous evidence for this early differentiation. The composition and subsequent evolution of the crust in turn has been greatly influenced by this early history. Accordingly, we begin with a discussion of these early events and follow with a broader evaluation of the composition and evolution of the martian crust in the next chapter.

5.1 The origin of Mars

Like all terrestrial planets, Mars was assembled from the rocky debris remaining behind in the inner nebula immediately after formation of the outer giant gas- and ice-rich planets. At that time, there was sufficient radial variation in nebular composition which, combined with narrow feeding zones (~0.1 AU), resulted in Mars accreting from a different population of planetesimals than Mercury, Venus or Earth. This is best reflected in the distinct oxygen-isotope composition compared to the Earth [3] and in the differences in their density (see below). The reason for the small size of Mars, no larger than perhaps a single or at most a few planetary embryos, is not entirely clear but it could be related to the chance circumstance of a planetary embryo that escaped being swept up into a larger terrestrial planet and acquired a stable orbit [4]. Possibly the Mars feeding zone, close to the asteroid belt, was depleted by the early formation of Jupiter.

Bodies up to the size of Mercury and Mars formed early and rapidly once the terrestrial planets began to accumulate. Within a few million years at most, they were common in the inner Solar System but most disappeared as they provided most of the mass to Earth and Venus in the form of the late-stage collisions (one of which gave rise to the Earth's Moon). Fairly large impacts occurred on Mars, for example, producing large basins and perhaps even the crustal hemispheric dichotomy [5]. On the other hand, unlike Mercury that is about a factor of two smaller, there is no evidence for any late-stage impacts that profoundly influenced the bulk composition of the planet.

5.1.1 A volatile-rich and oxidized planet

Later in this chapter we discuss in detail constraints on the bulk composition of Mars but a few comments are in order here. The planetesimals that accreted into terrestrial planets were basically dry and volatile-element depleted but there was also considerable variability both in the degree of depletion of moderately volatile elements, such as K and Rb, and the oxidation state (i.e., $Fe_{metal}/Fe_{(silicate+oxide)}$) within the inner nebula. The origin and causes of this variability were discussed in detail in Chapter 1. For Mars, there were several important consequences. The planet appears to have accumulated from a population of planetesimals that were enriched (by about a factor of two) in the moderately volatile elements, compared to their abundance in the Earth. Mars also appears to be relatively oxidized, such that the core is proportionately smaller than the Earth's while the primitive mantle appears to be enriched in iron, again by about a factor of two compared to the Earth [6]. In addition to influencing the subsequent evolution of the crust, these characteristics also influence the bulk physical properties, mineralogy and internal structure of Mars.

5.2 The interior of Mars

Mars has a mean radius of 3389.9 km leading to a volume of 1.6317×10^{11} km^3. The mass of Mars is well constrained at 6.4185×10^{23} kg leading to a compressed density of 3.934 g/cm^3 [7]. As discussed in Chapter 1, uncompressed density is an important constraint on planetary composition but is model dependent. For Mars, where the moment of inertia (MOI) factor is well known (see below) and pressure corrections are relatively modest, uncertainties in uncompressed density are due mainly to uncertainties in composition and thermal structure. The most recent estimate for the uncompressed density of Mars is 3.70 g/cm^3 or about 6.5% less than the Earth, consistent with Mars having accreted from a different population of planetesimals [8].

In the absence of seismic data, the internal structure of Mars is poorly constrained and largely model dependent [9]. One of the most important results of the Pathfinder mission was determination of a precise value for the polar MOI factor, I/MR^2, of 0.3650 ± 0.0012 [10]. This MOI factor is between those of the Earth and Moon. This suggests that Mars is largely differentiated with a substantial iron-rich metallic core and silicate mantle and crust but with a primitive mantle that is of higher density and/or proportionately larger than that of Earth.

Several workers have attempted to quantify the internal structure using MOI and other geophysical data and experimentally determined phase relations but uncertainties in crustal thickness, core/mantle/crust compositions and thermal structure lead to considerable scatter in the estimates. Various controversies exist and there are enough uncertainties in too many of the variables to permit any unique answer. For example, assuming a CI bulk planetary composition for refractory elements, the Fe/Si ratios (~1.3–1.6) derived from mass balance among the various proposed core and primitive mantle compositions (see below) cannot be reconciled the with the CI value of 1.71 [11]. A low Fe/Si ratio is consistent with the low uncompressed density for the bulk planet, compared to Earth, and comports with the smaller Mars not being of CI bulk composition for this elemental ratio. On the other hand, it is also possible that Mars has a roughly CI bulk Fe/Si ratio but that core, primitive mantle and crustal composition models are in error; in which case the low uncompressed density would be explained exclusively by the volatile-element rich and more oxidized nature of the planet.

Figure 5.1 provides the current best estimate of the internal structure of Mars [12] and below we discuss some of the uncertainties and caveats in adopting this or any other model.

5.2.1 Core

The physical state, composition and size of the core all are matters of debate. Several lines of evidence indicate that the core is at least partially and perhaps fully liquid [13]. Most workers consider the core to be composed of iron–nickel–cobalt metal with about 14–15% sulfur [14]. An important line of evidence in support of a sulfur-rich

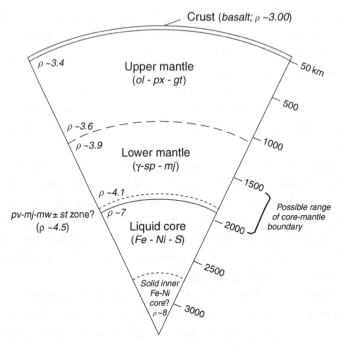

Fig. 5.1 Model for the internal structure of Mars [12]. Note that the uncertainty in the radius of the core–mantle boundary completely overlaps the perovskite-bearing zone of the lower mantle. Average densities (ρ) of the various layers are given in g/cm^3. Abbreviations as follows: ol – olivine; px – pyroxene; gt – garnet; γ-sp – gamma spinel; mj – majorite; pv – perovskite; mw – magnesiowüstite; st – stishovite.

core is depletion of chalcophile elements in martian meteorites, that suggests their partition into a sulfur-bearing core [6]. The presence of about 14–15% sulfur also allows for several hundred degrees depression of the melting point, again consistent with at least the outer part of the core being liquid [13].

Because of the uncertainties in the composition and physical state of the core and mantle, the radius of the core can only be constrained to be about 1600 ± 200 km, or about half of the radius of the planet [15]. A core of this size and composition represents roughly 20–25% of the mass of the planet. Note that this uncertainty in the size creates additional problems for evaluating the constitution of the lowermost martian mantle (see below).

5.2.2 Mantle

Bertka and Fei [16] evaluated the phase relationships for an iron-rich martian primitive mantle that has a bulk composition similar to that proposed by Heinrich Wänke and co-workers (Note 6 and see below). In this model, the mantle can be divided into either two or three regions (Fig. 5.2). The upper mantle consists

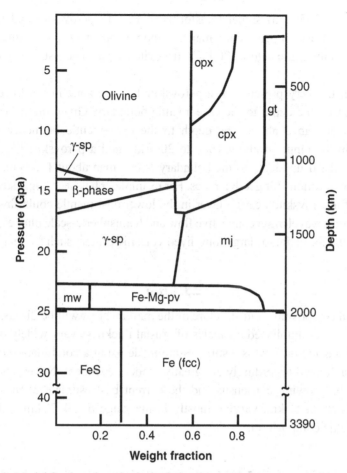

Fig. 5.2 Model for the mineralogical relationships of the martian primitive mantle. Redrawn from Bertka and Fei [16]. Note the uncertainty in the depth to the core–mantle boundary. Abbreviations in addition to those in Fig. 5.1 as follows: opx – orthopyroxene; cpx – clinopyroxene; β-phase – wadsleyite; Fe-Mg-pv – Fe–Mg–perovskite; Fe (fcc) – face-cubic-centered metallic iron.

of olivine, orthopyroxene, clinopyroxene and garnet to a pressure of 135 kbar (about 1050 km depth). Over the pressure range 20–95 kbar, orthopyroxene changes into clinopyroxene and an assemblage of olivine, clinopyroxene and garnet exists beneath about 90 kbar. The lower mantle is marked by the first appearance of γ-spinel at about 135 kbar. Because of the iron-rich nature of the mantle composition used in this model, the transformation sequence olivine–β-phase (wadsleyite)–γ-spinel is more complicated than in the Earth (Fig. 5.2), however, with either a little more or a little less iron, the nature of this transition could be quite different. Between about 120 kbar and 170 kbar, clinopyroxene changes into majorite garnet and the lower part of the lower mantle is characterized by γ-spinel and majorite.

At 225 kbar (~ 1850 km) spinel transforms to Fe–Mg perovskite and the lower mantle here consists of perovskite, majorite and perhaps magnesiowüstite. If lower mantle temperatures are below 1700 °C, stishovite may also be a stable phase in this lowermost region.

However, the very presence of a perovskite-bearing zone in the lower mantle is dependent on the depth to the core–mantle boundary. Given the uncertainty in core radius described above, the depth to the core–mantle boundary could be anywhere in the range of about 1600 to 2000 km and a perovskite-bearing zone would not exist if the depth to the boundary is less than about 1850 km. This has important implications for geodynamics. For example, it has been suggested that the presence of a perovskite-bearing layer in the lowermost mantle could favor development of long-wavelength convective flow and hemispheric-scale plumes, possibly explaining the formation of large long-lived volcanic centers such as Tharsis [17].

5.2.3 Crust

The detailed composition and structure of the martian crust will be discussed in the following chapter. Published estimates of crustal thickness vary widely due to the absence of seismic data. Most estimates are made using a combination of gravity and topography and accordingly are greatly model dependent. Nevertheless, there appears to be growing consensus and the current best estimate of mean crustal thickness is about 50 km, varying mostly in the range of 20–70 km and with an overall basaltic composition.

5.3 Martian stratigraphy

The geological record of Mars is divided into three major epochs: Noachian, Hesperian and Amazonian. In turn, these are subdivided into Early, Middle and Late except for the brief span of Hesperian that warrants only a two-fold subdivision. Absolute timescales are not well constrained due to difficulties in establishing well-dated and well-calibrated cratering histories on a planetary surface that has undergone complex and extended geological activity and for which there are few well-dated events. Constraints on absolute dating of the martian surface come from models based on the relatively well-calibrated impact record of the Moon [18]. Radiometric dating of martian meteorites (see below) provides additional calibration points that allow constraints to be placed on the oldest and youngest surfaces of the planet [19]. In contrast to the terrestrial geological record where temporal resolution decreases with increasing age, on Mars uncertainties in absolute ages, derived from counting craters, increase on younger surfaces until they may reach as much as a factor of 2 or 3 for the Amazonian.

Most of the geological record is compressed into the first billion years of the planet's history and so the boundary between the oldest Noachian and Hesperian epochs is estimated at about 3.5–3.7 Gyr, with 3.7 Gyr used here [20]. Thus the geologically rich Noachian of Mars is roughly equivalent to the Earth's Hadean that is characterized by a complete lack of a rock record. The most likely age of the Hesperian–Amazonian boundary is 2.9–3.3 Gyr, making the Amazonian equivalent to more than all of Post-Archean time on Earth. The current estimates of the ages of stratigraphic boundaries together with a summary of major geological activity are given in Table 5.1.

Table 5.1 *Martian stratigraphic subdivisions, with summaries of major geological events and approximate ages of boundaries [21]*

System	Major activity	Approximate age and duration
Late Amazonian	Glaciation and formation of polar layered deposits (PLD); major outflow channels; eolian activity; formation of shergottites	0.4 ± 0.2–0.0 Gyr
Middle Amazonian	Resurfacing of northern plains; formation of nakhlites and chassignites	1.5 ± 0.5–0.4 Gyr
Early Amazonian	Resurfacing of northern plains; volcanic activity at Elysium Mons; deep erosion of Valles Marineris	3.1 ± 0.2–1.5 Gyr
Hesperian–Amazonian boundary (3.3–2.9 Gyr)		
Late Hesperian	Numerous outflow channels; limited volcanic activity; fluvial infilling northern plains	3.4 ± 0.2–3.1 Gyr
Early Hesperian	Formation of outflow channels; volcanic activity; Tharsis volcanism continues; formation of Gusev plains volcanics	~3.7–3.4 Gyr
Noachian–Hesperian boundary (~3.7 Gyr)		
Late Noachian	End of heavy bombardment; formation of valley networks; Tharsis volcanism begins; layered volcanic and sedimentary rocks in Valles Marineris; deposition of Burns formation under acidic conditions; Columbia Hills alkaline volcanism	3.8 ± 0.1–~3.7 Gyr
Middle Noachian	Heavy bombardment; highland volcanism and formation of intercrater plains; late heavy bombardment; dynamo inactive	4.0 ± 0.2–3.8 Gyr
Early Noachian	Magma ocean; crust–mantle–core differentiation; formation of ALH84001; magnetic dynamo active; formation of crustal dichotomy; intense bombardment; loss of primary atmosphere	4.57–4.0 Gyr

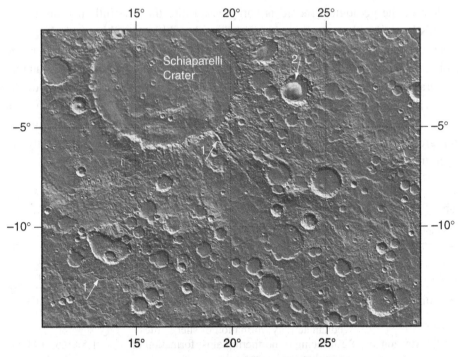

Fig. 5.3 Mars orbiter laser altimeter (MOLA) shaded relief map showing an example of the heavily cratered southern highlands immediately southeast of Schiaparelli Crater (460 km diameter). Note the degraded nature of crater rims, channels (arrow 1), remnants of crater fill deposits (arrow 2) and relatively smooth intercrater plains. Image taken from MOLA shaded relief quadrangle map MC20 [22].

5.4 Cratering record and the age of the martian surface

Impact processes dominate the visible surface of Mars, especially in the ancient southern highlands (Fig. 5.3). Within the highlands, there are also a number of well-defined multiringed basins with Hellas and Argyre being the largest at about 2300 km and 1800 km diameter, respectively. The bottom of Hellas is 8.2 km below the average martian surface and is the lowest topographical point on Mars, thus contributing nearly a third of the overall 29.5 km of topographic relief on the planet. Elsewhere on Mars, there are numerous other large basins either well exposed at the surface (e.g. Isidis at the boundary between highlands and lowlands) or buried and obscured beneath younger cover (e.g. Utopia in the northern plains).

The cratering record of Mars, notably that preserved in the southern highlands, differs in several important ways when compared to the record on the Moon or Mercury. These differences reflect a variety of dynamic geological processes that have occurred on the martian surface throughout its history. Craters in the southern highlands commonly are heavily degraded by erosional processes and cut by channels and valley networks (Fig. 5.3). Smooth plains are also common between these craters, due to younger volcanism. These craters are also the host of the extensive layered deposits of Noachian and Hesperian age, many of which have been interpreted to represent ancient indurated sedimentary rocks (see Chapter 6) [23].

5.4.1 Crustal dichotomy

The geomorphology of Mars readily divides into hemispheric terrains of roughly equal area: the heavily cratered southern highlands and the relatively smooth low-lying northern plains [24]. The contrast is so extreme that "It almost seems as if halves of two dissimilar planets have been fused together" [25]. In turn, this hemispheric dichotomy largely mirrors the age and geological history of the planet. Approximately, but not precisely, coinciding with the topographic break is a dichotomy in crustal thickness. There are a variety of other features that approximately coincide with the boundary, such as preservation of magnetized crust in the southern highlands (see below) and enrichment in iron content in the northern plains (see Chapter 6).

Thoughts on the origin of the dichotomy fall into two broad models: (1) mantle convection processes (or magma ocean overturn) either gave rise to the differences in crustal thickness or later modified the crust; (2) one or more giant impacts into the northern hemisphere produced crustal thinning thus giving rise to the dichotomy. Mantle convection is largely a default explanation (though well supported by modeling) due to perceived difficulties with the impact models, including: non-circularity of the dichotomy boundary, unlikelihood of multiple giant impacts on only one hemisphere, lack of detailed correlation between crustal thickness and other expressions of the dichotomy boundary, and so forth. Among the confounding complexities, however, is that the dichotomy has been greatly modified by volcanic, sedimentary, erosional and possibly convective processes, thus obscuring detailed correlations. Excavation of Hellas Basin also added significant topography to parts of the southern highlands. In any case, the origin of the dichotomy must be considered an unresolved problem.

5.4.2 Quasi-circular depressions

The age of the relatively lightly cratered smooth northern plains of Mars, which in turn constrains the age of the crustal dichotomy, had long been an unresolved question [26]. It was unclear whether the crust underlying this volcanic and sedimentary terrain was of Noachian age, similar to the southern highlands, or of Hesperian/Amazonian age, similar to the surficial deposits themselves. An important result from Mars Global Surveyor was to precisely map martian topography. From these maps, a number of near-circular features with subdued topography have been identified, termed "quasi-circular depressions" (QCD), which are interpreted to be the sites of large buried impact craters [27]. In the northern lowlands, there are more than 600 such depressions with diameters > 50 km and at least four basins > 1300 km. The age implied by this population of craters is model dependent but estimated to be ≥ 4.1 Gyr. Regardless of the exact age, the presence of numerous large impact structures in the northern lowlands, including some of the largest basins on Mars, clearly indicates that the northern crust is ancient, perhaps even older than the southern highland crust, and is covered by a relatively thin veneer (~ 1–2 km) of later volcanic and sedimentary material. Accordingly, the dichotomy must have formed within about the first 400–500 million years of martian history, well before the termination of the heavy bombardment.

5.4.3 Tharsis and Valles Marineris

Two dominating topographic structures are the Tharsis plateau (or "bulge") and Valles Marineris, features whose origins are related. Tharsis, consisting of two broad rises centered near the equator at about $250°$ E, covers about one-quarter of the martian surface. The canyon system of Valles Marineris extends some 4000 km eastward from the central region of Tharsis and is up to 600 km wide and 7 km deep. Tharsis is a broad topographic high with several superimposed large volcanoes, including Olympus Mons [28]. The plateau itself is about 10 km high, 8000 km across and provides enough mass to influence the moment of inertia and thus the obliquity of the planet. Tharsis is the site of a large positive gravity anomaly and a surrounding well-defined gravity low. This pattern typically is indicative of loading on an elastic lithosphere, that is, for something the size of Tharsis, about 100–150 km thick [29].

Surrounding Tharsis is a vast network of faults and fractures, the largest being the enormous canyon system of Valles Marineris, formed by tectonic processes that resulted from the loading of the Tharsis volcanic pile. Valles Marineris provides a unique window into martian geological history during and prior to the early development of Tharsis. The canyon walls display thick, layered deposits of volcanic and sedimentary origin [30]. One importance of these layered deposits is that they demonstrate that Tharsis is an ancient feature, forming mostly prior to about 3.5 Gyr (Late Noachian/Early Hesperian) [31].

Although Tharsis is mostly old, the most recently erupted volcanic flows on the large volcanoes of Tharsis and elsewhere are clearly very young since their surfaces are nearly devoid of impact craters (Fig. 5.4). If the young shergottites are derived from such regions (see below), then parts of their surfaces must be at least as young

Fig. 5.4 Mars Orbiter Camera (MOC) image of a region on the western flank of Tharsis, at about 220.0° W longitude and 6.0° N latitude (illumination from top left). The image, 4.5 km. across, shows dark lava flows on the west that embay an older, topographically higher and more heavily cratered terrain to the east (arrows). Note that only a very few small craters are present on the lavas, indicating a very young age. Part of image PIA08747 from the Planetary Photojournal [22].

as about 160 Myr. Some relatively young lava flows exposed on Olympus Mons show unusually well-defined flow features and crater counts suggest ages as young as ∼ 10 Myr [20]. It is likely that volcanism continues at a very low rate essentially through to the present day.

The origin of Tharsis has been the source of considerable debate. Any model must explain both the gravity–topography relationships and the long life of the plateau. Two commonly proposed mechanisms are (1) that the gravity anomalies are due to loading from long-lived volcanic outpourings and intrusions [29] and (2) that the anomalies are the result of a long-lived upwelling mantle plume [32]. In fact, both processes are likely to have contributed. Layered volcanic deposits in Valles Marineris coupled with ongoing volcanism suggest that Tharsis has been the site of a near-stationary plume for billions of years. There has been some success in modeling very large stationary plumes on Mars that are controlled by thermal barriers resulting from the phase transitions that give rise to mantle layering (Figs. 5.1, 5.2) [32]. The presence of a spinel–perovskite transition in the lowermost mantle appears to be especially important in promoting large plumes that are near stationary over long periods of time. Nevertheless, most modeling suggests that convective forces contribute only a fraction of the topography, with most resulting from the crustal loading of enormous volumes of volcanic material.

5.5 Early plate tectonics?

There is no doubt that for the past several billion years, Mars has been a one-plate planet in a stagnant lid convection regime. Although suggestions of plate tectonics on early Mars have a long history, the first serious evaluation of the possibility was carried out by Norm Sleep [33]. His model proposed that ancient thick highland crust was subducted during the Late Noachian to Early Hesperian and a younger thin crust was produced in the northern lowlands by sea-floor spreading, analogous to the formation of oceanic crust on the Earth.

However, it is now clear that the bulk of northern lowlands crust is at least as old as the southern highlands. The identification of numerous extremely large quasi-circular depressions, interpreted to be impact basins that are buried beneath a thin (< 1–2 km) veneer of younger lavas and sediment, indicates that the northern lowland crust is very ancient. Thus the crust in the northern plains does not result from any simple younger resurfacing process, analogous to production of the Earth's oceanic crust by sea-floor spreading; consequently this variant of plate-tectonic models has now been abandoned.

Undaunted by this negative result, more recent proponents of martian plate tectonics have looked instead to the southern highlands for evidence.

5.5.1 Crustal magnetization and plate tectonics

Mars does not have a global dipole magnetic field. However an important result from the Mars Global Surveyor magnetometer experiment was identification of remanent crustal magnetization anomalies, most notably in the form of an east–west trending striped pattern in the southern highlands [34]. These magnetized regions of the crust last cooled to below the Curie point prior to about 4 Gyr. Large ancient impact features, such as Hellas and Argyre clearly disrupt these features as do relatively younger volcanic regions such as Olympus Mons and Elysium (but note that the southernmost regions of Tharsis are magnetized). These relationships indicate that the magnetic field that produced them was ancient and short-lived, likely taking place entirely within the Early to Mid Noachian. Remanent magnetization preserved in magnetite and pyrrhotite within the martian meteorite ALH84001 also appears to have formed at approximately 4.1–3.9 Gyr [35].

In the absence of seismology and a firm understanding of the physical and chemical conditions on early Mars, "the problem of inferring or predicting the history of a martian dynamo is indeed formidable" [36]. Although it is feasible that an early and short-lived magnetic field could result from magma ocean processes, the most likely origin is from a short-lived core dynamo. Current debate centers on whether the convection required for an early dynamo was chemically or thermally induced. The former implies a solid iron–nickel inner core with a chemically distinct iron–nickel–sulfur liquid outer core (broadly comparable to the Earth) whereas the latter requires both a sulfur-rich core and an early period of exceptionally high heat flow. Although the origin of the martian magnetic field is by no means resolved, a thermally induced convection regime has several advantages. An early and short-lived history is expected for an initially hot and rapidly cooling planet, the model is consistent with a core containing about 14% S and the model readily allows for a mostly liquid core throughout martian history.

A variety of mechanisms have been proposed to explain the linear magnetic features:

i. magnetic striping formed by sea-floor spreading [37];
ii. lateral accretion of distinctive geological terranes by plate-tectonic processes [38];
iii. cooling of long networks of dike swarms [39];
iv. magnetic features are due to chemical remanent magnetization associated with large-scale hydrothermal activity [40].

All of these models are speculative; however, it is worth noting that in detail the linear magnetic features bear little resemblance to the magnetic striping so characteristic of sea-floor spreading on Earth. The size of the martian features, measured on the scale of hundreds of kilometers, is an order of magnitude larger than terrestrial magnetic stripes, observed on the scale of tens of kilometers. Both the overall strength of the martian remanent crustal magnetization and the amplitude of the striping are also an order of magnitude greater than seen in the terrestrial ocean floor [41]. Although the timescales of formation of the martian features are not well constrained, it is worth noting that magnetic stripes in the terrestrial ocean crust form on short time scales, typically 10^3–10^5 years.

On balance, neither the observational evidence nor theoretical modeling provides any compelling argument for the existence of plate tectonics at any time during martian history [42]. We conclude that Mars has always been a one-plate planet, like all of the other terrestrial planets and rocky satellites apart from the Earth.

5.6 Samples from Mars

5.6.1 Martian meteorites

It is now widely accepted that a subset of basaltic achondrites, termed SNC meteorites after Shergotty (S), Nakhla (N) and Chassigny (C), are derived from Mars [43]. These samples are central to understanding the evolution of Mars because sophisticated geochemical and isotopic techniques can be applied and thus Mars is one of only four bodies considered in this book for which there is direct sampling (others being the Earth, the Moon and the large asteroid 4 Vesta). The most persuasive evidence for their origin comes from the similarity in gas concentrations and isotopic ratios (e.g. $^{14}N/^{15}N$, $^{40}Ar/^{36}Ar$, $^{129}Xe/^{132}Xe$) trapped within SNC meteorite impact melt glasses with that measured in the martian atmosphere by the Viking landers and by remote sensing [44]. Additional lines of evidence in support of the martian origin are individually persuasive but non-definitive; however when taken together are conclusive [43]:

i. Great range (4.5–0.17 Gyr) but mostly young (≤1.3 Gyr) crystallization ages suggesting a parent body with surface geology unlike any other terrestrial planet or asteroid [45].
ii. Isotopic composition of trapped gases indicating atmospheric processing.
iii. Broad similarity in chemical composition and oxidation state between basaltic shergottites and martian soils and rocks measured *in situ*, but distinct from other potential parent bodies for which there are chemical data (e.g. 4 Vesta, the Moon).
iv. Rare earth element patterns consistent with relatively high-pressure equilibrium (i.e., garnet stability) indicating derivation from a body larger than asteroids or the Moon.
v. Low remanent magnetization in shergottites, consistent with the absence of a present-day martian magnetic field.
vi. Presence of "pre-terrestrial" secondary mineralogy (e.g. sulfates, halite, hydrous clays) consistent with near-surface aqueous activity.

At the time of writing, there are 37 recognized unique martian meteorites [46]. They can be divided into three broad groups on the basis of age, petrology, chemistry and isotopes (Tables 5.2 and 5.3):

i. A single sample of coarse-grained cumulate orthopyroxenite, ALH84001, with a crystallization age of 4.50 ± 0.13 Gyr and evidence of aqueous alteration at about 3.9 Gyr [48]. This is the only sample available from ancient Noachian crust and is characterized by incompatible-element-depleted and near-chondritic levels of refractory trace elements (Fig. 5.5).
ii. Coarse-grained ultramafic cumulate rocks with crystallization ages of ~1.3 Gyr, thus representing Middle Amazonian crust. These are further subdivided into clinopyroxenites (nakhlites) and dunites (chassignites) [49]. Nakhlites represent a related suite of cumulates, possibly cogenetic, derived from crystallization in basalt flows or shallow intrusions. They are enriched in incompatible elements (Fig. 5.5) but long-lived radiogenic isotope systems indicate derivation from ancient incompatible-element-depleted mantle sources. Nakhlite and chassignite chemistry are sufficiently different to suggest they are not cogenetic although they have similar radiogenic isotope characteristics and exposure ages and accordingly may be broadly related [50].

Table 5.2 *Summary of martian meteorite characteristics* [47]

	Basaltic shergottites	Olivine-phyric shergottites	Lherzolitic shergottites	Nakhlites/Chassignites	Orthopyroxenite
Crystallization ages	165–327 Myr	173–575 Myr	177–212 Myr	1.26–1.38 Gyr	4.50±0.13 Gyr
Major minerals	Pigeonite, augite maskelynite	Pigeonite, augite maskelynite	Orthopyroxene olivine, pigeonite	Augite, olivine	Orthopyroxene
Igneous textures	Melt veins, flow textures, cumulate	Porphyritic, xenocrysts, cumulate	Poikilitic, cumulate	Cumulate	Adcumulate
SiO_2 (wt%)	46–52	45–51	41–46	47–51 (Nakhlites) ~37 (Chassignites)	~53
FeO_T (wt%)	15–21	16–24	18–22	19–23 (Nakhlites) 20–27 (Chassignites)	~17
Mg #	23–52	59–68	~70	~51 (Nakhlites) ~68 (Chassignites)	~72
Samples*	Shergotty, Zagami, EETA79001B, Los Angeles, QUE94201, Dho378, NWA480, NWA856, NWA1669, NWA2975, NWA3171	EETA79001A, DaG476/489, SaU005/094/150, Dho019, NWA1068, NWA1195, NWA2046, NWA2626, Y980459	ALH77005, LEW88516, Y793605, GRV99027, GRV020090, NWA1950, NWA2646	Nakhla, Lafayette, Govornador Valadares, NWA817, NWA998, MIL03346, Y000593/749/802 Chassigny, NWA2737	ALH84001

Notes: * Under Nakhlites/Chassignites heading, Chassigny and NWA2737 are chassignites and the other samples are nakhlites. Samples DaG476/489, SaU005/094/150, Y000593/749/802 are paired meteorites. FeO_T is total iron as FeO_2.

Table 5.3 *Chemical composition of selected martian meteorites. These average values are obtained from data compiled in Mars Meteorite Compendium [46]*

Type:		Chassignite	Nakhlites		Lherz. shergottites		Ol-phyr. shergottites		Basaltic shergottites		
Sample:	ALH 84001	Chassigny	Nakhla	MIL 03346	ALH 77005	LEW 88516	EETA 79001A	Yam 980459	QUE 94201	Shergotty	Los Angeles
Lithology:	O'pxite	Dunite	C'pxite	C'pxite	Perid'ite	Perid'ite	Ol-Basalt	Ol-Basalt	Basalt	Basalt	Basalt
Oxide (wt%)											
SiO$_2$	52.8	37.4	48.6	49.5	42.4	45.5	50.2	49.9	47.0	50.7	48.9
TiO$_2$	0.20	0.10	0.34	0.68	0.41	0.42	0.75	0.53	1.80	0.85	1.20
Al$_2$O$_3$	1.21	0.70	1.68	4.09	2.8	3.31	5.70	5.20	10.8	6.87	10.9
FeO$_T$	17.3	27.3	20.6	19.1	20.0	19.9	18.4	17.3	18.7	19.2	21.2
MnO	0.47	0.53	0.49	0.46	0.45	0.48	0.48	0.48	0.45	0.54	0.46
MgO	24.7	31.8	12.1	9.26	28.1	24.9	16.4	18.7	6.20	9.14	3.73
CaO	1.82	0.80	14.7	14.4	3.30	4.24	7.22	6.80	11.0	9.98	9.92
Na$_2$O	0.15	0.12	0.46	0.96	0.44	0.55	0.85	0.65	1.40	1.36	2.20
K$_2$O	0.016	0.036	0.14	0.20	0.028	0.028	0.041	0.0156	0.045	0.168	0.30
P$_2$O$_5$	0.01	0.07	0.12	0.23	0.38	0.39	0.60	–	2.77	0.71	1.00
Σ	**98.7**	**98.9**	**99.2**	**98.9**	**98.3**	**99.7**	**100.6**	**99.6**	**100.2**	**99.5**	**99.8**
Element (ppm)											
Ba	4.0	7.6	30	57	4.5	4.9	5.3	1.5	–	30	47
Rb	0.83	0.52	3.5	4.4	0.76	0.20	1.2	–	0.60	6.55	12
Sr	4.5	7.2	60	121	13.9	15	20	–	46	53	81
La	0.150	0.39	2.26	3.89	0.34	0.31	0.45	0.166	0.40	2.29	3.97
Ce	0.430	1.12	6.09	11.3	0.91	0.87	1.28	0.426	1.47	5.54	9.84
Pr	–	0.13	–	1.78	0.13	–	0.19	0.084	–	0.86	1.43
Nd	0.265	0.54	3.59	8.04	0.95	0.82	1.17	0.567	2.20	4.50	7.07
Sm	0.104	0.11	0.835	1.83	0.49	0.47	0.81	0.466	2.30	1.37	2.64
Eu	0.032	0.038	0.263	0.52	0.22	0.23	0.36	0.254	1.04	0.564	2.02
Gd	0.140	0.11	0.818	1.86	0.92	–	1.61	1.13	4.30	2.20	4.28
Tb	–	–	–	0.30	0.17	0.16	0.34	0.244	0.87	0.44	0.79

Dy	0.240	0.12	0.766	1.66	1.08	1.1	2.21	1.70	5.80	2.94	5.04
Ho	0.068	–	–	0.32	0.25	0.24	0.47	0.379	1.19	0.56	1.03
Er	0.210	0.09	0.423	0.84	0.66	–	1.35	1.09	–	1.87	2.76
Tm	0.036	–	–	0.13	0.088	0.089	0.19	0.155	–	0.38	–
Yb	0.255	0.10	0.378	0.80	0.59	0.57	1.14	0.971	3.30	1.69	2.35
Lu	0.037	–	0.055	0.13	0.078	0.083	0.165	0.150	0.50	0.25	0.33
La_N/Yb_N	0.40	2.64	4.04	3.29	0.39	0.42	0.27	0.12	0.08	0.92	1.14
Y	1.6	0.6	3.3	8.5	6.2	–	11.6	–	31	14.3	29
Th	0.035	0.057	0.21	0.43	0.058	0.040	0.080	0.021	0.050	0.40	0.57
U	0.012	0.021	0.053	0.11	0.015	0.012	0.019	0.006	0.013	0.11	0.12
Zr	5.9	1.5	8.8	21.2	20	17	29	–	96	60	–
Hf	0.21	0.044	0.21	0.66	0.56	0.50	0.94	0.49	3.45	1.76	3.0
Nb	0.42	0.32	1.6	3.65	0.57	0.51	0.83	–	0.68	2.7	5.0
Ta	0.032	0.022	0.09	0.21	0.03	0.034	0.04	–	0.023	0.25	0.30
Ga	3.0	0.7	2.9	6.8	7.4	8.0	13	11	27	15.5	23
Ge	1.1	0.011	–	–	0.58	0.6	0.8	–	2.0	0.69	–
W	0.045	0.041	0.27	–	0.084	0.14	–	–	–	0.40	0.66
Cr	7700	5000	1800	1300	6700	6000	4200	4800	950	1400	200
V	205	38	192	210	160	185	220	188	110	305	–
Sc	13	5	55	52	21	25	36	35	48	53	42
Ni	6	500	90	58	330	270	150	220	–	83	30
Co	45	124	50	39	70	65	47	55	24	38	30
Cu	–	3	10	13	5	–	12	–	–	13	23
Zn	95	70	60	62	60	60	72	80	100	68	70

Abbreviations: Lherz. – Lherzolitic; Ol-phyr. – Olivine-phyric; O'pxite – orthopyroxenite; C'pxite – clinopyroxenite; Perid'ite – peridotite; Ol-basalt – Olivine-rich basalt.

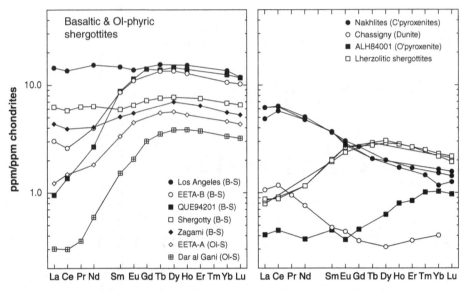

Fig. 5.5 Chondrite-normalized REE patterns for selected martian meteorites. See Notes 46 and 47 for sources.

iii. Shergottites are a suite of very young iron-rich mafic through ultramafic igneous rocks with crystallization ages between 575 and 165 Myr, thus sampling Late Amazonian crust. Shergottites are further divided on the basis of petrology into basaltic, olivine-phyric (or picritic) and lherzolitic (or peridotitic) shergottites. Lherzolitic and olivine-phyric shergottites are cumulates whereas the basaltic shergottites show varying degrees of cumulus (or possibly foreign) pyroxene crystals with only a few samples, at most, approaching true melt compositions. Shergottites show extreme but generally systematic variation in isotopic and trace element compositions (Fig. 5.5) suggesting derivation from ancient highly incompatible-element-depleted (and relatively reduced) sources that are mixed to varying degrees with more or less equally ancient incompatible-element-enriched (and relatively oxidized) sources. Whether the enriched component is the ancient martian crust or distinct enriched mantle reservoirs is the subject of debate and currently must be considered unresolved.

The age distribution of martian meteorites does not correlate with the age distribution of the surface, which is mostly older than 3.0 Gyr (Noachian and Hesperian). The young ages of all but one of the martian meteorites place their sources within Middle to Late Amazonian volcanic provinces, likely the broad region between Elysium, Amazonis Planitia and Tharsis. Intuition might suggest that sampling the martian surface by impact processes would be a random process and accordingly, the reason for the bias in sampling ages has been the subject of considerable discussion. One possibility is that only one or two ejection events are involved and so ejection processes are controlled by small number statistics. However, ejection ages [51] fall into at least four and most likely six or seven discernible events over the past 20 million years. Accordingly, the age bias is probably due to a combination of factors, including impact sites at high elevation that limit atmospheric filtering and low latitudes where increased angular velocity (~0.24 km/sec) can contribute to the velocity of ejected material. An additional factor is that only "strong" material may survive the ejection process, thus limiting the involvement of ancient heavily cratered southern highland terrains that have been highly disrupted by impact processes and contain extensive, presumably weak, sedimentary deposits [50, 52].

Many of the martian meteorites have also been affected by secondary aqueous alteration, including hydrothermal, weathering and evaporative processes, that are mostly associated with fractures and small veins. These features appear to be inherent to the samples and thus represent processes that occurred on Mars. All of the martian meteorite classes have been affected to some degree but nakhlites and ALH84001 appear to be most affected. The degree and type of alteration vary greatly from sample to sample [53].

5.6.2 Shergottite crystallization ages

The crystallization ages of ALH84001 (~4.5 Gyr), nakhlites and chassignites (both ~1.3 Gyr) are not in dispute. However, the interpretation of geochronological data for shergottites is contentious and unlike many disputes in geochronology, differences in opinion could not be less subtle. Well-established internal isochron techniques, using whole rocks, pristine minerals, leached mineral residues and leachates, give Rb–Sr, Sm–Nd, U–Pb and Lu–Hf ages of between 575 and 165 Myr with most clustering in the narrow range of 165–185 Myr. There is also good agreement among the different isotope systems [54]. However, $^{207}Pb/^{204}Pb - ^{206}Pb/^{204}Pb$ techniques on whole rocks and mineral separates indicate an age of about 4.0 Gyr leading to the suggestion that this is the true crystallization age with younger ages representing a resetting event, possibly due to impact processes and/or acid alteration that preferentially affects phosphate minerals, the main hosts for REE, Hf, Th and U (and thus radiogenic Pb) [55].

Resolving the question of the crystallization age of shergottites is of course critical for understanding the evolution of the martian crust–mantle system and, among other things, has important implications for constraining the compositional evolution of the crust (see Chapter 6) and for geodynamics. Young crystallization ages imply that the mantle reservoirs from which the shergottites were derived have been essentially isolated for some 4.5 Gyr in order to preserve isotope anomalies for the short-lived $^{182}Hf-^{182}W$ and $^{146}Sm-^{142}Nd$ isotope pairs (see below). Lack of mantle mixing that these anomalies imply would then argue against a convective regime in which there was large-scale mantle rehomogenization and certainly against any role for plate tectonics. On the other hand, if crystallization ages are ancient (~ 4.0 Gyr) then the anomalies in the short-lived isotopes would only need to have been isolated within about 0.5 Gyr of the formation of Mars. Significant mantle mixing after about 4.0 Gyr would be allowed, as this mantle would not be sampled by the martian meteorites, thus permitting later convective mixing of the mantle.

We consider young crystallization ages for shergottites to most likely be correct as discussed below and proceed from this conclusion. Young magmatic ages are by no means surprising for Mars as there is abundant independent evidence for young volcanism from geomorphological and crater-counting data (Fig. 5.4). Proponents of old ages have suggested that resetting of the Rb–Sr, Sm–Nd and Lu–Hf isotope systems has occurred through acid alteration that preferentially reset the phosphate minerals, that are the primary host of many trace elements. However the nature of acid alteration and aqueous conditions during the Amazonian is becoming increasingly well understood with the study of rock surfaces by the Spirit and Opportunity rovers and experiments. Although acid alteration of rock surfaces appears common, it only affects a few millimeters of the outer surface of the rocks. There is no evidence for the pervasive alteration (that should result in phosphate, olivine and iron–titanium oxide dissolution) that has been suggested for such a major resetting event. Indeed, it appears that young acid alteration likely proceeded under very low water/rock ratios with only incipient alteration patterns [56]. In any case, shergottites show no petrographic evidence for the type of alteration implied by such an isotopic resetting process.

Although the elemental budgets for REE, Hf, Th and U in shergottites are dominated by phosphates, the mineralogical controls on Rb and Sr are more complex and accordingly, it is difficult to understand the remarkable coherence in isochron ages given by the different isotope systems if they reflect a resetting event mostly involving one mineral class. Other independent isotope techniques, though preliminary, also appear to support a young crystallization age. $^{40}Ar/^{39}Ar$ dating of shergottites is greatly complicated by the addition of atmospheric argon during impact processes, however, recent laser probe $^{40}Ar/^{39}Ar$ dating of relatively pristine igneous minerals in NWA1950 gives an age of 382 ± 36 Myr, consistent with a young crystallization age and minor atmospheric argon addition [57]. In another recent study, ion microprobe analysis of baddeleyite in basaltic shergottite NWA856 gave a $^{206}Pb*/^{238}U$ age of 186 ± 12 Myr, essentially identical to whole rock Sm/Nd ages [58].

5.7 Early differentiation on Mars and magma oceans

A wide variety of both long-lived (e.g. $^{238}U–^{206}Pb$, $^{235}U–^{207}Pb$, $^{147}Sm–^{143}Nd$, $^{87}Rb–^{87}Sr$, $^{187}Re–^{187}Os$, $^{176}Lu–^{176}Hf$) and short-lived ($^{182}Hf–^{182}W$, $^{146}Sm–^{142}Nd$,

^{129}I–^{129}Xe) radiogenic isotope measurements on martian meteorites provide compelling evidence for a planet that differentiated into core, mantle and crust during approximately the first 30–50 million years of its history [59]. The fact that such evidence is so clearly preserved on Mars, in contrast to Earth where isotopic evidence for early differentiation processes is obscure at best, provides important constraints on the crustal and geodynamical histories of Mars.

The most direct and precise constraints on the timing of early differentiation come from the short-lived isotope systems of ^{182}Hf–^{182}W ($t_{1/2} = 9$ Myr) and ^{146}Sm–^{142}Nd ($t_{1/2} = 103$ Myr). Hafnium is a refractory lithophile element that remains in the silicate mantle whereas W is a refractory moderately siderophile element that preferentially partitions into metal during metal–silicate equilibrium. Both Sm and Nd are refractory lithophile elements. Accordingly, Hf/W fractionation is thought to take place due to metal–silicate separation during core formation; Sm/Nd ratios are unaffected by this process. During subsequent partial melting of the silicate mantle, these element ratios fractionate further. The lithophile elements Hf, Sm and Nd are all moderately incompatible during partial melting of mantle silicates whereas W is highly incompatible. Variations between the ^{182}W and ^{142}Nd isotope systems can thus be used to provide constraints on both the timing of core–mantle fractionation and the ^{182}W composition of the primitive mantle.

For reasonable estimates of Hf/W fractionation between core and mantle, model ages of core formation are mostly <30 Myr with the most often quoted estimate being $\sim 12 \pm 4$ Myr after the formation of the Solar System [60]. As discussed in Chapter 1, however, such an age actually records the average time that the silicate mantle acquired its current Hf/W ratio and this could be complicated by the delivery of large differentiated bodies (Moon-sized) during the accretion.

Early differentiation of the mantle is also clear from other isotope systems. For example, Fig. 5.6 shows that isotopic compositions for long-lived ^{147}Sm–^{143}Nd and short-lived ^{146}Sm–^{142}Nd systems are correlated in shergottites, suggesting that the silicate reservoirs sampled by these meteorites were established very early in martian history. The reference isochrons assume a simple two-stage model of early Sm/Nd fractionation followed by later melting (~ 165 Myr) to produce the shergottite magmas and thus represent an end-member case. For this simple model, an age of 4.525 Gyr is suggested for the formation of the source regions [61]. Osmium isotopic compositions of martian meteorites also correlate with the ^{182}W and ^{142}Nd isotope systems, consistent with very early differentiation [62].

Exactly how this early differentiation took place, whether by magma ocean processes or early recurrent volcanism, is less clear. Mars completed its accretion significantly earlier than Earth and if a large fraction of the accreted mass came from giant impacts (i.e., bodies about the size of the Moon or greater), there was probably sufficient energy to melt the planet at least to some considerable depth [36]. Isotopic

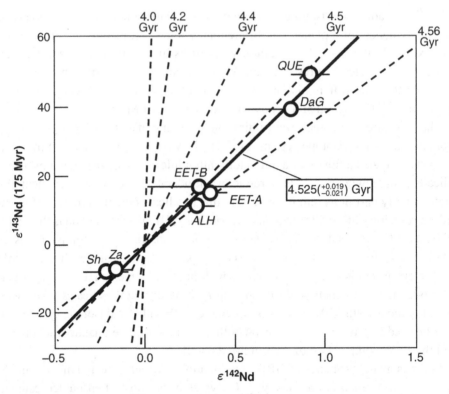

Fig. 5.6 Plot of initial ε^{143}Nd (175 Myr) vs. ε^{142}Nd indicating that the fractionation of Sm/Nd observed in shergottites occurred very early in martian history, at about 4.525 Gyr. ε_{Nd} represents deviation of the Nd isotopic composition from a chondritic reference in parts per ten thousand. Shergottites represent mixing between two distinct sources and so this trend could also be interpreted as a mixing line between two ancient sources. Dashed lines are model isochrons and the heavy line is a regression of the data. Sample abbreviations on the figure refer to meteorite names listed in Table 5.2. Diagram adapted from Borg *et al.* (2003) and Foley *et al.* (2005); see Note 61 for further details.

evidence that the timing of formation of core, known mantle reservoirs and perhaps even the earliest crust are essentially indistinguishable and within about 30 Myr of the formation of the Solar System seems most consistent with the formation of a magma ocean.

Several attempts have been made at modeling the evolution of martian magma oceans. Petrological and geochemical modeling suggests that crystallization of a deep magma ocean where majorite is stable at the base (> 1000 km) broadly (but not in detail) reproduces mantle reservoirs characteristic of the sources of basaltic shergottites [63]. Available geophysical modeling suggests that an early formed deep magma ocean would produce gravitationally unstable cumulus layering that would rapidly overturn, leading to a broadly layered mantle with the exact petrologic stratigraphy depending on various input parameters and assumptions [64]. In all cases, crystallization of such magma oceans results in an early formed crust of basaltic composition.

5.8 Multiple reservoirs and the age of the earliest crust

At least four isotopically distinct silicate reservoirs have been identified, all forming within ~30–50 Myr of the formation of the Solar System. These include the source regions for the 4.5 Gyr martian meteorite ALH84001, at least one source region for the *c*. 1.3 Gyr nakhlites and chassignites, and at least two distinct sources for the much younger (575–165 Myr) shergottites [60–63]. As described above, one of the shergottite source regions is characterized by long-term depletion of incompatible elements. However, the source of the second reservoir, characterized by long-term incompatible-element enrichment, is less certain. Most workers have suggested that this second reservoir is the martian crust but this interpretation fails to explain the lack of correlation between standard indices of differentiation (e.g. Mg-number, SiO_2 content) and degree of incompatible-element enrichment or isotopic composition [63]. Crystallization of a magma ocean gives rise to the possibility that the second component could be a KREEP-like mantle reservoir, analogous to that found on the Moon (see Chapters 2 and 3). The age constraints of these reservoirs are summarized in Fig. 5.7.

Trace element data are also consistent with multiple distinct geochemical reservoirs that in general coincide with those identified by radiogenic isotopes [65]. For example, REE patterns and ratios among various incompatible refractory elements (e.g. La/Th, Sm/Hf) are consistent with shergotittes being derived from mixing of two distinct geochemical reservoirs and with nakhlites/chassignites and ALH84001 also having distinct sources (Fig. 5.5, Fig. 5.8). However, it is unlikely that the entire crust–mantle system is characterized by the martian meteorites. For example, the Ba/La ratio of all martian meteorite classes is greater than the chondritic ratio of 9.3 (Fig. 5.8), suggesting that there must be an additional reservoir with subchondritic Ba/La ratios [66].

The geometry of these geochemical reservoirs is not known. A layered structure is a natural consequence of magma ocean processes although available modeling also predicts lateral compositional variations [64]. Thus various chemically layered mantle reservoir models have been proposed [67]. At least one author has pointed out that other geometries could facilitate much later melting to produce the shergottites [68]. Although a layered structure for major mantle reservoirs is appealing, it is worth keeping in mind a hard-learned lesson from the study of the terrestrial mantle; chemistry and isotopes, while very powerful for identifying and constraining the size of mantle reservoirs, provide essentially no direct information on geometry.

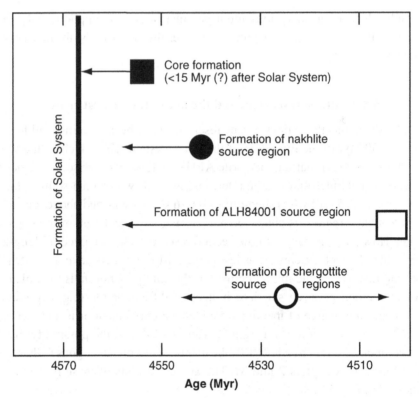

Fig. 5.7 Diagram showing age constraints (with uncertainties) on the timing of core formation and major silicate reservoirs on Mars. Adapted from Foley *et al.* [61].

Constraints on the formation of the earliest crust are more indirect. The combined $^{147}Sm–^{143}Nd$ and $^{146}Sm–^{142}Nd$ isotope systems place the differentiation of the silicate mantle reservoirs at 4.53 ± 0.02 Gyr, indistinguishable from the formation of the core within analytical uncertainty, but probably younger (Fig. 5.7). There is controversy about whether the earliest crust is directly reflected in these ages but most workers have concluded that formation of a significant part of the early crust coincided with this event. The oldest known rock from Mars is the martian meteorite, dated at 4.50 ± 0.13 Gyr and this cumulate orthopyroxenite may be the single example of the earliest formed crust of Mars.

5.9 The composition of Mars

There have been several models for the composition of Mars but few survive detailed scrutiny in light of martian meteorites, missions to Mars, experiments, and what we know about the formation of rocky planets and satellites [69]. The most reliable composition for the martian primitive mantle that meets most current constraints is

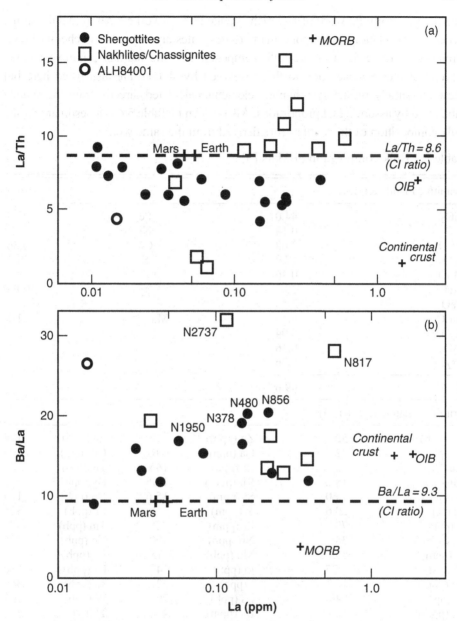

Fig. 5.8 Plots of (a) La/Th versus La and (b) Ba/La versus La for martian meteorites illustrating relationships among refractory incompatible lithophile elements. Shown for comparison (crosses) are the bulk compositions of the martian and terrestrial primitive mantles and compositions representing major terrestrial silicate reservoirs. Meteorites from desert environments, where terrestrial weathering may influence compositions, are labeled individually (prefix N in sample numbers refers to northwest Africa, or NWA in Table 5.2). These diagrams indicate that trace-element abundances in martian meteorites are consistent with multiple mantle (± crust?) reservoirs. Adapted from McLennan *et al.* [65] and Taylor *et al.* [66].

that derived from the work of Heinrich Wänke and co-workers based mainly on the geochemical relationships among martian meteorites and other cosmochemical constraints [6]. Table 5.4 lists the bulk composition of the martian primitive mantle. Several elements in addition to those reported by Wänke are also listed here but these are simply refractory lithophile elements calculated directly from the Wänke tabulation by assuming CI proportions. Also shown in Table 5.4 is an estimate for the bulk composition of the martian core derived from the same work.

Table 5.4 Bulk composition of Mars [6]

Primitive mantle oxides				Core
SiO_2	44.0		Fe	77.8
TiO_2	0.14		Ni	7.6
Al_2O_3	3.02		Co	0.36
FeO	17.9		S	14.2
MnO	0.46			
MgO	30.2		Σ	100.0
CaO	2.45			
Na_2O	0.50		Mass(%)	21.7
K_2O	0.04			
P_2O_5	0.16			
Cr_2O_3	0.76			
Σ	99.6			

Primitive mantle (elemental)						
Be (ppb)	52	Zn (ppm)	62	Eu (ppb)	114
F (ppm)	32	Ga (ppm)	6.6	Gd (ppb)	400
Na (%)	0.37	Br (ppb)	145	Tb (ppb)	76
Mg (%)	18.2	Rb (ppm)	1.06	Dy (ppb)	500
Al (%)	1.60	Sr (ppm)	15.6	Ho (ppb)	110
Si (%)	20.6	Y (ppm)	2.7	Er (ppb)	325
P (ppm)	700	Zr (ppm)	7.2	Tm (ppb)	47
Cl (ppm)	38	Nb (ppb)	490	Yb (ppb)	325
K (ppm)	305	Mo (ppb)	118	Lu (ppb)	50
Ca (%)	1.75	In (ppb)	14	Hf (ppb)	230
Sc (ppm)	11.3	I (ppb)	32	Ta (ppb)	34
Ti (ppm)	840	Cs (ppb)	70	W (ppb)	105
Cr (ppm)	5200	Ba (ppm)	4.5	Tl (ppb)	3.6
Mn (%)	0.36	La (ppb)	480	Th (ppb)	56
Fe (%)	13.9	Ce (ppb)	1250	U (ppb)	16
Co (ppm)	68	Pr (ppb)	180		
Ni (ppm)	400	Nd (ppb)	930		
Cu (ppm)	5.5	Sm (ppb)	300		

Model from H. Wänke and others [6]. Also listed for the primitive mantle are additional refractory lithophile elements, determined by assuming they are in CI relative proportions.

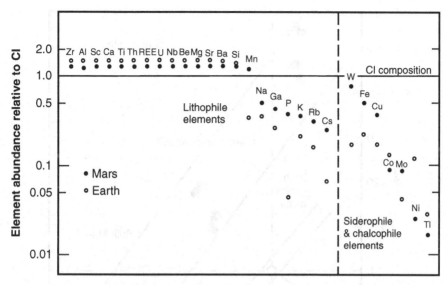

Fig. 5.9 CI-normalized abundances for the martian primitive mantle compared to that for the terrestrial primitive mantle (Chapter 8). Mars has lower abundances of refractory lithophile elements but the moderately volatile elements are less depleted than on Earth. Siderophile and chalcophile elements are also depleted in the martian primitive mantle, a consequence of their extraction into a sulfur-rich metal core, but comparisons to Earth are erratic, indicating that the mechanisms of metal–silicate equilibrium differ in detail.

This composition has a number of distinctive features that are relevant to this investigation (Fig. 5.9). Most notably, the martian primitive mantle is iron-rich with FeO at 17.9%, more than twice the value for the terrestrial primitive mantle. Thus, while the overall uncompressed density of Mars is less than that of the Earth, the amount of iron in the mantle is greater. An iron-rich primitive mantle is generally consistent with the higher martian MOI factor compared to the Earth but in detail there are difficulties (see below). A second feature of this composition is that the moderately volatile elements (e.g. K, Rb), while depleted compared to CI chondrites, are enriched by about a factor of two compared to the Earth. This observation is in line with the isotopic evidence that points, among other things, to a low μ ($^{238}U/^{204}Pb \sim 5$) and high Rb/Sr ratio for the mantle (Fig. 5.10) [70].

The composition is also notably depleted in siderophile and chalcophile elements, the latter being indicative of equilibrium with a sulfur-bearing core [71]. The content of highly volatile elements in the martian mantle is less clear. The above model calls for very low water content ($\sim 36\,ppm$) but relatively high levels of other highly volatile elements (e.g. halogens). Low water content has been explained by loss during oxidation of the mantle, thus giving rise to the high FeO content and loss of hydrogen to space. However there is also evidence for substantial early degassing

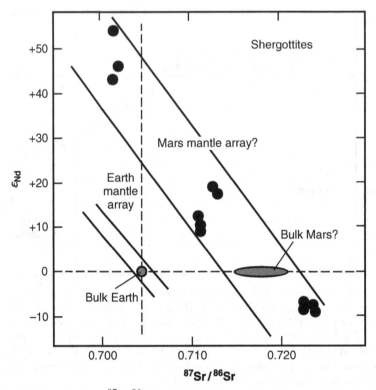

Fig. 5.10 Plot of ε_{Nd} vs. $^{87}Sr/^{86}Sr$ for martian meteorites. Shown for comparison is the terrestrial mantle array. Martian meteorites are offset to higher $^{87}Sr/^{86}Sr$, consistent with the primitive mantle being volatile-element rich (high Rb/Sr ratio).

and this leaves the volatile contents of much younger basalts, represented by the SNC meteorites, difficult to interpret [44].

5.9.1 A cautionary note

The above model is based fundamentally on geochemical relationships found among the SNC meteorites, most of which are cumulates or contain some proportion of cumulate minerals. These samples are at best representative of young martian magmatism and were derived largely from a mantle previously depleted by the formation of the early crust with a second component that could be either additional mantle reservoirs or assimilated ancient crust. In any case, not all of the major geochemical reservoirs are represented [66]. Thus, developing a primitive mantle composition from these rocks is comparable to trying to constrain the terrestrial primitive mantle by looking to a handful of young mid-ocean ridge basalts erupted through the Icelandic or Jan Mayen plateaus. Caution is warranted, especially when considering the details.

For example, on diagrams of Mg/Si versus Al/Si, most martian meteorites plot below terrestrial-mantle-derived nodules that are carefully filtered to be most representative of mantle compositions; this is commonly used to illustrate the distinctive nature of the martian primitive mantle. However, addition of cumulate pyroxene and olivine significantly lowers the Al/Si ratio and thus the distinctions on such diagrams may be related more to the petrological history of the meteorites rather than primitive mantle compositions [72].

The hallmark of this compositional model is its iron-rich nature, a characteristic reinforced by orbital gamma-ray mapping that shows the exposed crust is iron-rich ([73]; also see Chapter 6). However, it has long been recognized that it is difficult to reconcile this composition with the MOI factor, a relatively thin crust and a bulk planetary CI composition (i.e., Fe/Si = 1.71) [10,11]. Recent experimental data also suggest that parental melts to SNC meteorites may have iron contents as much as one-third lower than that suggested by the Wänke model (i.e., FeO $\sim$ 12–14%) but still significantly higher than the composition of the terrestrial primitive mantle (FeO = 8.0%) [74].

Synopsis

Mars, at one-tenth the mass of Earth, has an uncompressed density about 6.5% less. But curiously, the silicate portion is enriched in iron and moderately volatile elements by terrestrial standards. Accordingly, Mars accreted from a compositionally distinct relatively oxidized population of planetesimals. Its small size is likely a consequence of its nebular feeding zone being depleted of mass by the earlier formation of Jupiter. Indeed, Mars may be a good analog of late-stage planetary embryos most of which accreted to form the Earth and Venus.

Evidence from martian meteorites indicates the planet largely melted and differentiated early into a sulfur-rich core, iron-rich silicate mantle possessing several isotopically and geochemically distinct reservoirs, and basaltic crust. The ^{182}Hf–^{182}W isotope system suggests this occurred within $\sim$15 million years of T_{zero} but if Moon-sized planetesimals were delivered with metal cores "pre-packaged", this duration may be slightly underestimated. The ^{146}Sm–^{142}Nd system is more readily interpreted but less constraining and indicates differentiation within $\sim$ 30–50 million years of T_{zero}. Most geological activity occurred within the first billion years (Noachian and Hesperian epochs) but volcanism has persisted intermittently to the present. The basaltic shergottites represent products of young magmatism (575–165 Myr) despite recent suggestions that they are ancient.

Mars possesses a hemispheric dichotomy separating heavily cratered, high-standing thick crust in the south from low-lying thin crust, covered by smooth plains, to the north. Remnants of large impacts indicate the northern terrain is at least as old as the south but covered by a thin volcanic/sedimentary veneer. The origin of the dichotomy remains unresolved and may be related to a giant impactor or a mantle convection pattern established early in the planet's history, perhaps magnified by piling up debris from the excavation of Hellas. Most post-Noachian volcanism has been focused at just a few sites, probably the locations of long-lived plumes. The Tharsis plateau, the largest of these centers, has been the site of magmatism from a near-stationary plume for over 3.5 billion years. Much of this history is preserved within the volcanic and sedimentary layered deposits within Valles Marineris.

Suggestions of an early phase of plate tectonics persist, most recently driven by the recognition of broad Noachian-aged magnetic stripes in the southern highlands. However, the evidence is unconvincing and our judgement is that Mars has been a one-plate planet throughout its history.

The best estimate of the primitive mantle composition is one enriched in iron and moderately volatile elements compared to Earth, each by a factor of about two. However, this composition is derived largely from young martian meteorites which represent a biased sample. The composition is also difficult to reconcile with current understanding of internal structure, as dictated by the moment of inertia data. Accordingly, an iron- and moderately volatile-element-enriched primitive mantle appears secure but the degree of enrichment is rather less certain.

Notes and references

1. Urey, H. C. (1952) *The Planets: Their Origin and Development*, Yale University Press, p. 105.
2. The Viking missions of the mid 1970s were followed by a 20-year hiatus in Mars exploration (including several failed missions). The past decade (1996–2007) witnessed seven successful missions with plans for at least one launch at every opportunity (roughly every 26 months) for the foreseeable future. Recent missions include: 1996 Pathfinder (lander/rover ended 1997); 1996 Mars Global Surveyor (orbiter ended 2007); 1996 Mars 96 (failed launch); 1998 Nozomi (failed to reach Mars orbit); 1998 Mars Climate Orbiter (failed to reach Mars orbit); 1999 Mars Polar Lander (failed on landing); 1999 Deep Space-2 Penetrator (lost contact during landing); 2001 Mars Odyssey (operating orbiter); 2003 Mars Exploration Rover Spirit (ongoing mission); 2003 Mars Exploration Rover Opportunity (ongoing mission); 2003 Mars Express Orbiter (operating orbiter); 2003 Mars Express Beagle 2 Lander (failed on landing); 2005 Mars Reconnaissance Orbiter (operating orbiter). Approved NASA missions to Mars, at the time of writing, include 2007 Phoenix Lander (fixed lander) and 2009 Mars Science Laboratory (rover).
3. Clayton, R. N. (2003) Oxygen isotopes in the solar system. *Space Sci. Rev.* **106**, 19–32.
4. Chambers, J. E. (2004) Planetary accretion in the inner solar system. *EPSL* **223**, 241–52.
5. Watters, T. R. *et al.* (2007) Hemispheres apart: The crustal dichotomy on Mars. *Ann. Rev. Earth Planet. Sci.* **35**, 621–52.
6. The most influential model for the composition of Mars is that of Heinrich Wänke and co-workers: Wänke, H. (1981) Constitution of terrestrial planets. *Phil. Trans. Royal Soc. London* **A303**, 287–302. Dreibus, G. and Wänke, H. (1985) Mars: A volatile rich planet.

Meteoritics **20**, 367–82. Wänke, H. and Dreibus, G. (1988) Chemical composition and accretion history of terrestrial planets. *Phil. Trans. Royal Soc. London* **A349**, 285–93. Halliday, A. N. *et al.* (2001) The accretion, composition and early differentiation of Mars. *Space Sci. Rev.* **96**, 197–230.

7. Lodders, K. and Fegley, Jr., B. (1998) *The Planetary Scientist's Companion*, Oxford University Press. The mass of Mars is 10.7% and its volume 15.1% that of Earth.

8. Stacey, F. D. (2005) High pressure equations of state and planetary interiors. *Rep. Prog. Phys.* **68**, 341–83.

9. Only Viking 2 has landed an operational seismometer on Mars. Inadequate attachment with the surface resulted in seismic response being due to wind-generated noise. Although no unambiguous seismic activity was recorded, there is reason to believe that a seismic network would record significant seismicity resulting both from internal tectonics (marsquakes) and impact activity. Lognonné, P. (2005) Planetary seismology. *Ann. Rev. Earth Planet. Sci.* **33**, 571–604. Because efforts at Mars exploration have given priority to finding evidence for life, the curious situation has arisen where sub-millimeter-scale rock textures can be evaluated routinely on the surface but the internal structure of the planet, measured on scales of tens to thousands of kilometers is largely unknown.

10. In this term I is the moment of inertia, M is mass and R is radius of the planet. The ratio I/MR^2, the moment of inertia factor, depends on mass distribution. A sphere with an evenly distributed mass has a MOI factor of 0.40 and a sphere with mass highly concentrated near the center has a value of about 0.33. (For aficionados of H. G. Wells, the MOI factor of a hollow sphere is 0.67.) Thus the highly differentiated Earth with a large high-density metallic iron–nickel core, has a value of 0.3308 whereas the Moon with no more than a few percent metal core has a value of 0.394. Accordingly, in order to be useful, the MOI factor must be known with considerable accuracy and precision. The value for MOI factor first reported based on Pathfinder data was 0.3662 ± 0.0017. Folkner, W. M. *et al.* (1997) Internal structure and seasonal mass redistribution of Mars from radio tracking of Mars Pathfinder. *Science* **278**, 1749–52. Reanalysis of the data set resulted in the slightly lower value reported here. Yoder, C. F. *et al.* (2003) Fluid core size of Mars from detection of the solar tide. *Science* **300**, 299–303. This value represents the polar MOI but the mean MOI best constrains the interior structure of the planet, thus requiring correction for the planet's oblateness (~ 22 km difference between polar and equatorial radii). This mean planetary value has been estimated at 0.3635 ± 0.0012. Sohl, F. *et al.* (2005) Geophysical constraints on the composition and structure of the Martian interior. *JGR* **110**, doi: 10.1029/2005JE002520.

11. Bertka, C. M. and Fei, Y. (1998) Implications of Mars Pathfinder data for the accretion history of the terrestrial planets. *Nature* **281**, 1838–40. Spohn, T. *et al.* (2001) Geophysical constraints on the evolution of Mars. *Space Sci. Rev.* **96**, 231–62. Kavner, A. *et al.* (2001) Phase stability and density of FeS at high pressures and temperatures: Implications for the interior structure of Mars. *EPSL* **185**, 25–33. Sohl, F. *et al.* (2005) Geophysical constraints on the composition and structure of the Martian interior. *JGR* **110**, doi:10.1029/2005JE002520. Zharkov, V. N. and Gudkova, T. V. (2005) Construction of martian interior model. *Solar Sys. Res.* **39**, 343–73. These latter authors suggested the Fe/Si ratio would correspond to the CI value by adding about 50 mole % hydrogen to the core, unlikely for a body formed in the inner nebula. Adding silicon to the core also explains the discrepancy but makes it more difficult to keep the core molten. A sulfur-rich core makes most cosmochemical sense and has the added advantage of lowering the melting temperature thus allowing for a partially to fully liquid core. The question of the "light element" in the martian core is reminiscent of

arguments over the terrestrial core, and the various non-unique attempts to reconcile the Fe/Si ratio probably tell us as much about the lack of critical data as about the nature of the martian deep interior.

12. This model is based on the Wänke model for the bulk composition of Mars coupled with the experimentally determined mineralogical constitution of Bertka and Fei. The range in core radius is based on numerous estimates. See references in Notes 11 and 13.

13. Stevenson, D. J. (2001) Mars' core and magnetism. *Nature* **412**, 214–19. Yoder, C. F. *et al.* (2003) Fluid core size of Mars from detection of the solar tide. *Science* **300**, 299–303. Fei, Y. and Bertka, C. (2005) The interior of Mars. *Science* **308**, 1120–1. Stewart, A. J. *et al.* (2007) Mars: A new core-crystallization regime. *Science* **316**, 1323–5. Also see Stacey, F. D. (2005) High pressure equations of state and planetary interiors. *Rep. Prog. Phys.* **68**, 341–83.

14. See references in Notes 11 and 13. A sulfur-rich core is not universally accepted. For example, some siderophile-element partitioning experiments were used to suggest that the sulfur content of the core may be <1% thus resulting in a sulfur-depleted planet. Gaetani, G. A. and Grove, T. L. (1997) Partitioning of moderately siderophile elements among olivine, silicate melt and sulfide melt: Constraints on core formation in the Earth and Mars. *GCA* **61**, 1829–46. Such a model is difficult to reconcile with cosmochemistry and accordingly we consider it unlikely.

15. Published estimates for the core radius range from as low as 1370 km to as high as 1900 km. See references in Notes 11 and 13.

16. Bertka, C. and Fei, Y. (1997) Mineralogy of the Martian interior up to core-mantle boundary pressures. *JGR* **102**, 5251–64. Bertka, C. and Fei, Y. (1998) Density profile of an SNC model Martian interior and the moment-of-inertia factor of Mars. *EPSL* **157**, 79–88. This petrological model is based on the primitive mantle composition but note that the planet differentiated into multiple crust–mantle reservoirs early in its history. Accordingly, the present-day internal structure likely differs to some degree. For an alternative mineralogical model, based on mantle chemistry proposed by Morgan and Anders (1979), the reader is referred to Kamaya *et al.* (1993). Morgan, J. W. and Anders, E. (1979) Chemical composition of Mars. *GCA* **43**, 1601–10; Kamaya, N. *et al.* (1993) High pressure phase transitions in a homogeneous model martian mantle. *AGU Geophys. Monogr. Ser.* **74**, 19–26.

17. Breuer, D. *et al.* (1998) Three dimensional models of Martian mantle convection with phase transitions. *GRL* **25**, 229–32. van Thienen, P. *et al.* (2006) A top-down origin for martian mantle plumes. *Icarus* **185**, 197–210.

18. Neukum, G. *et al.* (2001) Cratering records in the inner solar system in relation to the Lunar reference system. *Space Sci. Rev.* **96**, 55–86. Ivanov, B. A. (2001) Mars/Moon cratering rate ratio estimates. *Space Sci. Rev.* **96**, 87–104.

19. Models for ejection of material into space suggest that martian meteorites were derived from the near subsurface – tens to hundreds of meters depth. The oldest meteorite ALH84001, at ~4.5 Gyr, provides a younger limit on the oldest surface of Mars and the youngest shergottites, at 165 Myr, provide an older limit on the youngest surface.

20. Chronology used in this book is mainly from Hartmann, W. K. (2005) Martian cratering 8: Isochron refinement and the chronology of Mars. *Icarus* **174**, 294–320.

21. Adapted from Ref. 20 and Head, J. W. *et al.* (2001) Geological processes and evolution. *Space Sci. Rev.* **96**, 263–92.

22. Images used in Chapters 5 and 6 are taken from a variety of publicly available World Wide Web sources including instrument team sites, the Planetary Photojournal (courtesy of NASA/JPL-Caltech) and the Planetary Data System (PDS).

23. Malin, M. C. and Edgett, K. S. (2000) Sedimentary rocks of early Mars. *Science* **290**, 1927–37. Malin, M. C. and Edgett, K. S. (2003) Evidence for persistent flow and aqueous sedimentation on early Mars. *Science* **302**, 1931–4.

24. Watters, T. R. *et al.* (2007) Hemispheres apart: The crustal dichotomy on Mars. *Ann. Rev. Earth Planet. Sci.* **35**, 621–52. Also see the special issue of *GRL* **33** (8): *The Hemispheric Dichotomy of Mars* (2006). The dividing line between hemispheres is an approximate great circle inclined at ~28° to the equator with highest and lowest latitudes at about 50° E and 230° E. The southern highlands cover approximately 60% of the surface but this is somewhat misleading because the original boundary beneath Tharsis and the effects of Tharsis loading are not fully known. There is a topographic offset, typically about 3–4 km across the dichotomy boundary and a difference in crustal thickness of about 25 km. In detail, however, the topographic break and other geological features do not correlate with changes in crustal thickness.

25. Carr, M. H. (1999) Mars, in *The New Solar System*, 4th edn. (eds. J. K. Beatty *et al.*), Sky Publishing, p. 147.

26. A good review of martian geology at about the time of the Viking missions can be found in Mutch, T. A. and Saunders, R. S. (1976) The geologic development of Mars: A review. *Space Sci. Rev.* **19**, 3–57.

27. Frey, H. V. *et al.* (2002) Ancient lowlands on Mars. *GRL* **29** (10) Art. #1384. Frey, H. V. (2006) Impact constraints on, and a chronology for, major events in early Mars history. *JGR* **111**, doi: 10.1029/2005JE002449. Radar sounding on the Mars Express Orbiter confirms the presence of large buried basins. Watters, T. R. *et al.* (2006) MARSIS evidence of buried impact features in the northern lowlands of Mars. *Nature* **444**, 905–8. The idea that large basins underlie the northern plains was suggested from Viking Orbiter data: e.g. McGill, G. E. (1989) Buried topography of Utopia, Mars: Persistence of a giant impact depression. *JGR* **94**, 2753–9. Confirmation of a large circular depression in Utopia that correlated with gravity anomalies confirmed McGill's suggestion that Utopia was a ~3300 km diameter impact basin.

28. Olympus Mons has an elevation of about 21.3 km, the highest on the planet, and is about 600 km in diameter. Other major volcanoes in Tharsis are Ascraeus Mons, Arsia Mons and Pavonis Mons. Alba Patera, north of Tharsis, and Elysium (14 km elevation), some 4000 km to the west, are the sites of the other major volcanic centers. Alba Patera and Olympus Mons are distinct from the main Tharsis gravity anomaly and thus likely have separate magma sources. It is worth noting for those interested in records that although Olympus Mons is the highest volcano and often referred to as the largest volcano in the Solar System, Alba Patera has the largest areal extent.

29. Phillips, R. J. *et al.* (2001) Ancient geodynamics and global-scale hydrology on Mars. *Science* **291**, 2587–91. Zhong, S. (2002) Effects of lithosphere on the long-wavelength gravity anomalies and their implications for the formation of the Tharsis rise on Mars. *JGR* **107**, doi: 10.1029/2001KE001589.

30. McEwen, A. S. *et al.* (1999) Voluminous volcanism on early Mars revealed in Valles Marineris. *Nature* **397**, 584–6. Also see references in Note 23.

31. Additional evidence for the ancient age for much of Tharsis is the presence of magnetic anomalies in the southern regions and orientation of Noachian tectonic features and valley networks consistent with being influenced by Tharsis crustal loading. Nimmo, F. and Tanaka, K. (2005) Early crustal evolution of Mars. *Ann. Rev. Earth Planet. Sci.* **33**, 133–61.

32. Harder, H. and Christensen, U. R. (1996) A one-plume model of martian mantle convection. *Nature* **380**, 507–9. Breuer, D. *et al.* (1998) Three dimensional models of

martian mantle convection with phase transitions. *GRL* **25**, 229–32. Defraigne, P. *et al.* (2001) Steady-state convection in Mars' mantle. *Planet. Space Sci.* **49**, 501–9.

33. Sleep, N. H. (1994) Martian plate tectonics. *JGR* **99**, 5639–55. Sleep, N. H. (2000) Evolution of the mode of convection within terrestrial planets. *JGR* **105**, 17563–78. For an earlier proposal, see Courtillot, V. E. *et al.* (1975) On the existence of lateral relative motions on Mars. *EPSL* **25**, 279–85.

34. Acuña, M. H. *et al.* (1999) Global distribution of crustal magnetization discovered by the Mars Global Surveyor MAG/ER experiment. *Science* **284**, 790–3. Connerney, J. E. P. *et al.* (1999) Magnetic lineations in the ancient crust of Mars. *Science* **284**, 794–8. Mitchell, D. L. *et al.* (2007) A global map of Mars' crustal magnetic field based on electron reflectometry. *JGR* **112**, doi: 10.1029/2005JE002564.

35. Weiss, B. P. *et al.* (2002) Records of an ancient Martian magnetic field in ALH84001. *EPSL* **201**, 449–63.

36. Stevenson, D. J. (2001) Mars' core and magnetism. *Nature* **412**, 216.

37. Connerney, J. E. P. *et al.* (2005) Tectonic implications of Mars crustal magnetism. *Proc. Natl. Acad. Sci.* **102**, 14970–5.

38. Fairién, A. G. *et al.* (2002) On origin for the linear magnetic anomalies on Mars through accretion of terranes: Implications for dynamo timing. *Icarus* **160**, 220–3. Fairién, A. G. and Dohm, J. M. (2004) Age and origin of the lowlands of Mars. *Icarus* **168**, 277–84.

39. Nimmo, F. (2000) Dike intrusion as a possible cause of linear martian magnetic anomalies. *Geology* **28**, 391–4.

40. Scott, E. R. D. and Fuller, M. (2004) A possible source for the martian crustal magnetic field. *EPSL* **220**, 83–90.

41. Martian magnetization intensity is estimated to be about 10–30 amperes/meter, about an order of magnitude greater than the terrestrial oceanic crust. The high values likely result from a thick magnetized basaltic crust. Note that the resolution of magnetic features on the surface is influenced by spacecraft height and small-scale features with dimensions comparable to magnetic lineations on the terrestrial ocean floor cannot be resolved on Mars.

42. Breuer, D. and Spohn, T. (2003) Early plate tectonics versus single-plate tectonics on Mars: Evidence from magnetic field history and crust evolution. *JGR* **108**, doi: 10.1029/2002JE001999.

43. McSween, H. Y. (1985) SNC meteorites: Clues to martian petrologic evolution? *Rev. Geophys.* **23**, 391–416. McSween, H. Y. and Treiman, A. H. (1998) Martian meteorites. *Rev. Mineral.* **36**, 6–1 to 6–53. Treiman, A. H. *et al.* (2000) The SNC meteorites are from Mars. *Planet. Space Sci.* **48**, 1213–30. McSween, H. Y. (2002) The rocks of Mars, from far and near. *MPS* **37**, 7–25. Bridges, J. C. and Warren, P. H. (2006) The SNC meteorites: Basaltic igneous processes on Mars. *J. Geol. Soc. London* **163**, 229–51.

44. Bogard, D. D. *et al.* (2001) Martian volatiles: Isotopic composition, origin, and evolution. *Space Sci. Rev.* **96**, 425–58.

45. Asteroids, the source of most basaltic achondrites, are small and thus differentiate and crystallize very early in their history due to rapid heat loss. Earth has lost its rock record prior to about 3.8 Gyr and what remains at the surface includes basalt, granite and sedimentary rock in very roughly equal proportions. Mercury and the Moon show no evidence for significant volcanism after about 3 Gyr and samples from the Moon bear no resemblance to martian meteorites. Venus appears to have been entirely resurfaced at about 0.75 Gyr.

46. The number of fragments is greater, but several are multiple fragments from the same meteorite. Shergottite EETA79001 has two compositional domains and is usually

considered as two samples (A and B). Several finds are made each year, mainly from Antarctica and the north African deserts. Most martian meteorites are finds. Shergotty, Zagami, Chassigny and Nakhla are falls and Lafayette, Governador Valadares and Los Angeles come from collections with uncertain origin. Sizes vary from < 10 g to about 18 kg. Small sizes (more than half are < 500 g) combined with coarse-grained textures present considerable analytical challenges for characterizing petrography and geochemistry. For example, published chemical data for many individual meteorites are highly variable, making it difficult to estimate overall bulk composition. A compilation of martian meteorite data is maintained and periodically updated on the World Wide Web: Meyers, C. (ed., 2006) *Mars Meteorite Compendium, Revision C,* JSC # 27672 (www-curator.jsc.nasa.gov/curator/..%5Cantmet/mmc/index.cfm).

47. Information in this table derived from compilations provided by McSween, H. Y. and Treiman, A. H. (1998) Martian meteorites. *Rev. Mineral.* **36,** 6–1 to 6–53. Borg, L. and Drake, M. J. (2005) A review of meteorite evidence for the timing of magmatism and of surface or near-surface liquid water on Mars. *JGR* **110,** doi: 10.1029/2005JE002402. Bridges, J. C. and Warren, P. H. (2006) The SNC meteorites: Basaltic igneous processes on Mars. *J. Geol. Soc. London* **163,** 229–51. See these publications for primary sources. McSween, H. Y. Jr. (2008) Martian meteorites as crustal samples, in *The Martian Surface: Composition, Mineralogy, and Physical Properties* (ed. J. F. Bell III), Cambridge University Press, pp. 383–95.

48. ALH84001is the martian meteorite that was thought to possess evidence for fossil biological activity within fractures. McKay, D. *et al.* (1996) Search for past life on Mars: Possible relic biogenic activity in martian meteorite ALH84001. *Science* **273,** 924–30. These features are now understood to be consistent with abiological origins and interpretations of biological activity are in serious, if not terminal, doubt. See for example: Zolotov, M. Y. and Shock, E. L. (2000) An abiotic origin for hydrocarbons in the Allan Hills 84001 martian meteorite through cooling of magmatic and impact-generated gases. *MPS* **35,** 629–38. Cady, S. L. *et al.* (2003) Morphological biosignatures and the search for life on Mars. *Astrobiol.* **3,** 351–68. Golden, D. C. *et al.* (2004) Evidence for exclusively inorganic formation of magnetite in martian meteorite ALH84001. *Amer. Mineral.* **89,** 681–95.

49. For an exhaustive review of nakhlites, see Treiman, A. H. (2005) The nakhlite meteorites: Augite-rich igneous rocks from Mars. *Chemie der Erde* **65,** 203–70.

50. Nyquist, L. E. *et al.* (2001) Ages and geologic histories of Martian meteorites. *Space Sci. Rev.* **96,** 105–64.

51. Ejection ages represent the sum of cosmic ray exposure ages, which dates the duration that the body spent in space, and terrestrial residence ages. For recent review of ejection ages and potential source regions of martian meteorites, see McSween, H. Y. (2008) Martian meteorites as crustal samples, in *The Martian Surface: Composition, Mineralogy, and Physical Properties* (ed. J. F. Bell III), Cambridge University Press, pp. 383–95.

52. The question is reviewed by Hartmann, W. K. and Barlow, N. G. (2006) Nature of the Martian uplands: Effect on Martian meteorite age distribution and secondary cratering. *MPS* **41,** 1453–67. One recent paper even suggested an influence of climate on atmospheric pressure and that occurrence of extensive glaciations could influence impact ejection. Fritz, J. *et al.* (2007) The Martian meteorite paradox: Climatic influence on impact ejection from Mars? *EPSL* **256,** 55–60.

53. Bridges, J. C. *et al.* (2001) Alteration assemblages in martian meteorites: Implications for near-surface processes. *Space Sci. Rev.* **96,** 365–92. Martian meteorites from desert environments commonly exhibit clear signs of terrestrial weathering that disturb

trace-element distributions. Crozaz, G. *et al.* (2003) Chemical alteration and REE mobilization in meteorites from hot and cold deserts. *GCA* **67**, 4727–41.

54. See reviews in Nyquist, L. E. *et al.* (2001) Ages and geologic histories of Martian meteorites. *Space Sci. Rev.*, **96**, 105–64. Borg, L. and Drake, M. J. (2005) A review of meteorite evidence for timing of magmatism and of surface or near-surface liquid water on Mars. *JGR* **110**, doi: 10.1029/2005JE002402. For an example where multiple isotope techniques give concordant results in a single shergottite, see Borg, L. E. *et al.* (2005) Constraints on the U–Pb isotopic systematics of Mars inferred from a combined U–Pb, Rb–Sr, and Sm–Nd isotopic study of the Martian meteorite Zagami. *GCA* **69**, 5819–30.

55. Blichert-Toft, J. *et al.* (1999) The Lu–Hf isotope geochemistry of shergottites and the evolution of the Martian mantle–crust system. *EPSL* **173**, 25–39. Bouvier, A. *et al.* (2005) The age of SNC meteorites and the antiquity of the Martian surface. *EPSL* **240**, 221–33.

56. Hurowitz, J. A. *et al.* (2005) *In situ* and experimental evidence for acidic weathering of rocks and soils on Mars. *JGR* **110**, E02S19. Hurowitz, J. A. and McLennan, S. M. (2007) A ~3.5 Ga record of water-limited, acidic weathering conditions on Mars. *EPSL* **260**, 432–43.

57. Walton, E. L. *et al.* (2007) A laser probe ^{40}Ar/^{39}Ar investigation of two martian lherzolitic basaltic shergottites. 70th Annual Meteorical Society Meeting, Tucson, AZ, August 13–17, 2007, Abst. # 5211.

58. Misawa, K. and Yamaguchi, A. (2007) U–Pb ages of NWA 856 baddeleyite. 70th Annual Meteorical Society Meeting, Tucson, AZ, August 13–17, 2007, Abst. #5228.

59. Halliday, A. N. *et al.* (2001) The accretion, composition and early differentiation of Mars. *Space Sci. Rev.* **96**, 197–230.

60. Lee, D.-C. and Halliday, A. N. (1997) Core formation on Mars and differentiated asteroids. *Nature* **388**, 854–7. Kleine, T. *et al.* (2004) ^{182}Hf–^{182}W isotope systematics of chondrites, eucrites, and martian meteorites: Chronology of core formation and early mantle differentiation in Vesta and Mars. *GCA* **68**, 2935–46. Foley, C. N. *et al.* (2005) The early differentiation history of Mars from ^{182}W–^{142}Nd isotope systematics in the SNC meteorites. *GCA* **69**, 4557–71.

61. Borg, L. E. *et al.* (2003) The age of Dar al Gani 476 and the differentiation history of the martian meteorites inferred from their radiogenic isotopic systematics. *GCA* **67**, 3519–36. Foley, C. N. *et al.* (2005) The early differentiation history of Mars from ^{182}W–^{142}Nd isotope systematics in the SNC meteorites. *GCA* **69**, 4557–71. Devaille, V. *et al.* (2007) Coupled ^{142}Nd–^{143}Nd evidence for a protracted magma ocean in Mars. *Nature* **450**, 525–8. This last paper suggests that the array is best explained by mixing of two reservoirs, a REE-depleted reservoir about 4.535 Gyr and a REE-enriched KREEP-like reservoir about 4.457 Gyr. Such distinctions in interpretation rely in part on subtle differences in initial bulk planetary Nd-isotopic compositions, at levels consistent with inhomogeneous distributions of nucleosynthetic products in the solar nebula.

62. Brandon, A. D. *et al.* (2000) Re–Os isotopic evidence for early differentiation of the martian mantle. *GCA* **64**, 4083–95.

63. Borg, L. E. and Draper, D. S. (2003) A petrogenetic model of the origin and compositional variation of the martian basaltic meteorites. *MPS* **38**, 1713–31.

64. Elkins-Tanton, L. T. *et al.* (2003) Magma ocean fractional crystallization and cumulate overturn in terrestrial planets: Implications for Mars. *MPS* **38**, 1753–71. Elkins-Tanton, L. T. *et al.* (2005) Early magnetic field and magmatic activity on Mars from magma ocean cumulate overturn. *EPSL* **236**, 1–12. Elkins-Tanton, L. T. *et al.* (2005) Possible formation of ancient crust on Mars through magma ocean processes. *JGR* **110**, doi:

10.1029/2005JE002480. It is worth noting that in these magma ocean overturn models, incompatible-element-enriched layers sink and are concentrated near the core–mantle boundary. Given the evidence for incompatible-element-enriched crust (Chapter 6) and possibly additional KREEP-like reservoirs sampled during shergottite magma ascent, they could present a serious overall mass balance problem.

65. Longhi, J. (1991) Complex magmatic processes on Mars: Inferences from the SNC meteorites. *LPSC* **21**, 695–709. McLennan, S. M. (2003) Large-ion lithophile element fractionation during the early differentiation of Mars and the composition of the martian primitive mantle. *MPS* **38**, 895–904.

66. Taylor, G. J. *et al.* (2008) Implications of observed primary lithologies, in *The Martian Surface: Composition, Mineralogy, and Physical Properties* (ed. J. F. Bell III), Cambridge University Press, pp. 501–18.

67. Jones, J. H. (2003) Constraints on the structure of the martian interior determined from the chemical and isotopic systematics of SNC meteorites. *MPS* **38**, 1807–14. Jones suggested an upper highly depleted mantle as the source of shergottites (mixed to varying degrees with enriched crust) and a less depleted lower mantle as the source for nakhlites. Models incorporating magma oceans have appealed to various cumulate layers and/or trapped liquids (analogous to lunar KREEP) as distinct reservoirs (see references in Notes 63–64). A stable layered mantle might provide a useful means of keeping reservoirs isolated over long timescales, as required by the isotopic data but may be more difficult to reconcile with geophysical evidence for long-lived stable convection originating from the lower mantle.

68. Kiefer, W. S. (2003) Melting in the martian mantle: Shergottite formation and implications for present-day mantle convection on Mars. *MPS* **38**, 1815–32. Kiefer points out the potential difficulty of later melting to form shergottites in a layered mantle due to lack of heat sources. He suggested that if highly depleted shergottite source regions are embedded within a more enriched (or at least less depleted) mantle later melting might be easier to accomplish.

69. An early model for the bulk composition of Mars (Morgan, J. W. and Anders, E. (1979) Chemical composition of Mars. *GCA* **43**, 1601–10) predates recognition of martian meteorites and relies on equilibrium condensation models for the terrestrial planets and unreliable Mars 5 gamma-ray spectroscopy data; it is now only of historical interest. A recent group of models derive the bulk composition by mixing various classes of chondritic meteorites to obtain the correct O-isotope composition. Volatile-element depletion occurs during accretion by giant impacts. Lodders, K. and Fegley, B. (1997) An oxygen isotope model for the composition of Mars. *Icarus* **126**, 373–94. Sanloup, C. *et al.* (1999) A simple chondritic model of Mars. *PEPI* **112**, 43–54. Lodders, K. (2000) An oxygen isotope mixing model for the accretion and composition of rocky planets. *Space Sci. Rev.* **92**, 341–54. These models suffer from the same difficulties as all models that attempt to build planets from a small subset of meteorite compositions (see Chapter 1), and also predict far higher levels of moderately volatile elements than indicated by modern gamma-ray spectroscopy.

70. Assuming that shergottites represent the martian mantle array, the intersection with $\varepsilon_{Nd}=0$ gives an $^{87}Sr/^{86}Sr$ value of about 0.719 for the primitive mantle. Assuming Rb/Sr volatile fractionation occurred at about 4.5 Gyr, this leads to a Rb/Sr ratio of about 0.1 which compares favorably to 0.07 for the Wänke model (Table 5.4).

71. As is often the case, siderophile-element distributions seem to shed as much heat as light. Moderately siderophile elements (e.g. Ni, Co) are highly depleted in this primitive-mantle composition and in martian meteorites in general but how this relates to conditions of metal–silicate equilibrium during core formation is less clear

with estimates ranging from low pressure/high temperature ($<$ 10 kbar; $>$ 2200 °C) to moderate pressure/low temperature ($\sim$ 80 kbar; 1625 °C). Highly siderophile elements in martian meteorites (Ir, Os, Ru, Pt, Au) are $\sim$ 0.01–0.001 $\times$ CI but not all in chondritic proportions and thus likely were added by a "late veneer" but subsequently differentiated within the mantle. Kong, P. *et al.* (1999) Siderophile elements in Martian meteorites and implications for core formation in Mars. *GCA* **63**, 1865–75. Righter, K. (2007) Early differentiation of Mars: Constraints from siderophile elements and oxygen fugacity (abst.). *Antarctic Meteorites* **XXXI**, 89–90 (Institute of Polar Research, Tokyo).

72. Filiberto, J. *et al.* (2006) The Mars/Earth dichotomy in Mg/Si and Al/Si ratios: Is it real? *Am. Mineral.* **91**, 471–4. The compositions of basaltic rocks found at Gusev Crater, that are interpreted to be dominated by melt, also fall near the terrestrial trend on the Mg/Si vs. Al/Si diagram. McSween, H. Y. *et al.* (2006) Characterization and petrologic interpretation of olivine-rich basalts at Gusev crater, Mars. *JGR* **111**, doi: 10.1029/2005JE002477.

73. Taylor, G. J. *et al.* (2006) Bulk composition and early differentiation of Mars. *JGR* **111**, doi: 10.1029/2005JE002645.

74. Borg, L. E. and Draper, D. S. (2003) A petrogenetic model for the origin and compositional variation of the martian basaltic meteorites. *MPS* **38**, 1713–31. Agee, C. B. and Draper, D. S. (2004) Experimental constraints on the origin of Martian meteorites and the composition of the Martian mantle. *EPSL* **224**, 415–29. But for a contrary view, see: Monders, A. G. *et al.* (2007) Phase equilibrium investigations of the Adirondack class basalts from the Gusev plains, Gusev crater, Mars. *MPS* **42**, 131–48.

6

Mars: crustal composition and evolution

Early science fiction portrayed Mars as totally alien and unfamiliar, but some aspects of the Martian surface would seem surprisingly recognizable to a human visitor.

(Bill Hartmann) [1]

Although Mars has been the focus of planetary exploration over the past three decades, most effort has centered on evaluating the distribution and history of near-surface liquid water as a marker for potential habitability – for ancient and extant life and for future exploration – and so much of the data are more relevant to questions of surficial processes over geological time. These findings are important, but not central to our investigation. Accordingly, we have resisted the temptation of focusing too much attention on these results, impressive as they are. Instead, we consider them where they address major questions of crustal evolution that are the subject of this enquiry. There are several up-to-date reviews of the recent findings from the Mars exploration programs for those so interested [2].

Martian crustal evolution represents a near-perfect intermediary between the simple and mostly ancient crustal histories of Mercury and the Moon, where primary crusts dominate, and the extended evolution of Earth. On Earth any primary crust that may have existed is long since lost from the geological record and both secondary and tertiary crusts formed, but at very different rates, over some four billion years [3].

6.1 Sampling martian crust

Mars presents unique challenges in obtaining representative sampling of the crust. It is one of few planetary bodies for which samples are available and thus is accessible to sophisticated analytical techniques in terrestrial laboratories. For the Earth and Moon, most samples have solid geological context; even for lunar meteorites we have some understanding of their geological setting because of extensive knowledge obtained from the lunar missions. But this is not the case for martian meteorites, delivered to Earth by young impacts. Indeed, martian meteorites are heavily biased towards Amazonian volcanic/plutonic terrains, ironically the time of least volcanic activity, and represent as few as four distinct locations, but even this is uncertain (see Chapter 5).

Five landed missions have studied the geochemistry of the martian surface and, for the last two missions have also evaluated the mineralogy of rocks and soils using visible, infrared and Mössbauer spectroscopy [4]. Of these, the rover Spirit has traversed across (at least) two distinct terrains (Gusev plains; Columbia Hills) and so effectively six distinct sites have been examined. Site selection is also biased due to various engineering constraints. These results can be placed into reasonable geological context but the scope of data is limited. Thus, major-element data are available but analytical quality is variable from mission to mission and, in some cases, difficult to interpret. Spirit and Opportunity analyzed several trace elements but coverage is limited (only Cr, Ni, Zn, Br were obtained routinely). Various spectroscopic observations produce incomplete and, in many cases, ambiguous mineralogical interpretations [5].

Orbital remote sensing is another way to sample the crust. Mars Odyssey gamma-ray spectroscopy (GRS) is especially relevant [6], providing the chemical composition of the surface with "footprints" ~ 300 km diameter and sampling depths on the order of a meter or less, depending on material properties of near-surface deposits. Element coverage is incomplete and effective spatial resolution highly variable among elements [7]. On the other hand, the heat-producing and incompatible elements K and Th, as gamma-ray emitters, are among the best determined as are several elements either important to petrologic modeling (e.g. Si, Fe, Ca, Al) or for understanding surficial processes (Cl, S). Thermal infrared spectroscopy constrains surface mineralogy and, in turn, chemical inferences can be made from modeled mineralogy but with great uncertainty. The thermal emission spectrometer (TES) instrument onboard Global Surveyor had a "footprint" of a few kilometers and mapped the entire martian surface but only to depths on the order of a few microns [8].

The important point here is that there is a wealth of high-quality chemical and spectral data available for the near-surface of Mars, but integrating various types of data into a coherent picture of the chemical and mineralogical constitution of the martian crust is more challenging.

6.2 Crustal dimensions

The mass of the martian crust scales directly with the degree to which incompatible elements (including heat-producing elements) have differentiated into the crust.

Mars has a surface area of 1.444×10^8 km^2, or equivalent to about 28.2% of the Earth's surface area, roughly two-thirds the area of terrestrial continental crust and almost identical (97.6%) to the area of the Earth's land mass. The planet has 29.4 km of surface relief with the bottom of the Hellas impact basin being the lowest (-8.2 km) and the top of Olympus Mons the highest ($+21.2$ km) [9].

Without seismic data, constraining crustal thickness must employ indirect and model-dependent methods. The best approach combines topography and gravity, which leads to an average value of about 50 km and with variations between about 3–92 km but mostly between 20–70 km. The crust beneath the southern highlands is about 25 km thicker than that beneath the northern plains [10]. Other approaches have included geochemical mass balance, moment of inertia data and viscous relaxation of topography, all of which include various assumptions (that could be avoided with some good seismic measurements). Wieczorek and Zuber [11] reviewed the various approaches and concluded that the average thickness of the crust was 50 ± 12 km, the value adopted here (Fig. 6.1).

Uncertainties in crustal thickness, volume of the mantle (see Chapter 5) and densities of both in turn place considerable uncertainties on the fraction of the planet and of the primitive mantle represented by the crust. Thus the volume of crust is about $7.11 \pm 1.68 \times 10^9$ km^3, which for a density of 3.00 g/cm^3 leads to a mass of $2.13 \pm 0.50 \times 10^{22}$ kg. These values translate into the crust representing $4.4 \pm 1.0\%$ of the volume of the planet. Of more interest to this study is what fraction the crust is of the primitive mantle. For the uncertainties in the volume of the martian mantle described in Chapter 5 ($\sim \pm 9\%$) and for a mean mantle density of 3.60 g/cm^3, the crust represents $4.9 \pm 1.3\%$ of the volume and $4.1 \pm 1.1\%$ of the mass of the primitive mantle. For comparison, these values are approximately six times the fraction of the terrestrial primitive mantle that is represented by the present-day terrestrial (continental + oceanic) crust.

6.2.1 Hypsometry

One manifestation of the terrestrial continental–oceanic crust dichotomy is bimodal hypsometry that corresponds to the crustal type (i.e., oceanic vs. continental). Martian hypsometry is also bimodal, broadly similar to Earth. The martian bimodal distribution corresponds largely to the hemispheric dichotomy separating ancient southern highlands from younger low-standing smooth plains to the north. On Earth, the bimodal distribution in crustal elevation reflects isostatic response to differing oceanic and continental crust and lithosphere compositions, raising the question that there may be analogous compositional differences on Mars, perhaps implying continental-like crust. However, when hypsometry is normalized to the planet's center of figure (as opposed to the center of mass) [13], the bimodality

(a)

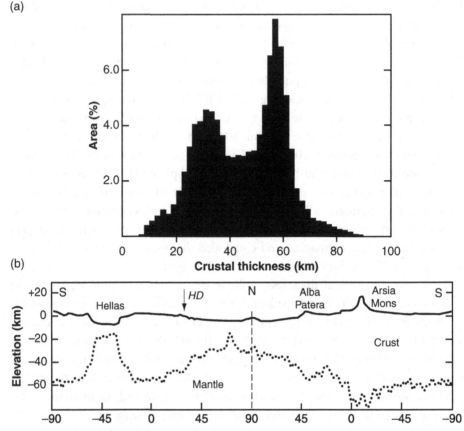

(b)

Fig. 6.1 (a) Histogram of crustal thickness on Mars. (b) Crustal thickness profile along longitudes 60° E–240° E. Note that the crust thins greatly beneath the Hellas impact basin and thickens beneath Tharsis (vicinity around Arsia Mons), where it obscures the hemispheric dichotomy boundary. On this transect, the dichotomy boundary is best resolved at about 60° north of the Hellas rim (denoted by arrow marked HD). Vertical exaggeration of 60:1. Adapted from Neumann *et al.* [12].

disappears and a unimodal distribution of elevation results. On Earth the offset between center of figure and center of mass is much smaller and when a similar correction is made, the bimodal hypsometry remains [14]. Accordingly, variation in crustal thickness is the dominant control on crustal elevation on Mars with no broader tectonic significance (Fig. 6.2).

6.3 Igneous diversity in a basaltic crust

It is widely understood that Mars is a "basaltic planet" [2, 3] with the crust dominated by basaltic lavas and their intrusive and volcaniclastic cousins. However, as with any remote region unavailable for sampling, there is a tendency to ascribe simplicity

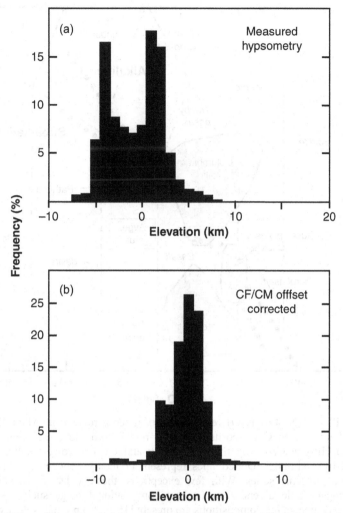

Fig. 6.2 Plots of martian (a) hypsometry and (b) hypsometry corrected for the 2.99 km CF/CM offset. Note that the bimodal distribution of hypsometry disappears when the effect of the CF/CM offset is taken into account [14].

in the place of ignorance. Thus until quite recently, martian meteorites dominated our thinking about the crust in spite of their obvious biases. As the surface of Mars has become more accessible and studied, our understanding of the complexity has increased greatly. While Mars remains a "basaltic planet", the crust in fact appears to have had a complex magmatic history (Fig. 6.3), the details of which are only beginning to be unraveled. A review of the various igneous lithologies that have been documented can be found in Taylor *et al.* [15]. With the possible exception of ALH84001, it is likely that all martian meteorites and rocks studied on the surface by rovers are representative of young secondary crust-forming processes.

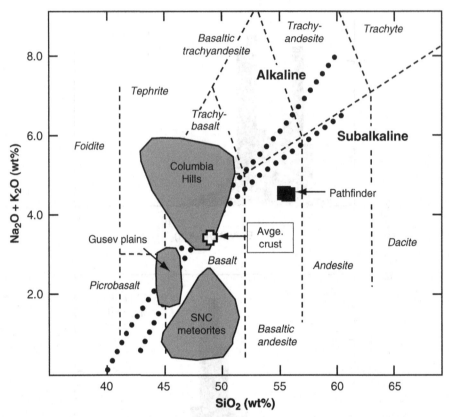

Fig. 6.3 Plot of ($Na_2O + K_2O$) versus SiO_2 for igneous rocks from Mars. Plotted are the fields for SNC meteorites, igneous rocks from the Gusev plains and Columbia Hills analyzed by the Spirit rover and the two lowest sulfur rocks analyzed by Pathfinder. Dotted lines represent boundary between alkaline and subalkaline volcanic series. With few exceptions, these rocks are basaltic but exhibit considerable diversity within that compositional range such that both alkaline and subalkaline compositions are present [16]. Also plotted is the estimate for the bulk composition of the martian crust (see Section 6.5).

6.3.1 SNC meteorites and crustal contamination

The importance of martian meteorites for understanding the composition and early differentiation of the mantle was provided in Chapter 5. These rocks, representing volcanic activity at six or seven distinct sites (but perhaps as few as four), were derived from the near surface during impacts and thus provide direct information about the crust [17]. Apart from the single Noachian-aged sample ALH84001, all other martian meteorites (SNC) sample Middle- to Late Amazonian crust and thus provide information about young crustal additions, which in terms of overall crustal mass are minor (see below).

Although there are significant petrological distinctions among the different varieties of shergottites, they all share some common traits. All represent lavas or cumulates from silica-saturated low-K basalts characterized by high FeO_T, low Al_2O_3 and low Mg-numbers. These features have long been interpreted to reflect derivation from a residual mantle (i.e., from which an aluminium-rich crust had been extracted) enriched in iron compared to the Earth (see Chapter 5 for further discussion and caveats). Among other notable characteristics are correlated variations in trace element and isotopic compositions and oxidation states indicative of mixing between two sources. This is clearly seen in REE distributions (Fig. 5.5), isotope variations (e.g. Fig. 5.10), and fO_2 [18]. One source is a long-term highly incompatible-element (e.g. LREE) depleted, relatively reduced ($\sim$ QFM -4; where QFM is the quartz–fayalite–magnetite oxygen fugacity buffer) mantle source. The other is relatively oxidized ($\sim$ QFM -1) and characterized by incompatible-element enrichment.

This enriched component may be of special importance to understanding crustal evolution but its origin is the subject of considerable debate. The standard model is that this component represents contamination from ancient crust either directly mixed into the magma or by more complex assimilation – fractional-crystallization-type processes. If so, the composition of the crustal component can be constrained as oxidized and modestly enriched in incompatible elements (La/Yb $\sim$3–4) with present-day $\varepsilon_{Nd} \sim -20$ [19]. However, the lack of correlation between apparent degree of crustal contamination and other indices of differentiation (e.g. Mg #) suggests that the second component could be a long-term enriched mantle source, possibly analogous to lunar KREEP [18, 20]. Although shergottites are related in general terms, they represent a range in crystallization age and geological settings (i.e., ejection ages). Accordingly, they no doubt represent a variety of petrogenetic histories and so care is required in interpreting the absence of correlations between trace elements, isotopes and other petrological indices. Nevertheless, the difficulty in distinguishing ancient large-ion lithophile (LIL)-enriched mantle components from crustal contamination has also plagued the study of terrestrial basalts [21] and so this controversy is unlikely to be resolved any time soon.

The 1.3 Gyr cumulate nakhlites (clinopyroxenites) and chassignites (dunites) have similar crystallization ages, isotopic compositions and ejection ages and may be derived from a single impact event $\sim$ 11 million years ago (see reviews in Refs. 15, 22, 23). The basalts from which they were derived were broadly similar to shergottites, enriched in Fe and depleted in Al. In contrast to shergottites, nakhlites and chassignites are incompatible-element enriched (e.g. La/Yb $\sim$ 6) but Nd isotopic evidence indicates derivation from long-term LREE-depleted sources (ε_{Nd} (1.3 Gyr) $\sim +15–+17$), generally reminiscent of many terrestrial plume volcanics. There are sufficient differences in composition between nakhlites and chassignites to indicate their petrogenetic relationship is complex. They likely represent zones of mineral accumulation at the base of lava flows or sills. Why only cumulate zones of the lava flows would be sampled by the ejection process is not understood.

Recent experiments based on chassignite olivine melt inclusions suggest fractionation assemblages similar to those observed in terrestrial silica-saturated alkaline lavas (e.g. hawaiites), such as those found at Ascension Island and the Azores [24]. These results add to the growing body of evidence that alkaline magmatism may have been more common on Mars than previously appreciated (discussed further below in Section 6.3.4). Alkaline suites on the Earth commonly show a broad range in composition, including small volumes of highly differentiated rocks such as dacites and rhyolites.

6.3.2 Hemispheric dichotomy, Surface Types 1 and 2 and martian andesites

Global mapping of TES defined two distinctive regions within low dust areas, termed Surface Types 1 and 2 (ST1, ST2), in the low-albedo regions of the southern and northern hemispheres, respectively [25]. Spectral analysis suggests that ST1 is best modeled as clinopyroxene–plagioclase basalt but interpretation of ST2 is controversial. Initially modeled as igneous rocks enriched in plagioclase and volcanic glass and interpreted to be andesitic, the spectrum can also be modeled as altered basalt with components of clays and/or amorphous silica [26]. The chemical compositions of ST1 and ST2 have been estimated by mixing mineral end members derived from spectral deconvolution; however, elements such as Fe, Mg, Ca, Na and K can vary by more than a factor of two depending on the minerals used in the spectral deconvolution and mixing calculations [27, 28]. This uncertainty makes it difficult to use such approaches to estimate the chemical composition of the martian crust in a robust manner.

Gamma-ray spectroscopy mapping provides additional insight on global-scale compositional variations. The GRS has a "footprint" about two orders of magnitude larger (~ 300 vs. ~ 3 km) and a sampling depth about four orders of magnitude greater (~ 50 cm vs. < 50 μm) than TES and so comparing results requires caution. When ST1 and ST2 regions are compared using GRS data, the only significant differences are elevated K and Th (but not Si) in ST2 [29]. Since TES data are far more sensitive to thin surface coatings, where aqueous alteration effects may dominate (see below) the discrepancy is not surprising and appears to support the alteration model.

An intriguing characteristic of GRS data is that correlations with major geological provinces, such as Tharsis or the dichotomy boundary for the most part are subtle [30]. Among the factors that may explain this result are the relatively large footprint, limited chemical variation on a "basaltic planet", mixing associated with impact processes and the ubiquitous presence of soils and dust that blunt regional igneous

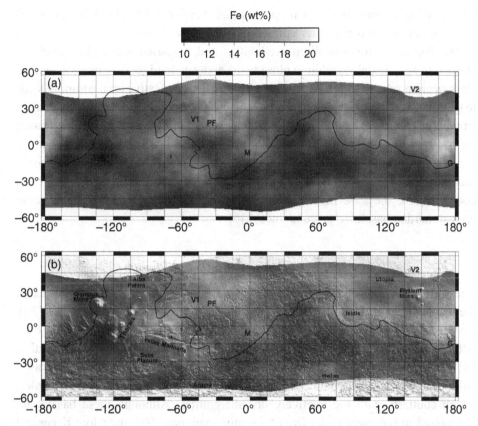

Fig. 6.4 Global maps of iron content in the martian near surface mapped by gamma-ray spectroscopy. (a) Map of iron content shown in grayscale. Black line is the 0 km elevation contour. Landing sites labeled: V1, V2 – Viking 1 and 2; PF – Pathfinder; M – Opportunity at Meridiani Planum; G – Spirit at Gusev Crater. Cut-off of data in north and south corresponds to water-rich polar regions ("H-mask"). The northern hemisphere surface has enriched iron compared to the south. (b) Iron content with shaded-relief overlay showing relationships between composition and major geomorphological features. Maps courtesy of Buck Janes and Bill Boynton.

compositional variations. A notable exception is a hemispheric contrast in iron content with the northern hemisphere being enriched by about 2–3% FeO_T, which likely reflects the composition of the younger resurfacing of the northern plains (Fig. 6.4).

Two proposed explanations are (1) the effect is due to secondary alteration where iron is leached from rocks in the south by acidic weathering and transported by surface waters to the northern hemisphere or (2) the effect is fundamentally igneous with younger surface lavas in the north being more iron-rich than the older crust exposed in the south [31, 32]. Lack of correlation between Fe and Cl [33] favors the latter explanation and thus it appears that there may be secular variations in crustal composition due to later volcanism (see below).

The Pathfinder mission discovered SiO_2- and K_2O-rich rocks [34], likely of Hesperian age, that plot within or close to the andesite field on various geochemical classification schemes (Table 6.1; Fig. 6.3) [35]. This evocative term naturally leads to speculation of a plate-tectonic origin. However, such rocks are chemically distinct from terrestrial andesites, for example; their content of FeO (~ 15%) is about twice the levels typically seen for calc-alkaline andesites, and so any such comparison is superficial. Assuming these rocks are indeed igneous (a question by no means settled) and that the measured composition is not strongly influenced by surface coatings [35], they are most likely late-stage differentiates derived from more typical martian basalts (e.g. icelandites) [36].

6.3.3 Gusev plains and Meridiani Planum

The Spirit rover analyzed three volcanic cobbles, thought to be of Hesperian age, on the plains of Gusev Crater [37, 38]. This lithology, termed Adirondack Class [39], represents a low-K (but high-Na) olivine-bearing picritic basalt [40] (Table 6.1). Although much older, they bear some resemblance to olivine-phyric shergottites but with lower SiO_2 and higher Al_2O_3 and Na_2O (reflecting more plagioclase). Adirondack rocks also contain vugs and vesicles suggesting magmatic volatiles, which contrasts with the relatively dry shergottites. Although these basalts are interpreted to represent melts from "primitive magmas" [40] their low K content (~ 250–900 ppm), comparable to the most LREE-depleted shergottites, suggests their sources were incompatible-element depleted.

Several basalt pebbles have been analyzed on the Meridiani Plains, the best-characterized being Bounce Rock [41]. This rock is chemically similar to basaltic shergottites, notably EETA79001B. Other basalt cobbles were too small to abrade with the rock abrasion tool and so the composition includes an altered surface (i.e., high S and Cl) but compared to Bounce Rock, these rocks are likely characterized by higher FeO_T, MgO, K_2O, P_2O_5, Cr, Ni and lower CaO. The ages of the pebbles are not known as they are likely ejecta but if they were derived from rocks underlying or in the vicinity of Meridiani Planum, they are likely much older than the shergottites, perhaps Late Noachian to Hesperian.

6.3.4 Alkaline volcanism and the Columbia Hills

The Noachian to Early Hesperian Columbia Hills, including the region termed Home Plate, in Gusev Crater [38] appear to be an alkaline volcanic province [42]. Rocks preserved within this terrain appear to be largely of pyroclastic origin and

Table 6.1 *Chemical composition of selected volcanic and pyroclastic rocks from the martian surface [37]*

	Gusev plains		Columbia Hills				Home Plate	Meridiani Planum	Pathfinder	
	Adirondack		Wishstone		Backstay		Fastball	Bounce Rock	Shark	
	1	2	1	2	1	2			1	2
SiO_2	45.7	46.0	43.8	44.6	49.5	50.1	45.3	51.6	54.3	55.1
TiO_2	0.48	0.48	2.59	2.64	0.93	0.94	0.67	0.74	0.6	0.6
Al_2O_3	10.87	10.94	15.03	15.3	13.25	13.40	7.85	10.48	10.3	10.5
FeO_T	18.8	18.9	11.6	11.8	13.0	13.1	17.8	14.4	15.2	15.4
Cr_2O_3	0.61	0.61	n.d.	–	0.15	0.15	0.49	0.11	0.1	0.1
MnO	0.41	0.41	0.22	0.22	0.24	0.24	0.47	0.40	0.4	0.4
MgO	10.83	10.90	4.50	4.58	8.31	8.40	12.0	6.84	3.7	3.8
CaO	7.75	7.80	8.89	9.05	6.04	6.11	5.80	12.09	8.2	8.3
Na_2O	2.4	2.4	5.0	5.1	4.15	4.20	2.35	1.66	3.5	3.6
K_2O	0.07	0.07	0.57	0.58	1.07	1.08	0.23	0.11	1.0	1.0
P_2O_5	0.52	0.52	5.19	5.28	1.39	1.41	0.79	0.92	0.4	0.4
SO_3	1.23	0.75	2.20	0.75	1.52	0.75	4.63	0.56	1.8	0.75
Cl	0.20	–	0.35	–	0.35	–	1.57	0.35	0.45	–
Σ	99.9	99.8	99.9	99.9	99.9	99.9	100.0	100.3	100.0	100.0
Cr	4175	4200	n.d.	–	1025	1035	3350	750	700	710
Ni	165	166	67	68	191	193	352	81	–	–
Zn	81	82	64	65	269	272	415	38	–	–
Br	14	19	22	–	26	–	370	39	–	–
Ge	–	–	–	–	–	–	70	–	–	–
	Abraded		Abraded		Abraded		As is	Abraded	As is	

Notes: For samples listed twice, column 1 represents the best analysis available and column 2 represents the recalculated values assuming $SO_3 = 0.75\%$, the highest values observed in martian meteorites. This approach is after McSween *et al.* [42]. For sample Fastball, the best available analysis is on an "as is" surface and the high SO_3 content indicates soil contamination and surface alteration and so the simple correction applied to other samples is inadequate. Bounce Rock is an abraded surface and the SO_3 content is <0.75% so no corrections are necessary. The Shark analysis is on an "as is" surface but the low SO_3 content suggests that soil contamination and alteration are relatively slight and accordingly, the simple correction is applied (also see Note 35). Major elements in wt% ; trace elements in ppm. **Data sources**: See Notes 34–37, 40–42. **Abbreviations**: FeO_T, total iron as FeO; n.d., not detected; –, no analysis available.

include volcanic, tuffaceous and volcaniclastic rocks, in places reworked by eolian processes. Because the Spirit rover lost its ability to abrade rock surfaces shortly after beginning the ascent of the Columbia Hills, exact primary chemical compositions are less secure. Nevertheless, the high levels of ($Na_2O + K_2O$) clearly point to mildly alkaline compositions. Indeed, McSween *et al.* [42] demonstrated that it is possible to derive Columbia Hills magmatic rocks by fractional crystallization from Adirondack-like volcanic compositions and accordingly the entire magmatic suite within Gusev Crater may be part of an alkaline province.

The global extent of such terrains is not well known. With increasing resolution, orbital spectroscopy has begun to identify small regions with highly evolved compositions, even including several small regions of quartz-bearing volcanic rocks [43]. However, orbital spectroscopy has generally been interpreted to suggest that the martian surface is dominated by subalkalic basaltic compositions but this could be due in part to inadequate mineral libraries; considerably more work is needed to evaluate this issue [28].

6.4 The sedimentary rock cycle on Mars

The terrestrial sedimentary record provides insight into the origin, composition and evolution of continental crust (see Chapters 10 and 11). Accordingly, existence of a sedimentary record on Mars promises to provide comparable information [44]. There are also many fundamental differences expected (and observed) for the sedimentary records of the two planets. On Earth, forces associated with plate tectonics are the main cause of uplift, subsidence, basin formation and the large-scale architecture of sedimentary deposits. On Mars, plate tectonics never occurred and associated large-scale thermal subsidence appears to be mostly absent. Instead, volcanic loading and impact processes provide the differential elevation, subsidence and basins that are required for sediment formation, transport, accumulation and preservation. Most sediment on Earth is derived from the "granitic" upper continental crust and the sedimentary record mostly reflects continental evolution. In contrast, Mars is a "basaltic planet" and accordingly, the sedimentary record responds very differently.

6.4.1 Water, wind and ice

The present-day inventory of water on or near the surface of Mars is likely only a small fraction of that on Earth both in terms of absolute amounts and concentration [45]. The volume of polar deposits is about $4–5 \times 10^6$ km^3 but the ratio of dust to water is not well constrained. Even assuming they are completely dominated by water ice, they only represent 0.3% of the terrestrial hydrosphere. It has long been understood that subsurface ice likely exists in pore spaces of the regolith and that this is the main reservoir of near-surface water [46]. Neutron mapping suggests that on average, the top meter contains about 14% water [47]. At high latitudes where near-surface ice is stable, this probably represents water ice. However, at mid latitudes where near-surface ice is thought to be unstable the situation is less clear. One possibility is that low-latitude subsurface water is in the form of ice that represents remnants of a previous climatic regime. A more likely possibility is that a significant fraction of low-latitude water is a structural component held by the various hydrated minerals that have been detected in soils and outcrops [48].

Mars has long been known to possess a dynamic sedimentary cycle [49]. In addition to extensive impact and volcaniclastic deposits, a variety of additional sedimentary processes have been documented, some of which implicate subaqueous conditions during the planet's early history. The canals of Lowell have long since given way to erosional channels of widely varying scale, branching valley networks and gullies, all suggesting the flow of liquid water at the surface. There is evidence that gullies continue to form today [50] but mostly the evidence for flowing water points to ancient processes (although some large channels may be as young as ~ 100 Myr). Valley networks are mostly restricted to the heavily cratered highlands, suggesting they are equally ancient, mostly Noachian with lesser occurrences during the Hesperian.

Extensive and thick-layered deposits, in some cases exhibiting clearly defined unconformities, are found in many craters in the ancient southern highlands (Fig. 6.5). Layered deposits likely formed by a variety of volcanic and impact processes in addition to eolian and subaqueous transport. However, in some of these craters geomorphological evidence suggests both moving and standing water, in the form of deltas, terraces and outlet channels. A popular proposal, but one for which convincing evidence remains elusive, is that the northern lowlands were the site of Noachian oceans, perhaps as much as 3 km deep.

Layered deposits are also exposed in the walls of the canyon systems of Valles Marineris, although uncertainty persists as to whether all of these deposits are part of the Noachian country rock and thus predate formation of Valles Marineris (Fig. 6.5), or formed predominantly after development of the canyons. Although much of this layering may be related to impact and volcanic processes, recent identification of sulfates and phyllosilicates in the layered deposits (see below) suggests that additional sedimentary processes, involving aqueous alteration and chemical sedimentation, were also active.

In spite of a thin atmosphere [51] the present-day surface contains abundant evidence for ongoing eolian processes in the form of dunes, eolian ripples, yardangs and wind streaks. Dust-devils and active wind transport have been observed by the Mars rovers and massive dust storms, sometimes of global extent, are an annual

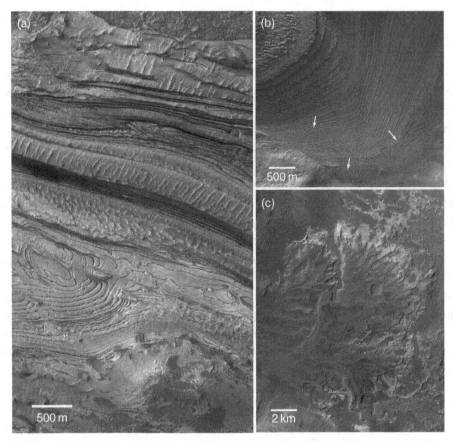

Fig. 6.5 Orbital images showing sedimentary layering in ancient terrains of Mars.
(a) Mars Orbiter Camera (MOC) image showing layered sedimentary rocks exposed
in the walls of Candor Chasma within Valles Marineris (NASA Photojournal
PIA07352). (b) Mosaic of MOC images showing sedimentary layering within
Galle Crater (52.3° S, 30.1° W). Note the numerous unconformities in the lower
part of the image (NASA Photojournal PIA08544). (c) Mosaic of MOC images
showing fan-shaped deposits (Eberswalde Fan) near Holden Crater, interpreted to
represent remnants of an ancient delta. NASA Photojournal PIA04869.

occurrence. Ancient sedimentary rock deposits, where studied in detail, also show
evidence for deposition by eolian processes (see below).

Glacial and periglacial processes have been important, perhaps throughout
much of geological time. Both poles are covered by up to ~ 3 km of rhythmically
layered deposits consisting of water ice and fine-grained debris likely dominated by
eolian dust (polar layered deposits, hereafter PLD). In turn, PLD are overlain by less
extensive, thin (~ 1–10 m) residual ice caps comprised mainly of water ice in the
north and water ice covered by CO_2 ice in the south. A very thin seasonally cyclical
veneer of CO_2 ice covers these deposits and at its maximum extends to latitudes as

low as 50° in the winter. Although mostly concentrated within the 80° latitude circle, thin PLD remnants extend equatorially to nearly 70° latitude. Also surrounding the main polar deposits and extending further towards the equator are periglacial features, including dune fields (termed the polar erg), patterned ground, and possible drumlins, eskers, thermokarst features and ice polygons, suggesting more extensive glacial and periglacial activity than is currently present. Of some interest is the occurrence of a 7500 km² gypsum-bearing erg in the north polar region, for which there is no obvious nearby source [52].

Polar deposits are young. Northern deposits have surfaces on the order of hundreds of thousands of years and southern deposits are two orders of magnitude older, ~5–10 Myr [53]. Thus, southern deposits are much older than timescales of obliquity variation and no doubt preserve a record of recent climate change [54]. Indeed, Mars may be emerging from an "ice age". Peripheral to the southern PLD lies the Hesperian-aged Dorsa Argentea Formation. Initially interpreted to represent volcanic or eolian deposits, landforms that look like eskers and features consistent with basal melting have led to suggestions that this unit may also be ice-rich. If so, it would be consistent with glacial activity occurring, at least periodically, over a significant part of martian history [55].

6.4.2 Surficial processes

Most of the sediment observed on Mars is composed of basaltic debris. However, there is increasing evidence that both ancient sedimentary rocks and surface soils contain abundant chemical constituents in the form of sulfates, amorphous silica, possibly chlorides, but only rarely carbonates (preserved within fractures in martian meteorites). Accordingly, it is clear that aqueous weathering/alteration has played a role in the sedimentary cycle. In detail, weathering patterns differ considerably from those seen on Earth (Fig. 6.6). Iron is relatively mobile (in spite of oxidizing surface conditions) suggesting acidic conditions. On the other hand, little evidence exists for large-scale aluminium mobility in spite of aluminium being orders of magnitude more soluble than ferric iron under similar low pH conditions. These observations suggest that highly soluble minerals, such as olivine and perhaps iron–titanium oxides, are involved in chemical weathering, but less soluble plagioclase is not. The weight of evidence thus points to surface weathering taking place dominantly under acidic conditions in aqueous systems that are rock-dominated rather than fluid-dominated (i.e., low water/rock ratios) [56].

Unlike the terrestrial sedimentary record where the carbon cycle dominates surficial processes, on Mars it appears that the sulfur cycle has dominated during much of the planet's history [44]. The atmosphere consists mainly of CO_2 but there is very little of it and apart from minor occurrences in martian meteorites, carbonate

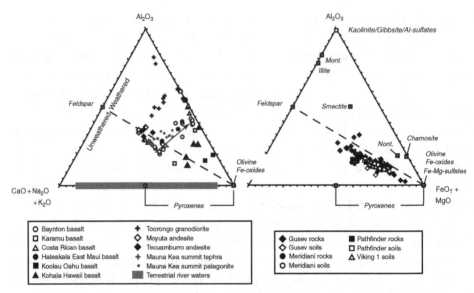

Fig. 6.6 Ternary diagrams of mole fraction Al_2O_3–$(CaO + Na_2O + K_2O)$–$(FeO_T + MgO)$ comparing terrestrial and martian alteration patterns. Terrestrial data are taken from a variety of weathering and alteration profiles. Mars data include all published results for martian meteorites and *in situ* analyses of all rocks and soils by Viking, Pathfinder and the Mars Exploration rovers (see Hurowitz and McLennan [56] for data sources). On Mars, iron is highly mobile during surficial processes due to low pH conditions and dissolution of the most soluble mafic minerals such as olivine. However, lack of aluminium-enrichments suggests that less soluble plagioclase is relatively unaffected, consistent with rock-dominated (low water/rock ratio) aqueous systems [56].

minerals have yet to be unambiguously identified at the surface. Instead, sulfates (± chlorides) dominate the chemically precipitated constituents in surficial deposits. Mineralogical associations suggest that they form mainly as evaporites and their diagenetically oxidized equivalents [57]. Sulfate minerals have also been identified from orbit and spectra suggest the presence of gypsum, kieserite and unspecified polyhydrated sulfates within ancient layered deposits and in the polar erg suggesting they are widespread in both space and time [52, 58].

Orbital infrared measurements have identified vibrational patterns consistent with the limited occurrences of clay minerals [59]. Although the exact mineralogy is not known, data appear most consistent with iron-rich clays (e.g. nontronites) and lesser occurrences of aluminous clays. Such minerals likely formed in the presence of near-neutral to slightly alkaline conditions. Bibring *et al.* [60] suggested that phyllosilicates are restricted to Early Noachian terrains and that there was an evolution in pH conditions from water-rich, near-neutral during Early Noachian to water-rich and acidic during the Late Noachian–Early Hesperian to water-limited and acidic during Late Hesperian/Amazonian. Although there certainly appears to have been a reduction in the amount of water involved in sedimentary processes over martian history, the idea that this was accompanied by a secular variation in pH is less secure.

6.4.3 Soils and dust

Much of the martian surface is covered by unconsolidated soils derived from a variety of impact, eolian and other sedimentary processes [61]. Accordingly, the origin of soils is complex and they are composed of a mixture of several components including: (1) impact ejecta of highly variable grain size that forms the planetary regolith [62]; (2) sand-sized and finer sedimentary deposits produced mainly by eolian transport; (3) chemical constituents consisting mainly of magnesium, calcium and iron sulfates (and possibly chlorides) and amorphous silica; (4) very fine grained, typically bright-toned dust.

Martian dust has long been thought to be a representative sample of the crust [63], analogous to terrestrial loess, although its exact chemical composition has proven elusive [64]. Dust may have been derived substantially from impact processes early in Mars' history and subsequently recycled at the surface. Indeed, if anything, dust may be biased towards sampling the more ancient southern highlands rather than less eroded younger volcanic terrains [65]. Magnet experiments on the Mars rovers isolated fine dust and analyses on opposite sides of the planet are remarkably similar [66] confirming the globally homogeneous nature. Table 6.2 lists the composition of soils that appear to be dominated by bright dust analyzed by the Spirit and Opportunity rovers. Accordingly, our judgement is that the dust, though possibly having a slight age bias, probably provides a good measure of the average composition of the exposed crust.

The idea that average sediment can be equated to upper crust, once components added during weathering (e.g. CO_2, SO_3, Cl) are accounted for, is well established on Earth [67]. Terrestrial sedimentary rocks are highly differentiated by lithology because chemical constituents are stored for long periods of time in oceans and deposited separately where conditions are favorable. Thus, clastics, carbonates, evaporites and siliceous sediments tend to form distinct lithologies and compositions for major elements and the more soluble trace elements are highly variable among these lithologies. On Mars, however, the situation appears to be simpler. Where studied, soils commonly appear to be intimate mixtures of both chemical and non-chemical components and lithological differentiation is less pronounced. Whether this is due to sedimentary mixing processes or to *in situ* alteration is not always clear and indeed both processes may be important.

The origin of sulfur and chlorine at the surface is ultimately due to volcanic activity, though its subsequent history is complex and not well understood. On

Table 6.2 *Chemical composition of selected soils and dust-rich soils from Mars*

	Gusev plains		Columbia Hills	Meridiani Planum			Pathfinder soil	Viking		Average
	Basaltic soil (n=15)	Dust-rich soil (n=4)	Basaltic soil (n=9)	Basaltic soil (n=8)	Dust-rich soil (n=4)	Hematitic soil (n=13)	(n=7)	Utopia soil (n=8)	Chryse soil (n=9)	Martian soil
SiO_2	46.2	45.9	46.3	46.4	45.5	39.6	42.1	47	47	45.41
TiO_2	0.86	0.89	0.87	1.02	1.05	0.75	0.87	0.59	0.69	0.90
Al_2O_3	10.1	9.83	10.3	9.46	9.22	7.86	9.5	–	7.9	9.71
FeO_T	16.3	16.1	15.5	18.3	17.9	29.5	21.6	17.5	17.7	16.73
Cr_2O_3	0.38	0.31	0.30	0.40	0.39	0.29	0.29	–	–	0.36
MnO	0.33	0.32	0.31	0.37	0.36	0.29	0.31	–	–	0.33
MgO	8.64	8.33	8.67	7.29	7.63	6.74	7.78	6.3	6.5	8.35
CaO	6.45	6.27	6.20	7.07	6.68	5.42	6.37	6.3	6.2	6.37
Na_2O	2.89	2.98	3.21	2.22	2.25	2.20	2.84	–	–	2.73
K_2O	0.43	0.47	0.44	0.49	0.48	0.39	0.60	n.d.	n.d.	0.44
P_2O_5	0.75	0.86	0.92	0.83	0.87	0.82	0.74	–	–	0.83
SO_3	5.66	6.85	6.18	5.45	6.81	5.22	6.27	8.9	8.3	6.16
Cl	0.69	0.82	0.73	0.63	0.73	0.70	0.76	0.5	0.85	0.68
Σ	**99.7**	**99.9**	**99.9**	**99.9**	**99.9**	**99.8**	**100.0**	–	–	**99.0**
Cr	2600	2120	2050	2740	2670	1980	1980	–	–	2460
Ni	424	615	466	445	516	845	–	–	–	490
Zn	264	385	274	273	382	332	–	–	–	286
Br	52	15	68	65	113	55	–	–	–	61

Notes: Analyses represent average Viking 1 and 2 soils, average Pathfinder soil, average dark "basaltic" soil from Gusev plains, Columbia Hills and Meridiani Planum, average dust-rich soils from Meridiani and Gusev and average hematitic soils from Meridiani Planum. Soil classifications into Basaltic, dust-rich and Hematitic from Yen *et al.* [68, 75]. Major elements in wt%; trace elements in ppm. **Data sources:** See Notes 34, 68, 75. Viking data compiled from Clark, B. C. *et al.* (1982) Chemical composition of martian fines. *JGR* **87**, 10,059–67. Average martian soil from Note 76. **Abbreviations:** FeO_T, total iron as FeO; n.d., not detected; –, no analysis available; *n*, sample numbers.

Earth, such constituents find their way into the ocean and marine sediments; a northern ocean on Mars, if it existed, should have provided a similar sink. Global GRS mapping provides no support for this notion.

It is also clear that soils contain a significant local influence on their compositions. Thus the chemical compositions of Pathfinder soils form mixing lines with local Si- and K-rich rocks as one end member, some Meridiani soils contain abundant hematitic concretions derived from the underlying bedrock (Table 6.2) and some soils from the Columbia Hills have inherited elevated phosphorus from nearby bedrock [34, 64, 68]. Nevertheless, these effects appear to be of second order and once corrected for local influences, soils are remarkably uniform in composition. It is unlikely that soils are globally mixed but rather that they sample large enough regions to effectively provide a reasonable sampling of the exposed, globally basaltic crust for both major and trace elements. Table 6.2 also lists the compositions of typical martian soils and they compare favorably to the composition of dust-rich soils.

6.4.4 Sedimentary rocks on Mars

The existence of ancient layered deposits has been known since Mariner 9. However, recent imaging shows the remarkable extent of such deposits, especially in Noachian and Hesperian terrains and geomorphological evidence points to many of these deposits being sedimentary rocks [69]. Most sedimentary deposits are relatively flat-lying with "layer cake" stratigraphy or in the case of impact-related and pyroclastic deposits may be draped over pre-existing topography. Depositional environments are difficult to establish from orbital imaging but in a few cases, such as at the Holden Northeast crater, apparent drainage patterns and fan-shaped geometry suggest an ancient deltaic or alluvial fan system [69, 70].

At the surface, both Spirit and Opportunity have identified ancient sedimentary rocks on opposite sides of the planet. At Meridani Planum, the Burns formation represents Late Noachian to Early Hesperian indurated eolian sandstones composed of sulfate-rich sand grains derived from altered basaltic debris, perhaps from a desiccating playa lake. Sands were deposited in a "wetting upward" dune–interdune environment and cemented by evaporitic sulfates during repeated acidic, very high ionic strength groundwater recharge that occasionally breached the surface for short periods of time (Fig. 6.7) [71]. In the Columbia Hills, Spirit encountered a Noachian layered sequence that includes sedimentary rocks. A variety of lithologies and depositional environments are represented but the overall sequence has a dominantly pyroclastic origin (Fig. 6.8) [72]. In the vicinity of Home Plate, associated deposits, enriched in amorphous silica, are consistent with hydrothermal processes that commonly accompany pyroclastic activity.

Fig. 6.7 Sedimentary rocks from the Burns formation at Meridiani Planum. (a) Meter-scale eolian cross-beds (arrow) of the Burns formation exposed on Cape St. Vincent in Victoria crater, Meridiani Planum. Cross-beds of this scale and geometry are likely deposited by eolian processes (NASA Photojournal PIA09695. Image stretched to highlight bedding features). (b) Centimeter-scale festoon-shaped ripple cross-laminations preserved in the rock Cornville in Erebus crater, Meridiani Planum. Small-scale cross-laminations with this geometry are diagnostic of subaqueous deposition and this facies is thought to represent part of an interdune deposit (adapted from Grotzinger et al. (2006) [71]).

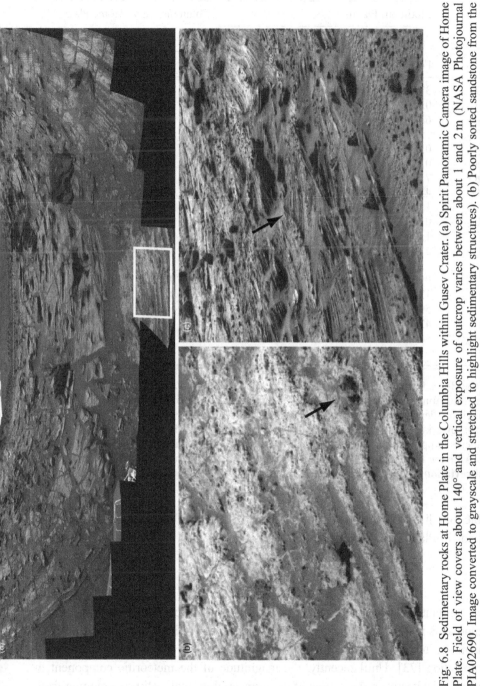

Fig. 6.8 Sedimentary rocks at Home Plate in the Columbia Hills within Gusev Crater. (a) Spirit Panoramic Camera image of Home Plate. Field of view covers about 140° and vertical exposure of outcrop varies between about 1 and 2 m (NASA Photojournal PIA02690. Image converted to grayscale and stretched to highlight sedimentary structures). (b) Poorly sorted sandstone from the base of the sequence containing a bomb-sag (arrow), a feature characteristic of pyroclastic deposition. Approximate location in (a) shown by white box (NASA Photojournal PIA08063. Image converted to grayscale and stretched to highlight sedimentary structures). (c) High angle cross-bedded, relatively well-sorted sandstone from the upper part of the sequence suggestive of eolian reworking. Scale across image approximately 1.5–2 m (NASA Photojournal PIA02055. Image converted to grayscale and stretched to highlight sedimentary structures). Adapted from Squyres *et al.* (2007) [72].

Table 6.3 *Chemical composition of selected sedimentary rocks from Mars*

	Meridiani Planum			Columbia Hills		Home Plate	
	Average Burns Formation	Upper Burns (Guadalupe)	Lower Burns (Mackenzie)	Peace	Alligator	Posey	Crawfords
SiO$_2$	37.1	36.2	43.0	37.3	41.8	45.8	46.6
TiO$_2$	0.74	0.65	0.86	0.45	0.53	1.02	1.11
Al$_2$O$_3$	6.40	5.85	7.27	2.24	5.49	9.39	9.98
FeO$_T$	15.6	14.8	15.6	20.4	18.3	15.5	15.4
Cr$_2$O$_3$	0.20	0.17	0.20	0.75	0.63	0.32	0.34
MnO	0.32	0.30	0.32	0.47	0.33	0.32	0.29
MgO	7.89	8.45	5.43	21.53	16.27	9.56	10.3
CaO	4.98	4.91	4.60	4.90	4.72	6.71	6.74
Na$_2$O	1.76	1.66	1.93	n.d.	1.6	3.53	3.36
K$_2$O	0.56	0.53	0.69	n.d.	0.19	0.42	0.32
P$_2$O$_5$	1.05	0.97	1.15	0.49	0.29	1.38	1.27
SO$_3$	22.18	24.91	17.01	10.6	8.48	4.85	2.91
Cl	1.00	0.50	1.90	0.72	1.26	1.96	1.35
Σ	**99.8**	**99.9**	**100.0**	**99.9**	**99.9**	**100.8**	**100.0**
Cr	1370	1160	1370	5130	4310	2190	2330
Ni	590	589	546	774	506	379	297
Zn	423	324	447	64	205	407	314
Br	100	30	9	71	217	181	91
Ge	–	–	–	–	–	30	30

Notes: Burns formation, Peace and Alligator analyses all from abraded surfaces; Posey and Crawford from brushed surfaces. Posey and Crawford are from the upper part of a pyroclastic deposit found at Home Plate. Because these rocks were likely derived by eolian reworking of the pyroclastics, they are included here. Major elements in wt%, trace elements in ppm. **Data sources**: Burns formation from Planetary Data System; Peace and Alligator from Gellert, R. *et al.* [37]; Home Plate from Squyres *et al.* [72]. **Abbreviations**: FeO$_T$, total iron as FeO; n.d., not detected; –, no analysis available.

Table 6.3 lists the chemical composition of selected sedimentary rocks from Meridiani Planum and the Columbia Hills.

6.4.5 Meteoritic components

An important issue to consider for ancient planetary surfaces is the role of meteoritic components in soils and other sedimentary deposits. On the Moon, meteoritic components in lunar soils are reasonably well constrained to about 2%, of CI average composition [73]. Until recently, the magnitude of the meteoritic component in martian soils has been the source of great uncertainty with estimates ranging from zero to 40% [74]. Some workers postulated a substantial meteoritic component due to the proximity to the asteroid belt (where meteoritic flux should be about one-third greater than at Earth), the scant atmosphere and size of Mars. The source of uncertainty comes from the fact that trace-siderophile data have not been available and major-element differences between martian crust and meteorites are too small to provide robust constraints.

The issue appears now to have been resolved with the sensitive APXS instrument on the Mars rovers that measures nickel concentrations at the ppm level. Yen *et al.* [75] identified nickel-enrichment in a wide variety of soils and ancient sedimentary rocks at the Meridiani and Gusev landing sites, relative to potential upper-crustal igneous sources. The differences are best explained by a meteoritic contribution of between 1 and 3%, similar to the amount found in soils on the Moon. However, comparison between the Moon and Mars is perhaps misleading. Although the martian surface is mostly ancient, it is younger on average than the lunar soils and accordingly, similar levels of meteoritic components are likely consistent with the greater meteoritic flux for Mars.

6.5 Bulk composition of the crust

An estimate for the bulk composition of the martian crust is given in Table 6.4 and is compiled from several sources [76]. Major element, and Cr, Ni and Zn abundances are taken from average S- and Cl-free soil and dust compositions after correction for a 2% meteoritic component assumed to be of CI composition. These values are broadly consistent with GRS data. Potassium and Th were derived directly from the global average GRS data but increased by a factor of 1.12 to account for H_2O, Cl and S in near-surface deposits. Uranium was calculated from $Th/U = 3.8$.

Estimates of additional elements fall closer to the realm of speculation but some constraints are possible. Using the chondritic La/Th ratio of 8.6 gives a La concentration of 6.0 ppm, which is probably an upper limit as most martian meteorites have subchondritic La/Th ratios [77] (see Fig. 5.8a) and in terrestrial settings petrologically evolved rocks also tend to have lower La/Th than primitive mantle [67]. Using the K/La ratio of martian meteorites leads to a La value of 5.5 ppm, but some caution is warranted since these meteorites are not representative of the bulk crust and the calculation assumes K and La are equally incompatible under a broad range of igneous conditions. Assuming that the enriched component of basaltic shergottites (see Fig. 5.5) is a crustal component rather than a second mantle reservoir, it is possible to estimate the REE pattern of this crustal component using mass balance with Nd isotopes and leads to the following values: $Nd = 8.5$ ppm, $Sm/Nd \sim 0.26$ and $La/Yb \sim 3$ [19]. Thus the most likely REE pattern for the average martian crust is one that is slightly LREE-enriched with abundances that vary between about $15 \times CI$ for La to about $7 \times CI$ for Yb. Rubidium is estimated from $K/Rb = 300$ and Ba

Table 6.4 *Estimate of the bulk composition of the martian crust*

Crust			
SiO_2	49.3		
TiO_2	0.98		
Al_2O_3	10.5		
FeO	18.2		
Cr_2O_3	0.38		
MnO	0.36		
MgO	9.06		
CaO	6.92		
Na_2O	2.97		
K_2O	0.45		
P_2O_5	0.90		
Σ	**100.0**		
Na (wt%)	2.20	La (ppm)	5.5
Mg (wt%)	5.46	Ce (ppm)	13.9
Al (wt%)	5.56	Pr (ppm)	1.9
Si (wt%)	23.0	Nd (ppm)	9.4
P (ppm)	3930	Sm (ppm)	2.7
K (ppm)	3740	Eu (ppm)	0.95
Ca (wt%)	4.95	Gd (ppm)	3.1
Ti (ppm)	5880	Tb (ppm)	0.55
Cr (ppm)	2600	Dy (ppm)	3.4
Mn (ppm)	2790	Ho (ppm)	0.70
Fe (wt%)	14.1	Er (ppm)	1.9
Ni (ppm)	337	Tm (ppm)	0.25
Zn (ppm)	320	Yb (ppm)	1.7
Rb (ppm)	12.5	Lu (ppm)	0.26
Y (ppm)	18	Th (ppm)	0.70
Ba (ppm)	55	U (ppm)	0.18

by assuming Ba and Th are in chondritic proportions. Similar results are obtained by assuming these LIL elements vary in their level of enrichment relative to the primitive mantle as a smooth function of ionic radius.

Since there appears to be no systematic pattern of Eu anomalies in the martian meteorites, we further conclude that neither average crust nor average depleted mantle possess a significant Eu anomaly. This is in stark contrast to the Moon where crystallization of the magma ocean has strongly fractionated Eu, during formation of the plagioclase-rich lunar highland crust on that dry and highly reduced body (see Chapter 2 for further discussion).

A crust of this composition that is 4% of a primitive mantle with a composition presented in Chapter 5 (Table 5.4) implies that ~50% of the most incompatible elements have been extracted from the mantle into the crust. Since the younger secondary crust of Mars appears dominated by incompatible-element-depleted

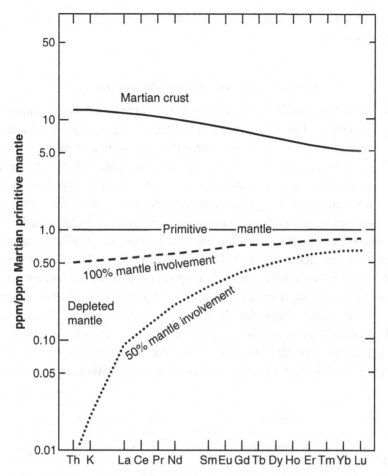

Fig. 6.9 Primitive-mantle normalized plot of selected trace elements in the martian crust and depleted mantle. Two depleted mantle models are shown, one assuming the entire mantle is involved in crust formation and the second assuming only half the mantle is involved.

basalts and related rocks, the vast majority of this enrichment is likely due to the early crust formed during the primary differentiation of Mars. This level of differentiation is nearly a factor of two greater than observed on Earth despite the highly differentiated nature of the continental crust.

Figure 6.9 plots the trace-element composition of the martian crust normalized to the primitive-mantle composition. Also shown are two compositions for the reciprocal "depleted" mantle, one where the crust has been extracted from the entire primitive mantle (i.e., deep magma ocean) and the second where only 50% of the primitive mantle is involved (i.e., shallow magma ocean). Note that if there is an additional reservoir of incompatible-element-enriched material in the mantle, analogous to lunar KREEP, then the depleted portions of the mantle would be far more

LIL-depleted. This appears to be consistent with the ultra-depleted sources implied for the shergottite data.

6.5.1 Compositional evolution of the martian surface

The mostly LREE-depleted nature of shergottites and their distinctive compositions (e.g. Mg/Si) compared to the more ancient basalts that have been analyzed by the Mars rovers (Ref. 40, Fig. 6.3) suggests that the composition of the youngest lavas may differ significantly from the earlier crust (also compare Tables 5.3 and 6.4). Is there any further evidence to support this concept? The age of the martian surface estimated from crater counting is not a particularly robust measure of crustal ages and only provides a crude estimate. Although compositional variations on the martian surface measured by GRS are subtle, there are some statistically significant age variations for a few elements (Fig. 6.10) [78]. Slight secular decreases are observed for K and Th that likely reflect crustal compositional evolution processes. A very small but statistically significant increase in Fe abundances ($\sim 1\%$ FeO$_T$) also may reflect crustal evolution supporting the idea that younger lavas on average are more iron-rich, especially in the northern hemisphere. Other elements show no variation (Si), are related to secondary processes (Cl) or currently have too poor a resolution to evaluate (Al, Ca). These secular variations in composition are consistent with

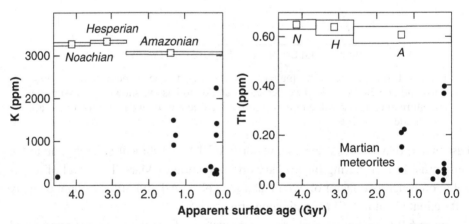

Fig. 6.10 Plots of potassium and thorium abundances on the martian surface, measured by GRS, as a function of apparent surface age compared to abundances in martian meteorites. Boxes represent standard error of estimate interval over the age range considered. Note that younger terrains have lower K and Th consistent with greater contributions from an incompatible-element-depleted crust during Amazonian times. Iron abundances (not shown) exhibit a small but statistically significant increase between Noachian and Hesperian/Amazonian times. Adapted from Ref. 78.

incorporation of Amazonian crustal components that are incompatible-element depleted and relatively iron-rich, similar to the SNC meteorites. Any such secular variations in crustal composition, observed from orbit, will of course be greatly moderated by the effects of abundant soils and dust, with relatively uniform compositions, at the surface.

6.6 Heat flow and crustal heat production

Heat flow is related to crustal abundances of heat-producing elements but no direct heat flow measurements have been made on Mars. However, it is possible to constrain heat flow indirectly to see if it is broadly consistent with compositional models. One approach uses gravity and topography to estimate the thickness of the mechanical lithosphere, the base of which is temperature dependent. Thermal gradients and heat flow are then estimated by assuming appropriate crust/mantle rheologies (e.g. diabase for crust, dry olivine for mantle) and thermal conductivities [79]. Another approach uses estimates of fault depths as a measure of the brittle–ductile transition, which is also temperature dependent. On Mars, wrinkle ridges may represent faults that localize horizontal lithospheric shortening with spacing being related to the depth of the brittle–ductile transition. Thermal gradients and heat flow are estimated as above [80].

McGovern *et al.* [79] compiled various heat flow estimates over geological time and, not surprisingly, concluded that heat flow declined rapidly during the Noachian from values of $\geq 50\,\text{mW/m}^2$ and then more slowly during the Amazonian to present-day values of $\leq 20\,\text{mW/m}^2$. By way of comparison, for a crust of 50 km average thickness, the crustal abundances of the heat-producing elements (discussed above) predict approximately $6.5\,\text{mW/m}^2$ of crustal contributions to present-day heat flow. This is consistent with the independent geophysical estimates and also consistent with the conclusion above that a very large fraction of the incompatible elements have been fractionated into the crust.

6.6.1 Compositional variation with depth

Does the martian crust possess large-scale compositional layering analogous to the upper and lower continental crust of Earth? If so, it would provide important additional constraints on the origin and evolution of the martian crust. For example, an incompatible-element-enriched upper crust might imply intra-crustal differentiation by partial melting or crystal fractionation processes; or might suggest substantial crustal growth from beneath by underplating. There is certainly some reason to believe that at least locally, the crust could have vertical variations in composition. Amazonian volcanic rocks, represented by the shergottites, have a significantly different composition than average crust and recognition of evolved alkaline igneous rocks may suggest differentiation in deep magma chambers [16, 24]. On the other hand, these younger magmas represent only a small fraction of the overall crust (see below). Although GRS mapping is at low resolution, there is no indication that ejecta from large basins, such as Hellas, has a distinct composition compared to surrounding terrains [30].

Ruiz *et al.* [81] suggested that the crust in the Hesperian-aged Solis Planum region may be differentiated into a heat-producing-element-enriched upper crust and depleted lower crust. Their modeling used average global surface heat production from GRS mapping (extrapolated to Hesperian age) combined with estimates of local heat flow and crustal thickness. Results suggested that the crustal component of heat flow could be explained by ~20–30 km of crust with K, Th and U abundances measured (or inferred in the case of U) at the martian surface, suggesting a lower 35–45 km of depleted crust. However, the most recent versions of ‚elemental mapping indicate that global average K and Th abundances are lower than those used by Ruiz *et al.* and that Solis Planum in particular has much lower surface heat production than the martian average [31]. Surface heat production in Solis Planum may thus be a factor of two or more less than the values used by Ruiz *et al.*, requiring much thicker crust of that composition to explain the same level of inferred heat flow. Accordingly, evidence for a strongly differentiated martian crust remains unconvincing.

6.7 Crustal evolution on Mars

How did the ancient crust of Mars form? In our judgement, large volumes of incompatible-element-enriched crust, produced within a few million years of accretion, are best interpreted as a primary basaltic crust. Partial melting of the mantle to produce LIL-enriched basalt is relatively inefficient; generating a large volume crust (~3–4%) from primitive mantle by partial melting within ~30–50 Myr of the accretion of the planet is a daunting prospect. In contrast, crystallization of a magma ocean will result in a significant volume of late-stage incompatible-element-enriched liquids, analogous to lunar KREEP but with a composition differing in detail (e.g. lacking Eu anomalies) owing to different pressure–temperature conditions, composition (e.g. water content) and crystallization history. Also, unlike the smaller bone-dry Moon, plagioclase will not form in abundance and float to form a cumulate crust. Once the magma ocean is nearly crystallized (~80–90%) such liquids can be tapped to form basaltic lavas (probably mixing with crystallized mantle) that extrude onto the surface to produce the early crust.

Both the younger age (and therefore differing thermal history) and distinct composition of the lunar magma ocean make it a poor analogue for Mars. On the other hand, evolution of 4 Vesta may provide some important insights (see Chapter 13). Although much larger, Mars, like Vesta, appears to have substantially melted and differentiated very shortly after accretion producing basaltic melts (analogous to eucrites in the case of Vesta). Heat sources included the short-lived isotopes like ^{26}Al and possibly ^{60}Fe, further aided by early impacts and the factor-of-two higher ^{40}K abundances compared to Earth. The depth of melting on Mars is difficult to constrain but available geophysical modeling suggests that at least the upper third and perhaps the entire mantle was involved [82]. Our judgement is that most of the planet was likely to have melted.

Martian meteorite ALH84001, dated at 4.50 ± 0.13 Gyr, is the only known sample of the Early Noachian crust. Does it represent a sample of the primary crust? Perhaps, but the evidence is slim. The sample is a cumulate orthopyroxenite, likely derived from a plutonic parent. The bulk sample is LREE depleted but the parent magma was most likely slightly LREE enriched (La/Yb ~ 3) [83]. For example, the initial ε_{Nd} is slightly positive but likely chondritic within uncertainty (e.g. pyroxene $T_{CHUR} = 4.57 \pm 0.03$ Gyr) [84]. In any case, this sample, in spite of the enormous attention it has garnered, provides only limited insight into the earliest crust.

Exactly when the formation of the early primary crust ended and partial melting of the martian interior to form the secondary basaltic crust began is unknown but was likely early and perhaps these processes overlapped. Later magmatic activity, best characterized by the SNC meteorites, was derived by partial melting of the residual mantle. The dominant source of these basalts was a long-term "ultra-depleted" mantle, characterized, for example, by the very high present-day ε_{Nd} (up to $+50$) seen in many SNC meteorites. This degree of depletion cannot be accomplished for the entire mantle by extraction of the primary crust (Table 6.4; Fig. 6.9) and so this source can only represent a fraction ($< 50\%$) of the present-day mantle. As discussed above, whether the second, long-term LIL-enriched component (present-day $\varepsilon_{Nd} < -10$) of SNC meteorites is of mantle or crustal origin has not been resolved. If the former, it is likely the signature of the KREEP-like component involved in the formation of the earliest crust [20]. Additional geochemically and isotopically distinct mantle reservoirs are almost certainly present [77] but there simply is insufficient sampling to characterize these fully.

The total volume of secondary crust is intermediate between the extreme values of the Earth and Moon. Lunar maria are about 1.7×10^7 km^3 ($< 1\%$ of lunar crust) whereas Earth's oceanic crust is 100 times that value ($\sim 19\%$ of the terrestrial crust) and more than 20 times that volume again has been produced over geological time (but mostly subducted). For Mars, the calculation is complicated because it is not possible to distinguish primary from secondary crust in much of the ancient cratered terrains and the role of plutonic rocks is difficult to estimate. Estimates of volcanism since ~ 4.0 Gyr suggest $\sim 10^8$ km^3 but the total amount of crust this represents could easily be 10 times this value if intrusive equivalents are accounted for [85]. Magmatic material at Tharsis alone is estimated at $3 \times 10^8 \, km^3$ [86]. These values do not take into account the interval between 4.5 and 4.0 Gyr. Interpretation of argon isotopic data is highly model dependent but appears to be broadly consistent with the formation of somewhere between $1–2 \times 10^9 \, km^3$ of basaltic crust since accretion and early differentiation [87]. Thus the best estimate is that about $20 \pm 10\%$ of the martian crust is secondary and so the mean age of the crust must be significantly greater than 3.5 Gyr. No mechanism for recycling such crust into the mantle has been identified and so this likely represents the total volume produced during the history of the planet.

Perhaps the most perplexing issue that remains unresolved is how short-lived isotopic anomalies (^{142}Nd, ^{182}W) can be maintained for ~ 4.5 Gyr within a planet that has been so volcanically active throughout its history. Processes leading to large-scale mantle convection and mixing are untenable. On the other hand, heat production in a mantle from which $\sim 50\%$ of the heat-producing elements have been extracted into a primary crust still generates sufficient radiogenic heat to sustain melting; accordingly some form of convective mantle flow seems likely [88]. A KREEP-like component may further concentrate heat production within a single layer or few localized regions within the mantle. Magmatic activity has been highly concentrated at just a few volcanic centers, the most important being Tharsis and Elysium; Tharsis being intermittently active for nearly four billion years. This has led to suggestions that mantle convection is highly focused with long-lived stable boundaries thus restricting mantle mixing [88, 89]. Modeling further suggests that the endothermic reactions associated with the presence of a lower-mantle perovskite zone could lead to such localized convection patterns [89].

6.7.1 Tertiary crusts on Mars?

Is there any evidence for tertiary crust formed by recycling of earlier-formed crust through the mantle? As discussed in Chapter 5, no convincing evidence exists for plate tectonics at any time during martian geological history and so such processes were not available to produce anything like continental crust. However, it is possible that some other mechanism of recycling has taken place, such as foundering at the base of the crust, to produce evolved tertiary crust, although no evidence exists for such processes. Several limited occurrences of evolved compositions have been found, including Pathfinder rocks, local occurrences of quartz-bearing rocks, and perhaps Surface Type 2. The question remains whether all of these compositions represent primary igneous rock compositions (e.g. Pathfinder), but at least some may do. In any case, the number and scale of these occurrences is limited and can readily be explained by late-stage differentiation of basaltic rocks. The recognition of alkaline series basalts is also notable in this regard because small volumes of highly differentiated rocks, including dacites and rhyolites, are common products on the Earth. Accordingly, we conclude that tertiary crusts are absent on Mars.

Synopsis

The surface of Mars has been the focus of intensive study by orbital remote sensing and surface examination by landers and rovers, thus complementing the detailed studies of martian meteorites. Although critical geophysical measurements, notably seismology, are regrettably absent, the crust is reasonably constrained to be about 50 km thick and sustains nearly 30 km of relief. Mars has a hemispheric dichotomy in crustal thickness and relief but, when corrected for a 3 km center of figure–center of mass offset, has a unimodal hypsometry, consistent with a one-plate planet.

Mars is a "basaltic planet" but the surface is characterized by magmatic diversity, and compositions range from ultramafic cumulates through quartz-bearing dacites/ rhyolites identified by orbital remote sensing. Most common are low-K, silica-saturated basalts (and related intrusives and volcaniclastics) but a significant Noachian- to Early Hesperian-aged alkaline volcanic province has been identified in Gusev Crater and alkaline volcanism may be more important than previously recognized. Identification of small areas of highly differentiated compositions (e.g. perhaps Pathfinder "andesites", quartz-bearing volcanics) has attracted considerable attention, however minor occurrences of evolved rocks represent a normal conse-quence of basalt differentiation, especially within the alkaline series.

Like the Earth, Mars has possessed a dynamic sedimentary rock cycle throughout most of its history, but one dominated by the sulfur cycle (and sulfates) rather than the carbon cycle (and carbonates). Weathering and groundwater diagenesis mostly take place at low pH and under water-limited conditions. The composition of the crust can thus be estimated from a combination of sedimentary products (soils, dust, sediments) and global surface mapping by gamma-ray spectrometry. Surficial deposits contain about 2% meteoritic components, similar to lunar soils. The derived composition is best characterized as a slightly incompatible-element-enriched (e.g. La/Yb ~3) subalkaline basalt. No compelling evidence exists for compositional variation with crustal depth and this composition integrated over the 50 km thick crust accounts for about 50% of the planet's incompatible-element budget and contributes about 6.5 mW/m^2 to present-day heat flow.

The earliest (mostly primary) crust of Mars dominates this composition. Like Mercury and the Moon, Mars differentiated early, probably through magma ocean processes, but with a primary crust that differed in composition. Differentiation of Mars predates that of the Moon by ~ 50 to 100 Myr, due to the fact that Mars never accreted much beyond the planetary embryo stage and avoided late highly disruptive giant impacts. The asteroid 4 Vesta may provide a useful analog, although on a much smaller scale, to early magma ocean processes that gave rise to incompatible-element-enriched (KREEP-like) melts which contributed to an early basaltic crust. Later magmatic additions (secondary crust) amount to about 20% of the crust. The bulk composition of the secondary crust is not well constrained but may be best characterized by the incompatible-element-depleted SNC meteorites. The overall importance of alkaline volcanism in the formation of the secondary crust is uncertain. Although Mars produced basaltic secondary crust throughout its history there is no evidence for a differentiated tertiary crust analogous to the Earth's continental crust.

Notes and references

1. Hartmann, W. K. (2003) *A Traveler's Guide to Mars*, Workman Publ., p 4.
2. The standard reference through the end of the Viking era but predating 1996 Mars Global Surveyor and Mars Pathfinder is: Kieffer, H. H. *et al.* (eds., 1992) *Mars*, University of Arizona Press. Books that consider the flood of recent data are: Carr, M. H. (2006) *The Surface of Mars*, Cambridge University Press; Chapman, M. (ed., 2007) *The Geology of Mars: Evidence from Earth-based Analogs*, Cambridge University Press; Bell III, J. F. (ed.) (2008) The *Martian Surface: Composition, Mineralogy, and Physical Properties*, Cambridge University Press.
3. Various aspects of martian crustal evolution are reviewed in several accounts. McSween, H. Y. (2003) Mars, in *Treatise on Geochemistry* (eds. H. D. Holland and K. K. Turekian) vol. 1, 601–21. McSween, H. Y. *et al.* (2003) Constraints on the composition and petrogenesis of the martian crust. *JGR* **108**, doi: 10.1029/2003JE002175. Nimmo, F. and Tanaka, K. (2005) Early crustal evolution of Mars. *Ann. Rev. Earth Planet. Sci.* **33**, 133–61. Solomon, S. C. *et al.* (2005) New perspectives on ancient Mars. *Science* **307**, 1214–20. McSween, H. Y. (2008) Martian meteorites as crustal samples, in *The Martian Surface: Composition, Mineralogy, and Physical Properties* (ed. J. F. Bell III), Cambridge University Press, pp. 383–95; Taylor, G. J. *et al.* (2008) Implications of observed primary lithologies, in *The Martian Surface: Composition, Mineralogy, and Physical Properties* (ed. J. F. Bell III), Cambridge University Press, 501–18.
4. Overviews of recent landed missions can be found in several journal special issues (of course hundreds of additional papers have been published since). Pathfinder: *Science* (1997) **278**, No. 5344; *JGR* (1999) **104** (E4); *JGR* (2000) **105** (E1); Spirit: *Science* (2004) **306**, No. 5685; *JGR* (2006) **111** (E2); Opportunity: *Science* (2004) **305**, No. 5702; *EPSL* (2005) **240** (1); *JGR* (2006) **111** (E12).
5. A serious complicating issue is variable sampling depth of surface analyses, that vary from a few microns for infrared (IR) and alpha particle X-ray spectrometry (APXS) to ~ 0.1–2 mm for Mössbauer (depending on material properties); this represents a significant source of uncertainty in comparing results. The APXS analyses provide a special challenge. Primary radiation "interrogates" samples to different depths depending on atomic number. Thus, most of the Na signal comes from the top $\sim 4\,\mu m$ whereas the Fe signal comes from the top $\sim 100\,\mu m$. This is not a major problem for samples that are homogeneous both laterally (over ~ 4 cm sampling diameter) and with depth but can present a serious issue for inhomogeneous samples. Reider, R. *et al.* (2003) The new Athena alpha particle X-ray spectrometer for the Mars Exploration Rovers. *JGR*

108, doi: 10.1029/2003JE002150. Gellert, R. *et al.* (2006) Alpha particle X-ray spectrometer (APXS): Results from Gusev crater and calibration report. *JGR* **111**, doi: 10.1029/2005JE002555.

6. Gamma-ray spectrometers were carried by Soviet Mars-5 (1974) and Phobos-2 (1989) orbiters but had very little measurement time before spacecraft loss and these data have been completely superseded by Odyssey GRS. For a summary of earlier results, see Surkov, Yu. A. *et al.* (1994) Phobos-2 data on martian surface geochemistry. *Geochem. Int.* **31**, 50–8.

7. For description of the GRS technique, see Boynton, W. V. *et al.* (2004) The Mars Odyssey gamma-ray spectrometer instrument suite. *Space Sci. Rev.* **110**, 37–83. At the time of writing, global maps are available for the elements Si, Fe, Al, Ca, K, Th, Cl, H. As counting statistics improve, other elements, such as S and U, will become available but at lower spatial resolution.

8. For example, Christensen, P. R. *et al.* (2001) Mars Global Surveyor Thermal Emission Spectrometer experiment: Investigation, description and surface science results. *JGR* **106**, 23,823–71. Christensen, P. R. *et al.* (2003) Morphology and composition of the surface of Mars: Mars Odyssey THEMIS results. *Science* **300**, 2056–61. Bibring, J. -P. *et al.* (2005) Mars surface diversity as revealed by the OMEGA/Mars Express observations. *Science* **307**, 1576–81. Pelkey, S. M. *et al.* (2007) CRISM multispectral summary products: Parameterizing mineral diversity on Mars from reflectance. *JGR* **112**, doi: 10.1029/2006JE002831. An example of extracting chemical composition from TES spectra can be found in Wyatt, M. B. *et al.* (2001) Analysis of terrestrial and Martian volcanic compositions using thermal emission spectroscopy: 1. Determination of mineralogy, chemistry, and classification strategies. *JGR* **106**, 14,711–32. The Thermal Emission Imaging System (THEMIS) on Mars Odyssey has much higher spatial resolution and has also produced geological maps, but its spectral resolution is very limited so mineralogical deconvolutions are more limited. Combinations of THEMIS and TES data are most useful in mapping the distributions of specific compositional units. Because thermal infrared spectroscopy only samples the uppermost few microns, it is compromised by dust and weathered rock surfaces.

9. In the absence of a sea-level reference, zero elevation on Mars was originally defined as lying at a level where mean atmospheric pressure is at the triple point of water (6.105 mbar, 273.16 K). With improved topographic and gravity data, elevations are now referenced to the equipotential surface (aeroid – jargon for martian geoid) with an average equatorial radius of 3396.00 km. Smith, D. E. *et al.* (1999) The global topography of Mars and implications for surface evolution. *Science* **284**, 1495–503. As pointed out in Chapter 2, this amount of relief is greater than that of the Earth and thus neither planetary size nor plate tectonics are required to produce it.

10. Zuber, M. T. *et al.* (2000) Internal structure and early thermal evolution of Mars from Mars Global Surveyor topography and gravity. *Science* **287**, 1788–93. Also see Smith, D. E. *et al.* (1999) The global topography of Mars and implications for surface evolution. *Science* **284**, 1495–503. The approach assumes that topography is compensated by crustal thickness and the crust has homogeneous density (Airy isostasy). In the simplest expression, $A_{tc} = T/(1 - \rho_c/\rho_m)$ where A_{tc} is crustal thickness amplitude, T is topographic amplitude and ρ_c, ρ_m are crust and upper mantle density. Spohn, T. *et al.* (2001) Geophysical constraints on the evolution of Mars. *Space Sci. Rev.* **96**, 231–62. By constraining crustal thickness in some place (e.g. a value near zero in Hellas, the deepest impact basin and thus assuming the mantle is not exposed) and assuming that some topographic level is equivalent to that of average crust (e.g. zero topographic level) it is possible to determine mean crustal thickness. Crust and mantle

densities must be estimated and these also have some uncertainties. This approach can be used on either a global or local scale with the former having the disadvantage of some unlikely assumptions (e.g. uniform crustal density).

11. Wieczorek, M. A. and Zuber, M. T. (2004) Thickness of the martian crust: Improved constraints from geoid-to-topography ratios. *JGR* **109**, doi: 10.1029/2003JE002153.

12. Neumann, G. A. *et al.* (2004) Crustal structure of Mars from gravity and topography. *JGR* **109**, doi: 10.1029/2004JE002262.

13. Compared to the center of mass (CM), the martian center of figure (CF) is displaced 2.986 km southwards and slightly toward the Tharsis bulge. For comparison, the CF/CM offset for the much larger Earth is only about 0.8 km. Yoder, C. F. (1995) Astrometric and geodetic properties of Earth and the solar system, in *Global Earth Physics, A Handbook of Physical Constants*, AGU Reference Shelf 1, pp. 1–31.

14. Smith, D. E. *et al.* (1999) The global topography of Mars and implications for surface evolution. *Science* **284**, 1495–503. Aharonson, O. *et al.* (2001) Statistics of Mars' topography from the Mars Orbiter Laser Altimeter: Slopes, correlations, and physical models. *JGR* **106**, 23,723–35.

15. Taylor, G. J. *et al.* (2008) Implications of observed primary lithologies, in *The Martian Surface: Composition, Mineralogy, and Physical Properties* (ed. J. F. Bell III), Cambridge University Press, pp. 501–18.

16. Adapted from McSween, H. Y. *et al.* (2006) Alkaline volcanic rocks from the Columbia Hills, Gusev crater, Mars. *JGR* **111**, doi: 10.1029/2006JE002698.

17. Deeply excavated material does not achieve the velocities necessary for escape. Shock effects in martian meteorites are consistent with an origin from near the surface during the impact process. Thus peak shock pressures are in the range of < 40–450 kbar and post-shock temperatures < 100–600 °C, which compares to peak pressures of 10 000 kbar in most impact craters. Although large shock pressures (> 1000 kbar) are required for escape, materials from near the surface cannot be highly compressed and are ejected at high velocity but experience relatively little shock pressure. Melosh, H. J. (1984) Impact ejection, spallation, and the origin of meteorites. *Icarus* **59**, 234–60. Melosh, H. J. (1985) Ejection of rock fragments from planetary bodies. *Geology* **13**, 144–8. Earlier notions that the craters must be large (≥ 100 km) have given way to those < 5 km being potential sources, far more palatable given the young meteorite ages. Identifying impact sites has proven challenging, despite considerable effort. Although craters of the appropriate size, age and spectral characteristics have been identified, much of the martian surface with appropriate age is covered with dust thus precluding spectral comparisons. McSween, H. Y. (2008) Martian meteorites as crustal samples, in *The Martian Surface: Composition, Mineralogy, and Physical Properties* (ed. J. F. Bell III), Cambridge University Press, pp. 383–95. Searching for launch sites in non-dusty regions is reminiscent of the proverbial drunk looking for keys beneath a street lamp and although potential sites can be identified, it is unlikely if any of the launch sites will be uniquely identified.

18. Herd, C. D. K. (2003) The oxygen fugacity of olivine-phyric martian basalts and the components within the mantle and crust of Mars. *MPS* **38**, 1793–805.

19. Norman, M. D. (1999) The composition and thickness of the crust of Mars estimated from rare earth elements and neodymium-isotopic compositions of Martian meteorites. *MPS* **34**, 439–49. The model was refined in Norman, M. D. (2002) Thickness and composition of the martian crust revisited: Implications of an ultradepleted mantle with a Nd isotopic composition like that of QUE94201. *LPSC* **33**, Abst. #1175.

20. Borg, L. E. and Draper, D. S. (2003) A petrogenetic model for the origin and compositional variation of the martian basaltic meteorites. *MPS* **38**, 1713–31.

21. Hofmann, A. W. (1997) Mantle geochemistry: The message from oceanic volcanism. *Nature* **385**, 219–29.
22. McSween, H. Y. (2008) Martian meteorites as crustal samples, in *The Martian Surface: Composition, Mineralogy, and Physical Properties* (ed. J. F. Bell III), Cambridge University Press, pp. 383–95.
23. Treiman, A. H. (2005) The nakhlite meteorites: Augite-rich igneous rocks from Mars. *Chemie der Erde* **65**, 203–70.
24. Nekvasil, H. *et al.* (2007) Alkalic parental magmas for chassignites? *MPS* **42**, 979–92.
25. Bandfield, J. L. *et al.* (2000) A global view of martian surface compositions from MGS-TES. *Science* **287**, 1626–30. Thermal emission spectroscopy covers the spectral range ~ 6–50 μm with a gap between ~ 12–20 μm due to CO_2 absorption; spatial resolution is ~ 3 km. High-albedo (bright) regions of Mars are covered by ubiquitous martian dust (~ 3 μm) and not amenable to mapping surface mineralogy and so TES mineralogical mapping is confined to low-albedo (dark) regions. For recent treatment of these data, see Rogers, A. D. and Christensen, P. R. (2007) Surface mineralogy of martian low-albedo regions from MGS-TES data: Implications for upper crustal evolution and surface alteration. *JGR* **112**, doi: 10.1029/2006JE002727. Rogers, A. D. *et al.* (2007) Global spectral classification of martian low-albedo regions with Mars Global Surveyor Thermal Emission spectrometer (MGS-TES) data. *JGR* **112**, doi: 10.1029/2006JE002726.
26. Wyatt, M. B. and McSween, H. Y. (2002) Spectral evidence for weathered basalt as an alternative to andesite in the northern lowlands of Mars. *Nature* **417**, 263–6. The difficulty is that there is considerable TES spectral overlap for high-Si volcanic glass, amorphous silica and certain clays. The Mars Express OMEGA instrument, which works at lower wavelengths, 1.2–2.6 μm, is better suited to identify clays but does not appear to have resolved the issue. Ruff, S. W. and Christensen, P. R. (2007) Basaltic andesite, altered basalt, and a TES-based search for smectite clay minerals on Mars. *GRL* **34**, doi: 10.1029/2007GL029602.
27. McSween, H. Y. *et al.* (2003) Constraints on the composition and petrogenesis of the martian crust. *JGR* **108**, doi: 10.1029/2003JE02175.
28. Dunn, T. L *et al.* (2007) Thermal emission spectra of terrestrial alkaline volcanic rocks: Applications to martian remote sensing. *JGR* **112**, doi: 10.1029/2006JE002766.
29. Karunatillake, S. *et al.* (2007) Composition of northern low-albedo regions of Mars: Insights from the Mars Odyssey gamma ray spectrometer. *JGR* **111**, doi: 10.1029/2006JE002675 (printed **112** (E3), 2007).
30. Boynton, W. V. *et al.* (2007) Concentration of H, Si, Cl, K, Fe, and Th in the low- and mid-latitude regions of Mars. *JGR* **112**, doi: 10.1029/2006JE002887. Although correlations with major geological features are subtle, they do occur. The hemispheric discontinuity in Fe and distinctions between ST1 and ST2 regions in K and Th [29] are discussed in the main text. Among the other geological features that appear distinct in their GRS signature are Medusae Fossae (mainly due to Cl) and Tharsis, that can be identified from the GRS signature using multivariate statistics.
31. Taylor, G. J. *et al.* (2006) Variations in K/Th on Mars. *JGR* **111**, doi: 10.1029/2006JE002676.
32. Hahn, B. C. *et al.* (2007) Mars Odyssey gamma-ray spectrometer elemental abundances and apparent relative surface age: Implications for martian crustal evolution. *JGR* **112**, doi: 10.1029/2006JE002821.
33. Keller, J. M. *et al.* (2006) Equatorial and midlatitude distribution of chlorine measured by Mars Odyssey GRS. *JGR* **111**, doi: 10.1029/2006JE002679.
34. Pathfinder soil and rock compositions are controversial because independent calibrations from Chicago and Mainz gave significantly different results for some elements.

Brückner, J. *et al.* (2003) Refined data of alpha proton X-ray spectrometer analyses of soils and rocks at the Mars Pathfinder site: Implications for surface chemistry. *JGR* **108**, E128094, doi: 10.1029/2003JE002060. Foley, C. N. *et al.* (2003) Calibration of the Mars Pathfinder alpha proton X-ray spectrometer. *JGR* **108**, doi: 10.1029/2002JE002018. Foley, C. N. *et al.* (2003) Final chemical results from the Mars Pathfinder alpha proton X-ray spectrometer. *JGR* **108**, doi: 10.1029/2002JE002019. For this work, data from the two groups are averaged, except for Na, P and Mn where the Foley *et al.* data are used because α-proton mode data are considered superior for Na and Brückner *et al.* noted calibration difficulties for Mn and P.

35. Pathfinder rocks have much higher S and Cl than expected for igneous rocks and are thought to reflect contamination by soils and dust, giving rise to the concept of "soil-free rock". An additional concern is that the andesitic compositions reflect surface alteration and that the primary composition of fresh rock is quite different. Experience from the Mars Exploration rover indicates that after soil and dust are removed by brushing, a sulfur- and chlorine-rich surface remains due to aqueous alteration and sulfate mineral precipitation. Hurowitz, J. A. *et al.* (2006) *In-situ* and experimental evidence for acidic weathering on Mars. *JGR* **111**, doi: 10.1029/2005JE002515. Accordingly, the soil-free rock probably can no longer be considered a more accurate estimate of the primary igneous composition than are the lowest S- and Cl-bearing rocks (Shark (A17) and Barnacle Bill (A3)). The SO_3 contents of these surfaces are about 1.8 and 2.4% respectively suggesting relatively little alteration and surface soil accumulation for these rocks.

36. McSween, H. Y. *et al.* (1999) Chemical, multispectral, and textural constraints on the composition and origin of rocks at the Mars Pathfinder landing site. *JGR* **104**, 8679–715.

37. Alpha particle X-ray spectrometer rock analyses fall into three categories. "As is" analyses are taken on undisturbed surfaces and include altered surface and any adhering dust and soil. All Pathfinder analyses are of this category. "Brushed" analyses are where a wire brush has swept off loose dust and soil and more closely represent the rock's altered surface. "Abraded" analyses are where the rock abrasion tool grinds about 1–5 mm into the rock thus exposing relatively fresh surfaces. The Spirit rover lost the ability to abrade rocks on sol 416 (sol = martian day) but retained the ability to brush. Data for Spirit through sol 470 are available in Gellert, R. *et al.* (2006) Alpha particle X-ray spectrometer (APXS): Results from Gusev Crater and calibration report. *JGR* **111**, doi: 10.1029/2005JE002555.

38. The Noachian-aged Gusev Crater, landing site of Spirit, is breached by the large channel Ma'adim Vallis and the crater floor is thought to be the potential site of an ancient lake deposit. Such deposits were not found and, if present, are buried by younger lavas that in turn were disrupted by impact processes. In any case, the plains within the crater are younger than the crater itself, likely Hesperian. The Columbia Hills, found within Gusev Crater and the site of most of Spirit's traverse, are embayed by the plains and probably older, perhaps as old as Noachian in age. Squyres, S. W. *et al.* (2004) The Spirit rover's Athena science investigation at Gusev crater, Mars. *Science* **305**, 794–9. Arvidson, R. E. *et al.* (2006) Overview of the Spirit Mars Exploration Rover Mission to Gusev Crater: Landing site to Backstay rock in the Columbia Hills. *JGR* **111**, E02S01, doi: 10.1029/2005JE002499.

39. Squyres, S. W. *et al.* (2006) Rocks of the Columbia Hills. *JGR* **111**, doi: 10.1029/ 2005JE002562.

40. McSween, H. Y. *et al.* (2006) Characterization and petrologic interpretation of olivine-rich basalts at Gusev Crater, Mars. *JGR* **111**, doi: 10.1029/2005JE002477.

41. Zipfel, J. *et al.* (2004) APXS analysis of Bounce Rock – The first shergottite on Mars (abst). *MPS* **39**, Suppl. A118.

42. McSween, H. Y. *et al.* (2006) Alkaline volcanic rocks from the Columbia Hills, Gusev Crater, Mars. *JGR* **111**, doi: 10.1029/2006JE002698. Squyres, S. W. *et al.* (2006) Pyroclastic activity at Home Plate in Gusev Crater, Mars. *Science* **316**, 738–42. McSween, H. Y. *et al.* (2008) Mineralogy of volcanic rocks in Gusev Crater, Mars: Reconciling Mössbauer, alpha-particle X-ray spectrometer, and mini-thermal emission spectrometer spectra. *JGR* **113**, doi: 10.1029/2007JE002970.

43. Christensen, P. R. *et al.* (2005) Evidence for magmatic evolution and diversity on Mars from infrared observations. *Nature* **436**, 504–9.

44. McLennan, S. M. and Grotzinger, J. P. (2008) The sedimentary rock cycle of Mars, in *The Martian Surface: Composition, Mineralogy, and Physical Properties* (ed. J. F. Bell III), Cambridge University Press, pp. 541–77.

45. Although model dependent, isotopic evidence suggests that about two-thirds of the total amount of water outgassed from Mars has been lost to space by various mechanisms. Jakosky, B. M. and Phillips, R. J. (2001) Mars' volatile and climate history. *Nature* **412**, 237–44. For comparison, the mass of the terrestrial hydrosphere is 1.66×10^{21} kg (1.6×10^9 km^3) or 279 ppm of the planetary mass and 413 ppm of the primitive mantle.

46. Squyres, S. W. *et al.* (1992) Ice in the martian regolith, in *Mars* (eds. H. H. Kieffer *et al.*), University of Arizona Press, pp. 523–54.

47. Feldman, W. C. *et al.* (2004) Global distribution of near-surface hydrogen on Mars. *JGR* **109**, doi: 10.1029/2003JE002160.

48. A variety of hydrous minerals, including magnesium sulfates, gypsum, ferric sulfates and amorphous silica, have been identified or inferred in martian soils. As little as 10–20 volume% of such minerals in near-surface soils could probably account for the near-surface water measured at mid latitudes.

49. The history is well known that the "canali" (Italian for channels) of Giovanni Schiaparelli (1877) were mistranslated and embellished into the canals of martian civilizations by Percival Lowell: *"Not everybody can see these delicate features at first sight, even when pointed out to them; and to perceive their more minute details takes a trained as well as an acute eye, observing under the best conditions. ... These are the Martian canals"* (see P. Lowell (1906) *Mars and its Canals*, Macmillan, p. 175). Of course none of these features – neither canali nor canals – survived high-resolution imaging of the martian surface. For an up-to-date review of physical sedimentary processes on which much of this section is based, see Carr, M. H. (2006) *The Surface of Mars*, Cambridge University Press.

50. Malin, M. C. and Edgett, K. S. (2000) Evidence for recent groundwater seepage and surface runoff on Mars. *Science* **288**, 2330–5. Malin, M. C. *et al.* (2006) Present-day impact cratering rate and contemporary gully activity on Mars. *Science* **314**, 1573–7.

51. The atmosphere on average has a pressure of 5.6 mbar at the surface, varying considerably over the martian year as the seasonal ice caps wax and wane, and consisting of 95.3% CO_2, 2.7% N_2, 1.6% Ar, 0.13% O_2, 0.07% CO, ~0.03% H_2O and trace levels of Ne (2.5 ppm), Kr (0.3 ppm) and Xe (0.08 ppm). Owens, T. (1992) The composition and early history of the atmosphere of Mars, in *Mars* (eds. H. H. Kieffer *et al.*), University of Arizona Press, pp. 818–34. On Mars, particles move mostly by a combination of surface creep (mostly particles > 1 mm), saltation (mostly 0.1–1 mm) and suspension (mostly < 0.1 mm).

52. Fishbaugh, K. E. *et al.* (2007) On the origin of gypsum in the Mars north polar region. *JGR* **112**, doi: 10.1029/2006JE002862.

53. Herkenhoff, K. E. *et al.* (2000) Surface ages and resurfacing rates of the polar layered deposits on Mars. *Icarus* **144**, 243–53. Using crater counts to date ice deposits is difficult and controversial because craters on ice are more prone to viscous relaxation and thus may avoid detection.

54. Orbital parameters of Mars differ greatly when compared to other terrestrial planets. Our large Moon has stabilized the obliquity variations of Earth and dissipation of solar tides has had similar effects on Venus and Mercury. Planetary asymmetry in mass caused by Tharsis further influences Mars' obliquity. Accordingly, the present martian obliquity is 25.2°, similar to Earth, but has varied by about ± 10° over the past few million years and by even larger amounts over longer timescales suggesting highly variable cyclical changes in climatic conditions. Timescales of orbital cycle periodicity on Mars are approximately as follows: obliquity: 120 000 yr; precession: 51 000 yr; eccentricity (currently 0.093 and varying between 0.0 and 0.12): 97 000 yr and 2.4 Myr. Laskar, J. *et al.* (2004) Long term evolution and chaotic diffusion of the insolation quantities of Mars. *Icarus* **170**, 343–64.

55. Carr, M. H. (2006) *The Surface of Mars*, Cambridge University Press.

56. Evidence for low pH conditions includes (1) ubiquitous sulfur- and chlorine-rich   .
soils suggesting acid-sulfate conditions; (2) absence of carbonates at the surface; (3) evidence for extensive iron mobility in spite of oxidizing conditions (including the occurrence of jarosite at one location); (4) millimeter-scale alteration profiles on rock surfaces exhibiting geochemical relationships consistent with experimental alteration studies at low pH conditions. Hurowitz, J. A. *et al.* (2006) *In situ* and experimental evidence for acidic weathering on Mars. *JGR* **111**, doi: 10.1029/2005JE002515. Low water/rock ratio conditions are suggested by widespread evidence for iron mobility but relatively little aluminium mobility on a global scale, indicating that water/rock ratios were sufficient to dissolve minerals such as olivine and iron–titanium oxides but not great enough to strongly affect less soluble plagioclase. Hurowitz, J. A. and McLennan, S. M. (2007) A ~3.5 Ga record of water-limited, acidic weathering conditions on Mars. *EPSL* **260**, 432–43.

57. Tosca, N. J. *et al.* (2005) Geochemical modeling of evaporation processes on Mars: Insights from the sedimentary record at Meridiani Planum. *EPSL* **240**, 122–48. Tosca, N. J. and McLennan, S. M. (2006) Chemical divides and evaporite assemblages on Mars. *EPSL* **241**, 21–31. Bibring, J.-P. *et al.* (2007) Coupled ferric oxides and sulfates on the martian surface. *Science* **317**, 1206–10. Tosca, N. J. *et al.* (2008) Fe-oxidation processes at Meridiani Planum and implications for secondary Fe-mineralogy on Mars. *JGR* **113**, doi: 10.1029/2007JE003019. Note that it may not be easy to distinguish evaporite minerals forming from freezing rather than by evaporation. But where sediments have been studied in detail, such as at Meridiani Planum, it appears, from the sedimentary associations, that true evaporitic processes dominate.

58. Gendrin, A. *et al.* (2005) Sulfates in martian layered terrains: The OMEGA/Mars Express view. *Science* **307**, 1587–91.

59. Poulet, F. *et al.* (2005) Phyllosilicates on Mars and implications for early martian climate. *Nature* **438**, 623–7. Loizeau, D. *et al.* (2007) Phyllosilicates in the Mawrth Vallis region of Mars. *JGR* **112**, doi: 10.1029/2006JE002877.

60. Bibring, J. P. *et al.* (2006) Global mineralogical and aqueous Mars history derived from OMEGA/Mars Express data. *Science* **312**, 400–4. Bibring *et al.* divide martian epochs of martian surficial processes into Phyllosian (clay-rich, Early Noachian), Theiikian (sulfate-rich, Late Noachian to Early Hesperian) and Siderikian (ferric oxide-rich, late Hesperian through Amazonian). It will be interesting to see whether this classification scheme becomes widely accepted.

61. The term soil, when used in a planetary context, refers to unconsolidated and poorly consolidated material at the surface and is not to be confused with terrestrial soils that typically have an origin involving organic activity. Those who are irritated by this usage will have to stand in line behind sedimentologists who take issue with metamorphic "facies" and ore deposit "reefs".

62. Over much of the near surface of Mars a regolith exists that was produced during the planet's impact history and in many places is interbedded with volcanic and sedimentary deposits. The amount of regolith depends on surface age (and thus cratering history) and has been estimated as follows: Amazonian, 0–18 m; Hesperian, 12–50 m; Noachian, 50–>> 100 m. Beneath the regolith lies an even thicker fractured zone. Hartmann, W. K. and Barlow, N. G. (2006) Nature of the Martian uplands: Effect on Martian meteorite age distribution and secondary cratering. *MPS* **41**, 1453–67.

63. As the well-documented massive annual dust storms attest, it is possible to entrain large masses of fine particles into the atmosphere. The lower g of Mars requires that lifting forces needed to cause particle motion, and drag forces, are smaller ($g_{Mars}/g_{Earth} = 0.4$) thus making it easier to initiate sediment transport, other factors being equal. Since the surface of Mars has likely been affected by at least several hundred million such events, dust is likely to be homogenized on a global scale.

64. McSween, H. Y. and Keil, K. (2000) Mixing relationships in the martian regolith and the composition of globally homogeneous dust. *GCA* **64**, 2155–66.

65. Taylor, S. R. (2001) *Solar System Evolution: A New Perspective*, 2nd edn., Cambridge University Press.

66. Goetz, W. *et al.* (2005) Indication of drier periods on Mars from the chemistry and mineralogy of atmospheric dust. *Nature* **436**, 62–5.

67. Taylor, S. R. and McLennan, S. M. (1985) *The Continental Crust: Its Composition and Evolution*, Blackwell.

68. Yen, A. S. *et al.* (2005) An integrated view of the chemistry and mineralogy of martian soils. *Nature* **436**, 49–54.

69. Malin, M. C. and Edgett, K. S. (2000) Sedimentary rocks of early Mars. *Science* **290**, 1927–37. Malin, M. C. and Edgett, K. S. (2003) Evidence for persistent flow and aqueous sedimentation on early Mars. *Science* **302**, 1931–4.

70. Moore, J. M. *et al.* (2003) Martian layered fluvial deposits: Implications for Noachian climate. *GRL* **20** (24), doi: 10.1029/2-003GL019002. Jerolmack, D. J. *et al.* (2004) A minimum time for the formation of Holden Northeast fan, Mars. *GRL* **31** (21), doi: 10.1029/2004GL021326.

71. Grotzinger, J. P. *et al.* (2005) Stratigraphy and sedimentology of a dry to wet eolian depositional system, Burns formation, Meridiani Planum, Mars. *EPSL* **240**, 11–72. Grotzinger, J. P. *et al.* (2006) Sedimentary textures formed by aqueous processes, Erebus crater, Meridiani Planum, Mars. *Geology* **34**, 1085–8. McLennan, S. M. *et al.* (2005) Provenance and diagenesis of the evaporite-bearing Burns formation, Meridiani Planum, Mars. *EPSL* **240**, 95–121.

72. Squyres, S. W. *et al.* (2007) Pyroclastic activity at Home Plate in Gusev crater, Mars. *Science* **316**, 738–42.

73. Taylor, S. R. (1982) *Planetary Science: A Lunar Perspective*, Lunar and Planetary Institute (Houston), p. 151.

74. Clark, B. C. and Baird, A. K. (1979) Is the martian lithosphere sulfur rich? *JGR* **84**, 8395–403. Flynn, G. J. and McKay, D. S. (1990) An assessment of the meteoritic contribution to the martian soil. *JGR* **95**, 14,497–509. Bland, P. A. and Smith, T. B. (2000) Meteorite accumulations on Mars. *Icarus* **144**, 21–6. Morris, R. V. *et al.* (2000) Mineralogy, composition, and alteration of Mars Pathfinder rocks and soils: Evidence

from multispectral, elemental, and magnetic data on terrestrial analogue, SNC meteorites, and Pathfinder samples. *JGR* **105**, 14,427–34.

75. Yen, A. *et al.* (2006) Nickel on Mars: Constraints on meteoritic material at the surface. *JGR* **111**, doi: 10.1029/2006JE002797. The most obvious evidence for significant meteoritic components is the remarkable discovery of iron meteorites on the martian surface by the Mars rovers. The Late Noachian evaporitic sandstones of the Burns formation of Meridiani Planum, a dune–interdune sequence, are also characterized by elevated nickel at a level consistent with as much as 5–6% meteoritic component (McLennan, S. M. *et al.* (2005) Provenance and diagenesis of the evaporite-bearing Burns formation, Meridiani Planum, Mars. *EPSL* **240**, 95–121). However, this value could be far less since nickel may be mobile under the low pH conditions that are proposed for surficial processes on Mars (see Sections 6.4.2 and 6.4.4).

76. Hahn, B. C. and McLennan, S. M. (2007) Evolution and geochemistry of the martian crust: Integrating mission datasets. 7th International Conference On Mars, Pasadena, CA, July 9–13, 2007, Abst. #3179, LPI contribution no. 1353.

77. McLennan, S. M. (2003) Large-ion lithophile element fractionation during the early differentiation of Mars and the composition of the martian primitive mantle. *MPS* **38**, 895–904.

78. Hahn, B. C. *et al.* (2007) Mars Odyssey gamma-ray spectrometer elemental abundances and apparent relative surface age: Implications for martian crustal evolution. *JGR* **112**, doi: 10.1029/2006JE002821.

79. McGovern, P. J. *et al.* (2002) Localized gravity/topography admittance and correlation spectra on Mars: Implications for regional and global evolution. *JGR* **107**, doi: 10.1029/2002JE001854; McGovern, P. J. *et al.* (2004) Correction to "Localized gravity/topography admittance and correlation spectra on Mars: Implications for regional and global evolution". *JGR* **109**, doi: 10.1029/2004JE002286.

80. Montési, L. G. J. and Zuber, M. T. (2003) Clues to the lithospheric structure of Mars from wrinkle ridge sets and localization instability. *JGR* **108**, doi: 10.1029/2002JE001974.

81. Ruiz, J. *et al.* (2006) Evidence for a differentiated crust in Solis Planum, Mars, from lithospheric strength and heat flow. *Icarus* **180**, 308–13.

82. See Section 5.7 in Chapter 5 and references therein for discussion of the likelihood and scale of magma ocean processes.

83. Mittlefehldt, D. W. (1994) A cumulate orthopyroxenite member of the martian meteorite clan. *Meteoritics* **29**, 214–21.

84. The parameter T_{CHUR} is the Nd-model age referenced to a chondritic uniform reservoir. Nyquist, L. E. (1995) "Martians" young and old: Zagami and ALH84001. *LPSC* **XXVI**, 1065–6.

85. Greeley, R. and Schneid, B. D. (1991) Magma generation on Mars: Amounts, rates, and comparisons with Earth, Moon, and Venus. *Science* **254**, 996–8: McEwen, A. S. *et al.* (1999) Voluminous volcanism on early Mars revealed in Valles Marineris. *Nature* **397**, 584–6.

86. Phillips, R. J. *et al.* (2001) Ancient geodynamics and global-scale hydrology on Mars. *Science* **291**, 2587–91.

87. Tajika, E. and Sasaki, S. (1996) Magma generation on Mars constrained from an ^{40}Ar degassing model. *JGR* **101**, 7543–54. Hutchins, K. S. and Jakosky, B. M. (1996) Evolution of martian atmospheric argon: Implications for sources of volatiles. *JGR* **101**, 14,933–49.

88. Kiefer, W. S. (2003) Melting in the martian mantle: Shergottite formation and implications for present-day mantle convection on Mars. *MPS* **38**, 1815–32.

89. Harder, H. and Christensen, U. R. (1996) A one-plume model of martian mantle convection. *Nature* **380**, 507–9.

7

Venus: a twin planet to Earth?

It was believed up until the 1960s that Venus might be Earth-like with respect to harbouring life and writers of science fiction endowed its surface with advanced civilizations.

(Henry S. F. Cooper) [1]

7.1 The enigma of Venus

Venus has historically been regarded as a "twin planet" to the Earth as amongst the planets, it is closest to the Earth in mass, density, size and in distance from the Sun. However it has, by terrestrial standards, extraordinary crustal features and a geological history that bears little resemblance to that of the Earth. In addition, it does not possess a satellite and has a retrograde rotation with a period of 243 days.

The planet clearly warrants closer study particularly as the differences between these twin planets emphasize the problems of building crusts or discovering habitable planets in other planetary systems. So it is useful to contrast crustal development on Venus with that of its twin planet Earth, that occupy the following five chapters [2].

The density of Venus ($5.24 \, \text{g/cm}^3$) is about 5% less than that of the Earth ($5.514 \, \text{g/cm}^3$). This difference is mostly due to the slightly lower internal pressures as the planetary radius is 320 km less than that of the Earth. But the uncompressed density of both planets is very close (Earth $3.96 \, \text{g/cm}^3$; Venus $3.9 \, \text{g/cm}^3$) [3]. The similar density of Venus to the Earth and the presence of a basaltic crust on the planet are the basis for assuming a broadly similar composition and internal structure. From analogy with the Earth, a metallic core is expected, but the lack of data for the moment of inertia precludes estimates of its size.

Although various bulk planetary compositions have appeared in the literature, these are model dependent with the status of inspired guesses [4]. Because of the density similarity, there is probably little real difference in the major element

181

composition (Si, Ti, Al, Fe, Mg, Ca) of Venus compared to that of the Earth. Until more data are available, the best guide to the bulk composition of Venus is that of the Earth (Chapter 8).

The abundances of the heat-producing elements, as discussed later, are probably similar in both planets. The MgO contents of the venusian basalts are around 10% and the Mg# is about 70. This indicates that mantle temperatures are similar to those of the Earth and are not hot enough in Venus to produce MgO-rich lavas analogous to terrestrial magnesium-rich komatiites (MgO $\approx$ 18%). An important caveat in all such comparisons is that Venus was resurfaced at about 750 Myr and accordingly surface geochemical measurements on Venus reflect igneous activity from that event or subsequently [5].

Measured surface temperatures, cooler at higher altitudes, range from 453 °C to 473 °C, hot enough to melt not only lead but also tellurium (m.p. 449.5 °C). The surface pressure of 95 bar is equivalent to the pressure on the Earth beneath about 1 km of seawater. The crustal thickness (the best estimate is about 30 km) appears to be relatively uniform in contrast to the varying crustal thicknesses observed on the Earth, Moon or Mars so that the venusian crust is similar in this respect, albeit thicker, to the uniform oceanic crust of the Earth. A uniform crustal thickness of 30 km represents approximately 1.5% of the volume of the planet.

However there are major differences between the crust of Venus and the Earth. Venus lacks the bimodal distribution of elevations of the terrestrial crust, while the venusian crust is strong and, in contrast to the Earth, displays a positive correlation between gravity and topography with gravity values reaching 70 mgal [6]. Compared with the Earth, the rocks are much stronger, a consequence of the effective absence of water [7].

7.2 Surface features of Venus

Because of the strength of the venusian crust, steep slopes on the surface of Venus apparently remain in place for millions of years, while large volcanoes are supported indefinitely by the crust [7]. The tectonic differences between the dry one-plate crust of Venus and the wet multiplate subduction environment on Earth mean that processes capable of producing the continental crust of the Earth are absent on Venus. "Like an armadillo, Venus has encased itself in a strong dry rigid shell of basalt" [8].

7.2.1 Plains

The surface is mostly of low relief, consisting of plains reminiscent of the terrestrial sea floor. Over 80% of Venus is covered with these flat, radar-smooth rolling plains.

Fig. 7.1 Smooth basaltic plains in the Lakshmi region, Venus showing cross-faulting. Image is 40 × 80 km. NASA JPL P36699.

They were probably emplaced by flood volcanism and appear to have been formed over large regions (10^7 km^2) on very short timescales although there is the usual divergence of views on this, as on almost every aspect of the interpretation of the venusian surface [9]. Although the closest terrestrial analogue is the basaltic floor of our ocean basins, there are significant differences. Thus the venusian plains show pervasive evidence of tectonic deformation, being typically crossed by sets of regularly spaced faults intersecting at high angles (Fig. 7.1). The amount of deformation is surprising and ubiquitous. This is in great contrast to flat-lying relatively undeformed basaltic plains on the Moon, Earth or Mars. On Earth, such intense deformation occurs only near plate boundaries.

Wrinkle ridges are more common on Venus than on the other terrestrial planets. They are typically about 1 km wide and spaced about 20 km apart in parallel arrays that often show sets of cross-cutting faults (Fig. 7.1). The causes of this widespread deformation appear to be linked to the dry rigid nature of the lithosphere, so that in the lack of a weak low-viscosity zone, mantle stresses are directly coupled to the crust [10].

7.2.2 Channels

One of the great lessons of comparative planetology is that caution is needed in applying terrestrial analogues when studying other planets. Thus there are at least 200 unique channels on the plains of Venus. They do not possess tributaries and appear to be only 1 or 2 km wide. They do not resemble lunar rilles, but meander in a form resembling that of terrestrial rivers. No boulders are apparent and they are very smooth down to centimeter scales. Some channels are compound and some deltas are present. Many are over 500 km in length. The longest is Baltis Vallis that extends for a distance of 6800 km, longer than the Nile (6695 km). But it is only about 1 km wide and 20 m deep and was cut into a surface where the temperature is around 460 °C. The only effective eroding agent that could form such a uniform lengthy channel would appear to have been low-viscosity lava, possibly carbonate-rich. This has led to the comment that on Venus, lavas flow like water, while on the outer icy planets, water ice behaves like terrestrial lavas [11].

7.2.3 Volcanoes

The presence of over a million small shield volcanoes that closely resemble terrestrial oceanic floor seamounts, reinforces the basaltic nature of venusian volcanism. There are many larger volcanic structures on Venus. Thus the edifice of Beta Regio rises nearly 10 km above the mean venusian datum. This structure is interpreted to be the result of volcanism and uplift over a mantle plume [12]. It contains a large volcano, Theia Mons. Maat Mons is the tallest volcano on Venus standing 8 km above the mean planetary radius (Fig. 7.2).

Fig. 7.2 Maat Mons, 8 km high, the tallest volcano on Venus located at 0.9° N and 194.5° E. The summit caldera is about 30 km in diameter. Vertical exaggeration × 23. NASA JPL PIA 00106.

This volcano has a large caldera, 31 × 28 km that contains several collapse craters up to 10 km in diameter. Large shield volcanoes that have flows extending over a diameter of 500 km, are thought to have formed over mantle hot spots.

There are 1500 volcanoes that are over 20 km in diameter, while another 150 are more than 100 km in diameter. However, these volcanoes are mostly low and broad and rarely exceed 2 km in elevation. This forms a contrast to the high volcanoes on the Earth and Mars and can be attributed to a combination of factors including high surface temperatures and pressures on Venus [13].

7.2.4 Coronae

Coronae are volcano-tectonic features on Venus that again have no terrestrial analog nor appear elsewhere in the Solar System [14]. These exotic structures consist of concentric rings of grooves and ridges, typically 150 to 1000 km in diameter but with extremes ranging from 60 km to the enormous corona, Artemis that is 2500 km in diameter [15] (Fig. 7.3).

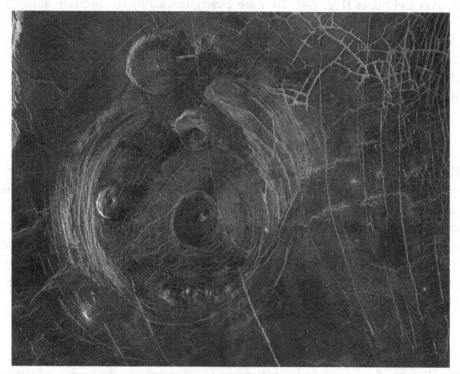

Fig. 7.3 Aine Corona (one of the smaller examples), 200 km in diameter, centered at 59° S and 164° E. A pancake dome, 35 km in diameter, occurs on the northern flank and others occur within the structure. NASA JPL FMIDR 59s164.

Coronae are probably the surface expression of hot mantle upwelling, perhaps analogous to terrestrial mantle plumes and so may be linked to the large volcanoes [16].

Like terrestrial mantle plumes, coronae have been the subject of controversy. At an early stage, they were thought to have formed in a short time period after the flooding of the regional plains but earlier than the formation of the great volcanic shields. Later studies have shown that they are not restricted to any particular time. This debate is a good example of the difficulties of establishing a venusian stratigraphy [17]. The problems of trying to account for the coronae and the search for terrestrial analogs provide yet another example of the difficulty in looking for some kind of patterns or processes that might have general applications to the other solid bodies of the Solar System. Such solutions are difficult to find in a planetary system dominated by stochastic processes [18].

7.2.5 Tesserae

Many other unique and unfamiliar features appear on the surface of Venus. These include tesserae, areas of deformed terrain that consist of closely packed sets of grooves and ridges (Fig. 7.4). They are clearly formed by compression although tesserae also contain extensional features. The tectonic structure of these ridged terrains resembles the complex tectonics of our terrestrial continents although the latter are of different composition. Because of their complex tectonics, tesserae have often been considered the oldest surfaces on Venus; but this opinion is based on lithological correlations rather than on precise dating.

7.2.6 Ishtar Terra and Aphrodite Terra

The small continent-sized areas such as Aphrodite Terra and Ishtar Terra are elevated with respect to the plains. The Maxwell Montes in Ishtar Terra reach 11 km. If this higher terrain is due to a lower density rather than to tectonic crumpling, it might be analogous in composition to our terrestrial continents. But examples of compressional tectonics are common (e.g. the banded terrain of Ishtar Terra) [19]. At the western end of Ishtar Terra, Lakshmi Planum (650° N, 335° E) reaches an elevation of about 3 km. Compressional features surround this plateau. Here the crust appears to have thickened due to the convergence of tectonic forces.

7.3 Impact craters and the age of the surface

The absence of the normal terrestrial processes of erosion on the surface of Venus have resulted in the preservation of 940 impact craters formed by meteorite, asteroidal or cometary impacts [20] (Fig. 7.5).

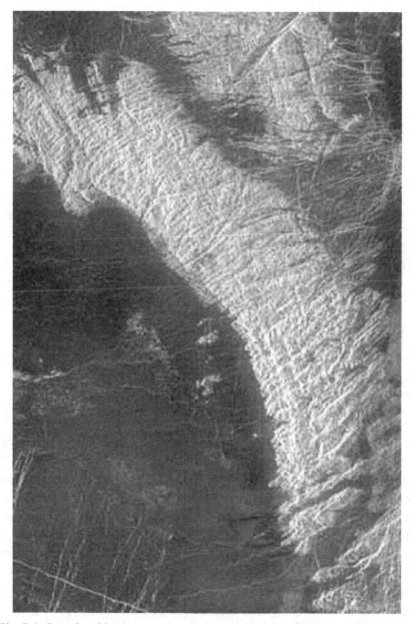

Fig. 7.4 Complex ridged tessera terrain rising from the lava plains of Leda Planitia at 41° N and 52° E. Image is 220 × 275 km. NASA JPL.

The thick atmosphere has acted as a filter to all but the larger impactors, so that there are few craters less than 30 km in diameter and almost none below 5 km. Seventy-two are large peak-ring craters that range in diameter from 31 to 109 km while there are four probable multiringed impact basins [21].

Fig. 7.5 Three impact craters with diameters from 37 to 50 km formed on fractured basaltic plains centered at 27° S and 339° E. NASA JPL PIA 00086.

About 16% of the craters are multiple, resulting from the break-up of the projectile in the atmosphere. Other impactors that almost penetrated through the atmosphere exploded close to the ground. The effects of those events are recorded on the surface as radar-dark splotches that are seen on the smooth terrains of the plains. These splotches are due to shock waves from the objects that exploded as they broke up in the atmosphere. The splotches are not preserved on the more rugged terrains, suggesting that they are superficial features. If evenly distributed over the venusian surface, they would number about 1100. The shock waves from these atmospheric explosions have the potential to pulverize the surface and perhaps are a partial cause of the fractured platy landscape seen in the Venera panoramas from the USSR landers [22].

A curious feature, unique to Venus, is that for some craters, sectors of the crater ejecta blankets are missing, probably due to atmospheric interaction during entry and impact (Fig. 7.6) [23].

Another unique feature of the formation of impact craters on Venus is the large-scale production of impact melt, some of which has flowed out to several crater diameters, well beyond the debris in the ejecta blankets. These flows must be

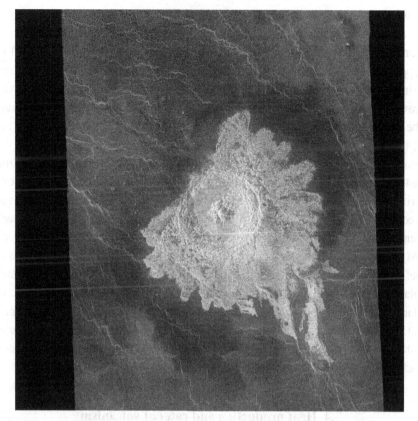

Fig. 7.6 Aurelia Crater, 32 km in diameter, showing a missing sector on the ejecta blanket. NASA JPL PIA 00239.

formed from low-viscosity melts [24]. Compared to terrestrial impacts, those on Venus create up to 25% more impact melt [25]. This increase in melt production on Venus is mostly a consequence of the higher surface temperature of Venus. Massive sheets of impact melt are also produced on Earth but are mostly removed by erosion, so that the efficiency of terrestrial geological processes has inhibited our recognition of the amount of impact-melted rock.

Among the many surprises that Venus has provided, the most significant in the present context is that the impact craters on Venus are mostly fresh. A few have been embayed by lavas or affected by tectonic processes but most are pristine and are randomly distributed on the venusian surface. It is worth contemplating the data. From the total of 940 craters, Sasha Basilevsky and co-workers conclude that, "743 appear pristine, 107 lightly fractured, 27 heavily fractured, 26 embayed by lavas from external sources and lightly fractured, 6 embayed by lavas from external volcanic sources and heavily fractured, 8 compressed, 2 compressed and lightly fractured and 5 craters are mantled by ejecta from other impact craters" [26].

However, although most craters retain a well-preserved rim and ejecta blanket, some of the older craters appear to be floored with dark basalt, recalling the appearance of craters such as Plato and Archimedes on the Moon. This fill does not appear to be impact-generated, as there are many large bright-floored craters. Here interpreted as lava, it is a few hundred meters thick at most. These observations indicate an extended period of volcanism as some lavas may also have been erupted on the plains where they are less readily distinguished [27]. However, like many interpretations of the geology of Venus, this is controversial [28].

Most workers agree that the crater distribution cannot readily be distinguished from that of a randomly distributed population [29]. Based on estimates of the cratering flux, the age of the surface on which the craters have been formed varies between 300 and 1000 Myr ago. In our opinion, the best estimate for the age of the surface of Venus on which the current population of craters was formed is 750 Myr [30]. The interesting conclusion is that the crust that we observe by the Magellan radar is geologically young and records only the last 15% of the history of the planet. The crust is thus not primordial, but secondary. It appears to have been emplaced relatively quickly, possibly in less than 100 Myr about 750 Myr ago but the rate of resurfacing is uncertain [31]. Only minor geological activity has occurred since that time, a period that on Earth extends back into the Late Proterozoic, before the rich fossil record of the Phanerozoic [32].

7.4 Heat production and rates of volcanism

The abundances of the heat-producing elements potassium, uranium and thorium in Venus are probably similar to those of the Earth. This conclusion follows from the similarity of the uncompressed densities of the two planets. For rocky terrestrial planets, it follows that the abundances of the refractory elements calcium, aluminium and titanium should be similar. But the abundances of the refractory elements, uranium and thorium are also correlated with those of the refractory major elements. Such elements are not separated by nebular processes and so these heat-producing elements are likely to be similar in abundance in both planets.

What about the abundance of potassium, whose ^{40}K isotope has been a significant contributor to the terrestrial heat-production budget? A value could be obtained from the few measured K/U ratios on the venusian surface that average about 7000 and that is somewhat lower than the terrestrial average. So the potassium abundance on Venus might be lower by as much as a factor of two compared to that of the Earth or might be similar. The accuracy of the USSR data does not allow us to reach a more definite conclusion. But even if the lower USSR value turns out to be correct, the heat production from ^{40}K is now much reduced from that of earlier times.

As on the Earth, 90% of the current heat production now comes from ^{238}U and ^{232}Th. It was only before about 2.5 Gyr ago that the heat production from ^{40}K would have contributed more than 30% to the total. The consequence is that the radiogenic heat production in both planets can now be expected to be similar and has been so since the beginning of the Proterozoic era on Earth. For these reasons, geophysicists have generally assumed that "direct scaling of terrestrial radiogenic concentrations to Venus is appropriate" [33].

According to geophysical calculations, the current heat loss on Venus is between 10 and 30 mW/m^2. This compares with the average heat flow of 100 mW/m^2 for the oceans, 48 mW/m^2 from continental areas and 84 mW/m^2 for the whole Earth [34]. So the venusian heat loss is low compared with that of the Earth, much of which is due to the operation of plate tectonics. The Earth loses most of its heat by the formation of 18 km^3 of lava per year at the mid-ocean ridges. Smaller amounts of heat are lost at island arcs or through hot-spot intra-plate volcanism. In comparison, Venus currently loses only a small amount of its planetary heat production by volcanic activity, that produces perhaps about only 1 km^3 of lava per year.

Although the basaltic surface of Venus bears some superficial resemblance to that of the terrestrial oceanic crust, it is much older than our ocean floors, which have a mean age of about 60 Myr. Although Venus and the Earth are of similar size, density, bulk composition and heat production, the production of basaltic magma clearly follows a different path. Since the last major resurfacing event 750 Myr ago, apparently less than 10% of the surface has been modified by more recent volcanic activity [35]. On the basis of such models, the heat flux is much less than that for the Earth. As the heat production is similar in both planets, the consequence is that the mantle of Venus is heating up. The prediction is that this will eventually lead to another major resurfacing [36].

The contrast between the two planets, that results from this difference in heat production–surface heat-flow relationships, is great. Given a similar heat production, the Earth has utilized this energy by producing the oceanic crust, recycling it into the mantle, and forming the continental crust. Venus in contrast has undergone episodic massive resurfacing. Such episodes, however, seem likely to have taken place over millions or perhaps tens of millions of years.

7.4.1 A one-plate planet

The terrestrial example of plate tectonics continues to fascinate modelers [37]. However, despite much searching, there appears to be no sign of crustal subduction or the operation of anything resembling plate tectonics on the surface of Venus. Neither are such processes expected. The near absence of water and the lack of oceans means that even if subduction was occurring, melting of the dry subducting basalt would be inhibited. The slightly lower pressures on Venus due to the smaller size of the planet, mean that the basalt to eclogite transition occurs at a depth of 65 km on Venus. So the change to a denser phase that would facilitate sinking and recycling of the crust is probably too deep for such processes to occur. Finally the dry nature of the mantle of Venus would inhibit the development of hydrous melts and so production of a weak asthenosphere (low seismic velocity zone) on which plates might slide.

Venus forms another example of a one-plate planet. In contrast to the Earth, there are no mid-ocean ridges on the rolling basaltic plains. Subduction zones are absent. The consequence is that the basaltic lavas that form the plains stay on the surface along with the erupted volatiles such as chlorine and sulfur. This is reminiscent of the situation on the martian surface. Hence lateral tectonics and basaltic volcanism are the major features that shape the surface of Venus.

7.5 Crustal composition

Although it might appear difficult to establish an overall composition for the crust of Venus, there are many clues. Much information is provided by the geomorphology of the surface as revealed by the Magellan radar. The smooth plains that dominate the surface appear to be volcanic. This is reinforced by the presence of a million basaltic-looking volcanoes. But the lessons of comparative planetology must engender caution and portraits may be deceptive. Fortunately the extraordinary USSR Venera and Vega missions produced several chemical analyses of the surface (see Fig. 7.7). The spacecraft variously carried XRF and gamma-ray detectors, sometimes both. All the locations gave similar results and the major-element compositions revealed by the Russian X-ray fluorescence (XRF) instruments (Table 7.1) are basaltic.

These data are of course subject to the usual caveats concerning spot analyses with additional potential biases imposed by the necessity for safe landing sites. Although a basaltic composition is consistent with the interpretation of the geomorphology of the surface topography, most of the missions landed on the plains [38].

There was no XRF instrument on the Venera 8 spacecraft, the first of the USSR landers to return data, so that no major element data are available for that site. Much interest was generated by the high values recorded of 4% K, 2.2 ppm U and 6.5 ppm Th. These are typical values for granitic rocks on the Earth and immediately raised the prospect of the presence of granites on Venus. Venera 9 and 10, also equipped only with gamma-ray detectors, landed on the flanks of Beta Regio, which resembles a large terrestrial hot-spot rise. Their data are consistent with basaltic compositions. But similar high values of 4% K to the Venera 8 data were recorded by the Venera 13 spacecraft XRF instrument at which site the overall major element composition was

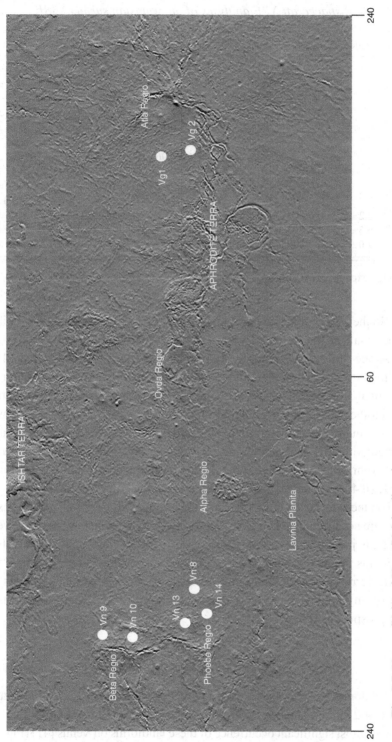

Fig. 7.7 Venera (Vn) and Vega (Vg) landing sites.

Table 7.1 *USSR spacecraft XRF analyses of the venusian surface [38]*

Constituent	Venera 8	Venera 9	Venera 10	Venera 13	Venera 14	Vega 1	Vega 2
SiO_2	–	–	–	45.1 ± 3.0	48.7 ± 3.6	–	45.6 ± 3.2
TiO_2	–	–	–	1.59 ± 0.45	1.25 ± 0.41	–	0.2 ± 0.2
Al_2O_3	–	–	–	15.8 ± 3.0	17.9 ± 2.6	–	16.0 ± 1.8
FeO	–	–	–	9.3 ± 2.2	8.8 ± 1.8	–	7.74 ± 1.1
MnO	–	–	–	0.2 ± 0.1	0.16 ± 0.08	–	0.14 ± 0.12
MgO	–	–	–	11.4 ± 6.2	8.1 ± 3.3	–	11.5 ± 3.7
CaO	–	–	–	7.1 ± 0.96	10.3 ± 1.2	–	7.5 ± 0.7
K_2O	4.8 ± 1.5	0.6 ± 0.1	0.4 ± 0.2	4.0 ± 0.63	0.2 ± 0.07	0.54 ± 0.27	0.48 ± 0.24
S	–	–	–	0.65 ± 0.4	0.35 ± 0.31	–	1.9 ± 0.6
Cl	–	–	–	<0.3	<0.4	–	<0.3
U (ppm)	2.2 ± 0.7	0.6 ± 0.2	0.5 ± 0.3	–	–	0.64 ± 0.47	0.68 ± 0.38
Th (ppm)	6.5 ± 0.2	3.7 ± 0.4	0.7 ± 0.3	–	–	1.5 ± 1.2	2.0 ± 1.0
K	4.0 ± 1.2	0.5 ± 0.07	0.33 ± 0.15	3.3 ± 0.5	0.17 ± 0.05	0.65 ± 0.2	0.40 ± 0.2

Data in wt% except where indicated in ppm.

basaltic. So the Venera 8 landing site, that seems to be on plains, might be similar to the alkali basalt at the Venera 13 site [38]. Alternatively Venera 8 might have landed on one of the small number of "pancake" features (one is located 140 km north of the estimated landing point but within the landing ellipse [38]) rather than on plains of alkali basalt. That Venera 8 encountered a small patch of differentiated rocks is another possibility, something that might be expected to occur occasionally in a dominantly basaltic terrain [39]. More light was shed by the Vega 2 lander that carried both XRF and gamma-ray instrumentation and landed on the eastern flanks of Aphrodite Terra. Here the XRF major element data and the gamma-ray data for K (0.40%), U and Th both indicated basaltic compositions.

Based on these few data, the surface of Venus appears to be dominated by basaltic rocks that are occasionally alkali-rich. Aphrodite Terra, for which we have the Vega 1 and 2 data, is judged to be crumpled-up basaltic crust. The composition of Ishtar Terra remains enigmatic but there is no reason to postulate any difference from Aphrodite Terra, except for the hazardous analogy with terrestrial continents. The absence of any sign of plate tectonics on the surface of Venus makes this an unlikely prospect [40].

7.5.1 Pancake domes: rhyolites on Venus?

Steep-sided domes or "pancakes" occur in a variety of sizes on Venus. Some are as small as 1 or 2 km in diameter, but the 175 larger domes (19 to 94 km in diameter) represent the most significant occurrence of these landforms on Venus [41] (Fig. 7.8).

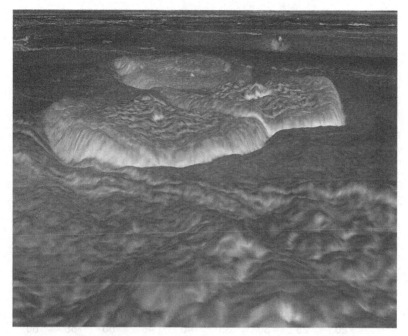

Fig. 7.8 Three-dimensional perspective view of a pancake, about 30 km in diameter on the eastern edge of Alpha Regio. Vertical exaggeration × 23. NASA JPL PIA00246.

The domes are steep-sided, mostly circular (10% are elongate or irregular) and flat-topped or concave. Mostly (90%) they occur close to other volcanic edifices or coronae. Commonly they occur in chains. Many (80%) of the domes have pits, analogous to such features on the rhyolite domes of the Mono Craters in California. So at first glance, the pancake domes resemble terrestrial rhyolite domes.

As there is little evidence for large areas of acidic rocks on Venus, these pancakes might represent silica-rich differentiates and so be the only occurrence of "granitic" rocks on the surface of a planet that is closely similar in bulk composition to the Earth. From our terrestrial experience, one might expect that some formation of acidic differentiates might occur during such massive outpourings of basalt that dominate the venusian surface. This happens on Earth at locations such as Iceland, although only rarely during the eruptions of plateau basalts, that seem a closer analogue to the basaltic plains of Venus.

Due to the high viscosity of silica-rich lavas, the crusts of terrestrial domes of rhyolitic composition break up into meter-size blocks so that the surface, virtually impassable, is amongst the roughest terrain encountered on this planet. But unlike the blocky surfaces of terrestrial silicic domes, the upper surfaces of the venusian pancakes appear smooth and unfragmented as measured by the Magellan radar [42] (Fig. 7.9). The radar images returned from the venusian pancakes resemble the radar

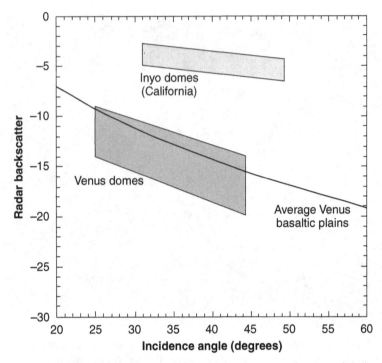

Fig. 7.9 Contrast between the radar reflections from rough terrestrial rhyolite domes and smooth venusian pancakes [42].

reflectivity observed from the smooth crusts of Hawaiian pahoehoe low-viscosity basaltic lavas.

At this point, it should be recalled that the extrusion of lavas on the surface of Venus takes place under pressures equivalent to about 1 km of seawater. Indeed, the close similarity of the pancakes to terrestrial seamounts has been noted [43]. It is our assessment of the current evidence that the pancakes are basaltic domes flattened under the 95 bar atmospheric pressure during extrusion. But any such speculation about a dry planet that is based on terrestrial analogs remains hazardous and in conflict with the hard-won experience of comparative planetology.

7.5.2 The differentiation of Venus

How much differentiation has occurred on Venus and what fraction of the incompatible elements have been concentrated in the crust? This is an inherently difficult question, on a planet where the crustal thickness is uncertain and where events occurring earlier than about one billion years ago, such as the possible formation of an early crust, are not represented by surface exposures. Melting in the mantle of Venus, even if producing mostly basaltic magmas, is likely to incorporate the heat-producing elements K, U, and Th, so that these will be concentrated over time in the crust and so deplete the mantle in incompatible elements independently of early crustal formation.

However, some insight can be gained by considering the amount of radiogenic argon in the venusian atmosphere, produced by decay of ^{40}K to ^{40}Ar, that has a half-life of 1.193×10^9 years [44]. As most of the radiogenic argon was produced early in the 4.5 Gyr history of Venus, this might provide us with a window into the missing 3.5 Gyr of geological history on the planet. Furthermore the atomic weight of ^{40}Ar is too high for it to escape from the atmosphere of Venus, once the planet has reached its present mass. The amount of ^{40}Ar in the venusian atmosphere at present amounts to that formed since accretion, less that retained in the mantle. On the Earth, approximately half of the ^{40}Ar generated is in the atmosphere. But the amount of ^{40}Ar in the atmosphere of Venus is a factor of four times less than in that of the Earth [44].

The amount of ^{40}Ar generated depends on the bulk potassium content of the planet. Our estimates depend on the K/U ratios as measured by the Russian landers (Table 7.1). However, these remarkable data have uncertain errors so that the best that one can say is that Venus has transferred somewhere between one-third and one-half as much potassium into the crust as has the Earth. Such a value is also broadly consistent with the very crude crust–mantle mass-balance estimates available for incompatible elements.

Assuming that the primitive mantle and internal structure of Venus are similar to the Earth, this degree of differentiation would imply bulk crustal levels of about ~1000–3000 ppm K (~0.4–1.6 ppm Th; ~0.1–0.4 ppm U) for a 30 km crust which is within the general magnitude of surface measurements (Table 7.1). Hence Venus does not seem to have been differentiated to the same degree as the Earth, by a factor of somewhere between a third and a half.

This has the further implication that the earlier hidden history of Venus was not very different from that which we infer to have occurred in the past one billion years. If an earlier Earth-like differentiation (plate tectonics?) had occurred, then the atmospheric content of ^{40}Ar might have approached that of the Earth, particularly as most of the ^{40}Ar was produced before one billion years ago.

7.6 The geological history of Venus

The apparently moribund geological history of Venus stands in great contrast to that of the Earth. There is no sign on Venus of rocks older than Late Proterozoic, and the preceding four billion years of geological history are missing. Thus a caveat that underpins all comments about the geological history of Venus is that over 80% of the record is currently and perhaps forever inaccessible. This allows room for even more speculation than that which surrounds the Hadean Eon on Earth (see Chapter 9).

However it has proven difficult to establish a relative timescale for the rock record that is exposed on Venus that might be applied on a planetary-wide scale and considerable controversy surrounds this topic. The stratigraphic information from magnetic signatures and paleontology that allows the dating of our ocean floors is not available on Venus.

Although Venus has 940 craters that allow estimates of the overall age of the surface to be established, it is one of the ironies of comparative planetology that there are not enough craters to be useful as a stratigraphic tool. This is a consequence of the deficiency of small craters due to the protection afforded by the thick atmosphere. Only areas greater than $5 \times 10^6 \, km^2$ contain a sufficient number of craters to be useful in establishing relative ages. Thus attempts to establish the relative ages of portions of the surface are hampered by a paucity of craters. Furthermore, volcanoes, coronae and other geological regions often contain no craters or too few to be useful. These dilemmas have led to differing versions of the geological history of the planet [45].

In an attempt to get around this problem, one stratigraphic scheme has been erected that is based on lithological correlations of areas lacking enough craters to be significant statistically. Synchronous formation of similar landforms such as coronae, tessarae and large volcanoes is assumed [46]. However some cautions are needed in the uncritical application of this concept.

One of the enduring geological truths based on long terrestrial experience, is that the correlation of strata on Earth using lithology is a blunt tool, fraught with the possibility of significant error. Although similar-looking rocks of differing ages are common, globally synchronous events are rare on the Earth. Localized observations of overlap or embayment on individual areas may have no planetary-wide significance.

Another subtle problem arises when studying the crater distribution on Venus. The larger volcanoes, for example, have fewer craters than are observed on the plains, and so might be considered to be younger. However these large volcanic edifices grow over significant periods of geological time, so that craters on earlier surfaces can be covered, leading to an apparently young age for the volcanic structures. The loss of craters in regions of greater tectonic activity may also produce areas that appear younger than they actually are [47].

An alternative viewpoint is that a stratigraphic sequence cannot be established from the available evidence and an essentially random sequence of geological processes has occurred on Venus [48]. Indeed it is unclear, bearing in mind the caveats noted above, that the major geological units such as coronae, plains units and volcanoes form an observable sequence [49] and in the opinion of one worker (among others) "the hypothesis of globally synchronous formation of landforms such as volcanoes or coronae cannot be validated with the available crater data" [50].

Moreover the major resurfacing on Venus seems unlikely to have been instantaneous and does appear to have occurred over an extended period of volcanism, lasting perhaps for 100 Myr rather than being a single catastrophic resurfacing event [49]. Coronae and massive volcanoes seem unlikely to be constructed on short timescales. Large shield volcanoes on Earth may take over a million years to reach full size, even on such an active planet. Indeed, many overlapping lava flows are observed on Venus, indicative of prolonged activity but on uncertain timescales that may simulate a stratigraphic sequence.

So the major resurfacing of the planet, the formation of the immense coronae, the building of the numerous large volcanoes and the lava flooding of the plains, all probably occurred on timescales more familiar to terrestrial geologists rather than in some apocalyptic Noachian flood of lava. Perhaps a venusian geologist might have found that period no more hazardous than that of current geologists investigating an active volcano on Earth.

Accordingly, we have a planet that underwent a major resurfacing around 750 Myr ago, an event that lasted for perhaps 100 Myr within our current ability to resolve ages. There are no older surfaces preserved. Tectonic compression forming tesserae, volcanism and formation of coronae seem to have occurred sporadically during that time. Compared to the Earth there is little resolvable geological history on Venus and there is no agreed sequence of events on the planet, the cornerstone to the establishment of a stratigraphic record.

7.6.1 Water on Venus

A basic question, relevant here, is why is there such a marked difference between the crustal development of the twin planets, Earth and Venus? They are so alike in size, mass and bulk density that they must be close in bulk composition. Also as noted, the abundance of the heat-producing elements, potassium, uranium and thorium, is likely to be similar on Earth and Venus, an observation that merely exacerbates the problem.

The difference in their subsequent evolution seems to be particularly a consequence of the low water content of Venus compared to the Earth. This is likely to lead to strong mechanical coupling between the dry crust and mantle. So unlike the Earth, a weak asthenosphere on which plates might slide is unlikely to develop. This lack of water also raises the temperature of remelting of a basaltic crust. In addition, the slightly smaller size of the planet may drive the basalt–eclogite (see Section 8.1) transition too deep to allow subduction to begin, even if other factors had been favorable. All these effects conspire and are the probable cause of the absence of any evidence for the operation of plate tectonics on this anhydrous planet. So the contrast between the tectonics of Venus and the Earth is mainly a consequence of the difference in water contents of the two planets, exacerbated by a slight difference in size.

The source and history of water in Venus as well as the Earth, continue to be debated and can be considered only briefly here. The present water budget on Venus is depleted relative to the primitive solar nebula (= solar) abundance by a factor about 2×10^{-9}. This is five orders of magnitude less than the water on Earth, that is already depleted relative to the primitive solar nebula abundance by a factor about 2×10^{-4}. The D/H ratio for Venus is about 0.025, a factor of 150 times that of the Earth [51]. Both are much higher than the value for the solar nebula, so ruling out a derivation from that source. Although comets were long regarded as a promising source for the Earth's water, the D/H ratio of the three comets measured is a factor of two higher than that of oceanic water, so that they can make only a minor contribution to the water budget of the Earth [52].

However the high D/H ratio on Venus has often been construed as evidence for an initially "wet" or "damp" Venus, in which an original complement of water was slowly depleted [53]. The evidence from the D/H ratios indicates that at best, Venus accreted only enough water for a global cover of a few meters [51] so that one might have difficulty in deciding between levels of dampness and dryness [54]. A runaway "greenhouse" that removed an early ocean is often invoked for Venus. In this concept, water was dissociated into H_2 and O, H_2 being lost. This led to the rise in the D/H ratio, while oxygen reacted with the surface rocks, leading to the present dry state of Venus. The problem with an Earth-like ocean on Venus is that removal of the oxygen by such a process would require oxidation of around 10^{25} g of Fe. Such an operation that would seem difficult to achieve, for example, on an already differentiated planet encased in a sheath of basalt that might hinder access to unoxidized material. Additionally, the evidence that overall planetary differentiation was much less efficient than on the Earth points to a very early loss of water if it was ever present. But perhaps the high D/H ratio of the minuscule amount of water (200 ppm including that present in H_2SO_4) in the dense venusian atmosphere might be due to late additions from comets.

In one view, the terrestrial planets originally were accreted from wet planetesimals and so initially Earth and Venus would contain similar concentrations of water [52]. This scenario, favored by experimental petrologists for obvious reasons, suffers from a major problem. Recall that the differentiation and degassing of Venus, as revealed by the atmospheric argon data, has been much less efficient than on the Earth. So if the accreting planetesimals were wet, rather than dry, the mantle of Venus could be expected to have retained significant amounts of water leading to the possibility of developing an asthenosphere, plate tectonics and perhaps felsic rocks.

We prefer the scenario in which the later supply of water for the Earth (about 500 ppm) seems mostly to have been derived from the arrival of a few planetesimals from the source region of the carbonaceous chondrites [54] or from later drift-back of icy bodies from the Jupiter region [55]. The basic question is if the accretion of water on the terrestrial planets was dominated by these few "hit or miss" events, there is no reason why Venus should have the same initial water content as the Earth.

The significance for the Earth is that just enough water was accreted to form surface oceans. Calcium, released by weathering of silicate rocks, was able to combine with the atmospheric CO_2 and precipitate as carbonates or be taken up by organisms forming calcite or aragonite shells. Both Venus and the Earth have similar levels of CO_2 (that is depleted at 3×10^{-5} solar) but their sources were probably distinct from those that supplied water [56]. The differing fates of the CO_2 budgets on these two planets emphasize once again the stochastic nature of rocky-planet formation and the difficulties of constructing habitable planets that resemble the Earth. Apart from the accidental incorporation of just enough water, that also allowed for the operation of plate tectonics and reduced the melting temperature of subducting basalt (so enabling the formation of continental crust), Earth might have resembled Venus. The conclusion in this very uncertain field is that the later evolution of both planets was governed largely by the stochastic nature of early planetary accretion.

Synopsis

Venus, with similar mass and uncompressed density to the Earth, is our "twin planet". It probably has a similar bulk composition. The best estimate of crustal thickness is 30 km. There is a strong correlation between gravity and topography: large volcanoes are supported by the crust and the rocks are strong due to the absence of water. There is no evidence for the operation of plate tectonics. The terrain is mostly rolling basaltic plains cut by some long meandering channels probably eroded by fluid lavas. There are a large variety of volcanoes, as well as coronae, probably the surface expression of hot spots or plumes. Two small continent-size regions (Ishtar and Aphrodite Terrae) are apparently uplifted by compressional tectonics, while there are highly deformed regions (tesserae).

Venus has a distinctive cratering record; for example, small impactors are effectively filtered out by the dense atmosphere. There are 940 mostly fresh impact craters, randomly distributed, that yield an age for the surface of 750 Myr. The craters are superimposed on a secondary basaltic crust that obliterated any pre-750 Myr terrain. The previous 85% of venusian geological history is completely obscured. This episodic event, possibly taking over 100 Myr, contrasts with the continuous recycling of the terrestrial plate-tectonic regime, although both planets likely have similar heat production with similar K, U and Th abundances.

The absence of plate tectonics on Venus is due to a combination of smaller size, that places the basalt–eclogite transition too deep and to the absence of an asthenosphere, due to lack of water. The crustal composition is basaltic, occasionally alkali-rich, high in K, U and Th. Minor occurrences of "pancake" domes may represent felsic differentiates or may merely be basaltic domes flattened under the 95 bar atmosphere. Venus appears much less differentiated than the Earth with less than 25% of the terrestrial atmospheric ^{40}Ar content (derived from ^{40}K).

Is there a decipherable geological sequence or history on the planet? Probably not. Despite the presence of 940 craters, they are not abundant enough to establish a relative sequence among the tesserae, plains, volcanoes or coronae. Attempts at constructing a geological chronology fall back on lithological correlations, a hazardous procedure. This twin of the Earth with similar heat production has had a totally different tectonic history due mostly to lack of water. The 200 ppm water in the venusian atmosphere has a high D/H ratio possibly indicating early loss, but perhaps was inherited from comets. Whether the deficiency is due to early loss or failure to accrete significant water, a conclusion that we favor, remains controversial.

Venus remains as a cautionary tale for seekers after "Earth-like" planets, with habitability depending on apparently trivial differences in mass and water content.

Notes and references

1. Cooper, H. S. F. (1992) *The Evening Star* Farrer Straus Giroux, p. 5
2. Crisp, D. *et al.* (2002) Divergent evolution among Earth-like planets: The case for Venus exploration, in *Astron. Soc. Pacific Conf. Proc.* **272**, 5–34. While there are strong possibilities of finding terrestrial-size rocky planets in extra-solar planetary systems, the presence of Venus casts a sobering shadow over the expectations of finding a habitable planet that resembles the Earth. The major reference is *Venus II – Geology, Geophysics, Atmosphere and Solar Wind Environment* (eds. S. W. Bougher *et al.*, 1997), University of Arizona Press. Like most similar multi-author books, its huge size does not guarantee easy access to information. The problem is exacerbated by the sometimes contradictory opinions of the various authors. Those looking for basic information about the planet will often be frustrated due to the absence of any real overview. Most authors seem content to leave this mundane but useful task to someone else. A review of the current state of knowledge about Venus is provided by *Exploring Venus as a Terrestrial Planet* (eds. L. W. Esposito *et al.*, 2007) AGU Geophysics Monograph 176, that includes speculations on astrobiology. See also Kaula, W. M. (1995) Venus reconsidered. *Science* **270**, 1460–4; Turcotte, D. L. (1996) Magellan and comparative planetology. *JGR* **101**, 4765–73.
3. Stacey, F. D. (2005) High pressure equations of state and planetary interiors. *Rep. Prog. Phys.* **68**, 341–83 (see discussion in Chapter 1).
4. e. g. Lodders, K. and Fegley, B. (1998) *The Planetary Scientists Companion*, Oxford University Press, pp. 123–4.
5. Nimmo, F. and McKenzie, D. (1998) Volcanism and tectonics on Venus. *Ann. Rev. Earth Planet. Sci.* **26**, 23–51.
6. Hansen, V. L. *et al.* (1997) Tectonic overview and synthesis, in *Venus II* (eds. S. W. Bougher *et al.*), University of Arizona Press, pp. 797–844; Schubert, G. *et al.* (1997) Mantle convection and the thermal evolution of Venus, in *Venus II* (eds. S. W. Bougher *et al.*), University of Arizona Press, pp. 1245–87; Phillips, R. J. *et al.* (1997) Lithospheric

mechanics and dynamics of Venus, in *Venus II* (eds. S. W. Bougher *et al.*), University of Arizona Press, pp. 1163–204.

7. Mackwell, S. J. *et al.* (1998) High temperature deformation of dry diabase with application to tectonics on Venus. *JGR* **103**, 975–84.

8. Taylor, S. R. (2001) *Solar System Evolution*, Cambridge University Press, p. 339.

9. Basilevsky, A. T. and Head, J. W. (1996) Evidence for rapid and widespread emplacement of volcanic plains on Venus: Stratigraphic studies in the Baltis Vallis region. *GRL* **23**, 1497–500.

10. Bannerdt, W. B. (1997) Plains tectonics on Venus, in *Venus II* (eds. S. W. Bougher *et al.*), University of Arizona Press, pp. 901–30.

11. Baker, V. R. *et al.* (1997) Channels and valleys, in *Venus II* (eds. S. W. Bougher *et al.*), University of Arizona Press, pp. 757–77.

12. Stofan, E. R. and Smrekar, S. E. (2005) Large topographic rises, coronae, large flow fields and large volcanoes on Venus: Evidence for mantle plumes? *Plates, Plumes and Paradigms* (eds. G. R. Foulger *et al.*), GSA Special Paper 388, 841–61.

13. Head, J. W. and Coffin, M. F. (1997) Large igneous provinces: A planetary perspective. *AGU Monograph* **100**, 411–38; Crumpler, L. S. *et al.* (1997) Volcanoes and centers of volcanism on Venus, in *Venus II* (eds. S. W. Bougher *et al.*), University of Arizona Press, pp. 697–756.

14. Stofan, E. R. (1997) Coronae on Venus: Morphology and origin, in *Venus II* (eds. S. W. Bougher *et al.*), University of Arizona Press, pp. 931–65.

15. Stofan, E. R. *et al.* (2001) Preliminary analysis of an expanded corona database for Venus. *GRL* **28**, 4267–70.

16. Grindrod, P. M. *et al.* (2006) The geological evolution of Altai Mons, Venus: A volcano-corona 'hybrid'. *J. Geol. Soc. London* **163**, 265–75.

17. Copp, D. L. (1998) New insights into coronae evolution: Mapping on Venus. *JGR* **103**, 19401–18; Guest, J. E. and Stofan, E. R. (1999) A new view of the stratigraphic history of Venus, *Icarus* **139**, 55–66. There are some large collapse calderas on Venus but these do not resemble, for example, the Paterae on Io; Radebaugh, J. *et al.* (2001) Paterae on Io: A new type of volcanic caldera? *JGR* **106**, 33,005–20.

18. Volcanic rises like Beta Regio may represent core–mantle boundary-type plumes and coronae might represent "plumelets" arising from shallower depths; see Stofan, E. R. and Smrekar, S. E. (2005) Large topographic rises, coronae, large flow fields and large volcanoes on Venus: Evidence for mantle plumes? *Plates, Plumes and Paradigms* (eds. G. R. Foulger *et al.*), GSA Special Paper 388, pp. 841–61.

19. Kaula, W. M. *et al.* (1997) Ishtar Terra, in *Venus II* (eds. S. W. Bougher *et al.*), University of Arizona Press, pp. 879–900.

20. Basilevsky, A. T. *et al.* (1997) The resurfacing history of Venus, in *Venus II* (eds. S. W. Bougher *et al.*), University of Arizona Press, p. 1078.

21. These are Klenova, Lise Meitner, Mead and Isabella, respectively 145, 150, 270 and 173 km in diameter. Alexopoulos, J. S. and McKinnon, W. B. (1994) Large impact craters and basins on Venus, with implications for ring mechanics on the terrestrial planets. Large *Meteorite Impacts and Planetary Evolution* (eds. B. O. Dressler *et al.*), *GSA Special Paper* 293, 29–50. Peak rings appear to form from the collapse of an unstable central peak. The rings of the multiring basins are inward facing fault scarps and do not necessarily follow the often-claimed √2 spacing. Spudis, P. D. (1993) *The Geology of Multi-ring Impact Basins*, Cambridge University Press.

22. McKinnon, W. B. *et al.* (1997) Cratering on Venus: Models and observations in *Venus II* (eds. S. W. Bougher *et al.*), University of Arizona Press, pp. 969–1014. The impactors causing venusian craters appear to belong to the post Late Heavy Bombardment

population or Population 2 of Strom *et al.* (2005); Strom, R. G. *et al.* (2005) The origin of planetary impactors in the inner Solar System. *Science* **309**, 1847–50.

23. These seem to have resulted from the out–thrown ejecta interacting with the atmosphere that was disturbed by the oblique passage of the incoming projectile. Some ejecta blankets resemble butterfly wings in plan, a consequence of impacts at oblique angles. Another difference, for example from the lunar craters, is the rarity of the ray systems again due to the damping effect of the thick atmosphere, that has blanketed the fine spray of particles from the impact. Strong east–west high-altitude winds occur on the planet and are responsible for the parabolic patterns of fine (1–2 cm) ejecta, seen, for example at the Adivar crater.

24. Herrick, R. R. *et al.* (1997) Morphology and morphometry of impact craters, in *Venus II* (eds. S. W. Bougher *et al.*), University of Arizona Press, p. 1033; Strom, R. G. *et al.* (2005) The origin of planetary impactors in the inner Solar System. *Science* **309**, 1847–50.

25. Grieve, R. A. F. and Cintala, M. J. (1995) Impact melting on Venus: Some considerations for the nature of the cratering record. *Icarus* **114**, 68–79.

26. Basilevsky, A. T. *et al.* (1997) The resurfacing history of Venus, in *Venus II* (eds. S. W. Bougher *et al.*), University of Arizona Press, p. 1078. For a different point of view see Herrick, R. R. and Sharpton, V. L. (2000) Implications from stereo-derived topography of Venusian impact craters. *JGR* **105**, 20,245–62.

27. Wichman, R. W. (1999) Internal crater modification on Venus. *JGR* **104**, 21,957–77.

28. See Kerr, R. A. (1999) Craters suggest how Venus lost her youth. *Science* **284**, 889.

29. e.g. Phillips, R. *et al.* (1992) Impact craters and Venus resurfacing history. *JGR* **97**, 15,923–84; Strom, R. G. *et al.* (1994) The global resurfacing of Venus. *JGR* **99**, 12,899–926; Campbell, B. A. (1999) Surface formation rates and impact crater densities on Venus. *JGR* **104**, 21,951–6.

30. McKinnon, W. B. *et al.* (1997) Cratering on Venus: Models and observations, in *Venus II* (eds. S. W. Bougher *et al.*), University of Arizona Press, pp. 980–5.

31. Campbell, B. A. (1999) Surface formation rates and impact crater densities on Venus. *JGR* **104**, 21,951–6.

32. Those workers studying terrestrial crater statistics and their possible relationship to mass extinctions might be well advised to study Venus. On this planet, the combination of the ~750 Myr resurfacing event coupled with the atmospheric filter on small impacts and absence of crater degradation has left a far better estimate of the cratering record of the Phanerozoic Earth than could ever be estimated from terrestrial impact-crater statistics. Thus over the past ~750 Myr, Venus has experienced one impact that resulted in a crater > 200 km in diameter, 4 craters >128 km, 17 craters > 90 km, and 84 craters >45 km diameter. Crater distribution is essentially random across the surface of Venus. The closer proximity of Earth to the asteroid belt and very slightly larger gravitational cross-section might result in a slightly greater impact rate. For reference, the Chicxulub crater, thought to be the remnant of the impact associated with the Cretaceous–Tertiary boundary on Earth is about 200 km in diameter. Strom, R. G. *et al.* (1994) The global resurfacing of Venus. *JGR* **99**, 10,899–926; Basilevsky, A. T. and Head, J. W. (2003) The surface of Venus. *Rep. Prog. Phys.* **66**, 1699–734.

33. Schubert, G. *et al.* (1997) Mantle convection and the thermal evolution of Venus, in *Venus II* (eds. S. W. Bougher *et al.*), University of Arizona Press, p. 1247; Phillips, R. J. *et al.* (1997) Lithospheric mechanics and dynamics of Venus, in *Venus II* (eds. S. W. Bougher *et al.*), University of Arizona Press, p. 1196.

34. Turcotte, D. L. (1995) How does Venus lose heat? *JGR* **100**, 16,931–40.

35. Strom, R. G. *et al.* (1994) The global resurfacing of Venus. *JGR* **99**, 10,899–926. But see also Stofan, E. R. *et al.* (2005) Resurfacing styles and rates on Venus: Assessment of 18 venusian quadrangles. *Icarus* **173**, 312–21.

36. Nimmo, F. and McKenzie, D. (1998) Volcanism and tectonics on Venus. *Ann. Rev. Earth Planet. Sci.* **26**, 23–51.

37. Turcotte, D. L. *et al.* (1999) Catastrophic resurfacing and episodic subduction on Venus. *Icarus* **139**, 49–54; van Thienen, P. *et al.* (2004) Plate tectonics on the terrestrial planets. *PEPI* **142**, 61–74.

38. Grimm, R. E. and Hess, P. C. (1997) The crust of Venus, in *Venus II* (eds. S. W. Bougher *et al.*), University of Arizona Press, pp. 1205–44; Basilevsky, A. T. *et al.* (1992) Geology of the Venera 8 landing site. *JGR* **97**, E10, 16,315–35; Weiss, C. M. and Basilevsky, A. T. (1993) Magellan observations of the Venera and Vega landing sites. *JGR* **98**, E9, 17,069–98.

39. For example there are substantial volumes of rhyolites and dacites in the Etendeka–Parana flood basalt province of southern Africa and South America. Marsh, J. S. *et al.* (2001) The Etendeka Igneous Province. *Bull. Volcan.* **62**, 464–86.

40. Geochemists have, like their geophysical colleagues, not hesitated to erect models using the USSR analytical data, despite its poor precision and the possibility of systematic error. Some of these models are based on close analogies with terrestrial examples, such as MORB, a hazardous procedure in our judgement; e.g. Nikolaeva, O. V. and Ariskin, A. A. (1999) Geochemical constraints on petrogenetic processes on Venus. *JGR* **104**, E8, 18,889–97.

41. Stofan, E. R. *et al.* (2000) Emplacement and composition of steep-sided domes on Venus. *JGR* **105**, 26,757–71.

42. The radar wavelength is 12.5 cm. Surfaces rougher on scales exceeding 12.5 cm strongly reflect the radar beam and appear bright; smoother surfaces are dark; Plaut, J. J. *et al.* (2004) The unique radar properties of silicic lava domes. *JGR* **109**, doi: 10.1029/2002JE002017.

43. Bridges, N. T. (1995) Submarine analogs to Venusian pancake domes. *GRL* **22**, 2781–4; Bridges, N. T. (1997) Ambient effects on basalt and rhyolite lavas under venusian, subaerial and subaqueous conditions. *JGR* **102**, 9243–55. The possibility has been suggested that an earlier high-temperature phase on Venus, induced by the massive resurfacing event might allow domes of rhyolitic composition to be extruded with smooth, rather than blocky, upper surfaces. Bullock, M. A. and Grinspoon, D. H. (1996) The stability of climate on Venus. *JGR* **101**, 7521–9. However the domes are superimposed on the basaltic plains and hence postdate the major resurfacing of the planet. Although extrusion under higher temperatures might result in the formation of smoother, rather than rugged blocky surfaces, the exceedingly dry nature of venusian rocks means that rhyolitic magmas on Venus are likely to be even more viscous than their terrestrial counterparts, so offsetting the effects of a potentially hotter atmosphere. Thus the identification of the venusian pancakes as having a silica-rich composition and so be analogs of terrestrial rhyolites is non-unique. Sakimoto, S. E. H. and Zuber, M. T. (1995) The spreading of variable-viscosity axisymmetric radial gravity currents: Applications to the emplacement of Venusian "pancake" domes. *J. Fluid Mech.* **301**, 65–77. In addition, the production of differentiates in the dry basaltic magma chambers of Venus is likely to be hindered in the effective absence of water.

44. Kaula, W. M. (1999) Constraints on Venus evolution from radiogenic argon. *Icarus* **139**, 32–9. Bill Kaula has given an exhaustive account of the various factors and conjectures affecting the conclusions reached from the abundance of ^{40}Ar in the

venusian atmosphere. See also Allègre, C. J. *et al.* (1996) The argon constraints on mantle structure. *GRL* **23**, 3555–7.

45. As with much of the interpretation of the Magellan data, there are some contrary views. Thus Hauck *et al.* (1998) maintain that the resurfacing of Venus could have extended over 500 Myr. Geological mapping programs have shown that more craters are embayed than previously thought. Hauck, S. A. *et al.* (1998) Venus: Crater distribution and plains resurfacing models. *JGR* **103**, 13,635–42. But what most workers seem to agree about the cratering record is that the global-resurfacing episode occupied a brief period (about 100 Myr?) and that any later resurfacing has not affected more than 20% of the planet. Stofan, E. R. *et al.* (2005) Resurfacing styles and rates on Venus. *Icarus* **173**, 312–21.

46. e.g. Basilevsky, A. T. *et al.* (1997) The resurfacing history of Venus, in *Venus II* (eds. S. W. Bougher *et al.*), University of Arizona Press, p. 1078; Price, M. H. *et al.* (1996) Dating volcanism and rifting on Venus using impact crater densities. *JGR* **101**, 4657–72; Basilevsky, A. T. and Head, J. W. (1995) Regional and global stratigraphy of Venus: A preliminary assessment and implications for the geologic history of Venus. *Planet. Space Sci.* **43**, 1523–53.

47. Campbell, B. A. (1999) Surface formation rates and impact crater densities on Venus. *JGR* **104**, 21,951–5. Major resurfacing events, albeit on a smaller scale, occur on Earth with the eruption of plateau or flood basalts. Thus between five and ten million cubic kilometers of lava were erupted at 56 Myr within about one million years during the opening of the Atlantic Ocean. Storey, M. *et al.* (2007) Paleocene–Eocene thermal maximum and the opening of the northeast Atlantic. *Science* **316**, 587–9.

48. Guest, J. E. and Stofan, E. R. (1999) A new view of the stratigraphic history of Venus. *Icarus* **139**, 55–66.

49. Stofan, E. R. *et al.* (2005) Resurfacing styles and rates on Venus. *Icarus* **173**, 312–21.

50. Campbell, B. A. (1999) Surface formation rates and impact crater densities on Venus. *JGR* **104**, 21,951.

51. Bertaux, J.-L. *et al.* (2007) A warm layer in Venus' cryosphere and high altitude measurements of HF, HCl, H_2O and HDO. *Nature* **450**, 646–9.

52. Drake, M. J. and Righter, K. (2002) Determining the composition of the Earth. *Nature* **416**, 39–44; Ikoma, M. and Genda, H. (2006) Constraints on the mass of a habitable planet with water of nebular origin. *Astrophys. J.* **648**, 696–706.

53. Donahue, T. M. *et al.* (1982) Venus was wet: A measurement of the ratio of deuterium to hydrogen. *Science* **216**, 630–3; Robert, F. (2006) Solar system deuterium/hydrogen ratio, in *Meteorites and the Early Solar System II* (eds. D. S. Lauretta and H. Y. McSween), University of Arizona Press, pp. 341–51.

54. Morbidelli, A. *et al.* (2000) Source regions and time scales for the delivery of water to Earth. *MPS* **35**, 1309–20; Prinn, R. G. and Fegley, B. (1987) The atmospheres of Venus, Earth and Mars: A critical comparison. *Ann. Rev. Earth Planet. Sci.* **15**, 171–212.

55. Lunine, J. L. (2006) Origin of water ice in the solar system, in *Meteorites and the Early Solar System II* (eds. D. S. Lauretta and H. Y. McSween), University of Arizona Press, pp. 309–19; Robert, F. (2006) Solar system deuterium/hydrogen ratio, in *Meteorites and the Early Solar System II* (eds. D. S. Lauretta and H. Y. McSween), University of Arizona Press, pp. 341–51; Morbidelli, A. *et al.* (2000) Source regions and time scales for the delivery of water to Earth. *MPS* **35**, 1309–20; Raymond, S. N. *et al.* (2007) High-resolution simulations of the final assembly of Earth-like planets 2: Water delivery and planetary habitability. *Astrobiol.* **7**, 66–84.

56. Lunine, J. L. (2006) Physical conditions on the early Earth. *Phil. Trans. Royal Soc.* **B361**, 1721–31.

8

The oceanic crust of the Earth

If the great ocean were our domain, instead of the narrow limits of the land, our difficulties would be considerably lessened ... an amphibious being, who should possess our faculties, would still more easily arrive at sound theoretical opinions in geology

(Charles Lyell) [1]

The next five chapters deal with the formation of crusts on the Earth. These occupy a significant fraction of this book, partly on account of their intrinsic importance to us, but also because we know so much about them. We begin by considering the oceanic crust, both because it forms a good example of a secondary crust and because the continental crust, discussed in the succeeding four chapters is effectively derived from it.

8.1 The sea floor and plate tectonics

The oceanic crust differs significantly in composition from the continental crust, a fact that has been known only for the past half-century. Before that time, the ocean floors were commonly thought to be underlain by sunken continental crust. Land bridges were invoked to explain puzzling cross-ocean similarities in fossil faunas. But in the 1950s, it was established that the oceanic crust, in great contrast to the continental crust, was both more dense and only a few kilometers thick. Thus it was most likely to be composed of dense basalt, or "sima" in the jargon of the time, that contrasted with the less dense continental granitic crust or "sial" [2].

The oceanic crust consists mostly of basaltic lavas that are formed by partial melting due to decompression of the mantle as it rises by convection beneath the mid-ocean ridges. The hot lavas are of lower density than the mantle peridotites and so are buoyant. The stability of the crust so formed depends on its lower density compared to the underlying mantle material. As it cools it loses buoyancy (or reaches the interesting oxymoronic condition of "negative buoyancy") and returns back down into the mantle at the subduction zones along plate margins. The driving mechanism for the operation of plate tectonics at present is the drag of the cool down-going slabs [3]. This occurs at subduction zones, driven by the phase transition from basalt to denser eclogite, that occurs typically at depths in the mantle around 30–40 km. Slab-pull provides over 90% of the forces driving plate tectonics on the Earth, with ridge-push, more accurately described as gravitational sliding on the asthenosphere providing an order of magnitude less [4].

It was the observation of the symmetry of the magnetic anomalies on either side of the mid-ocean ridges that informed us that the sea floor was spreading away from the ridges. This evidence, coupled with the observation that the outline of the Mid-Atlantic Ridge paralleled that of the coastlines of Africa and America, was the key to establishing the plate-tectonic paradigm [5].

Along the way from the mid-ocean ridge to its descent into the mantle, the basaltic crust is increasingly covered with a veneer of sediment that comes both from continental weathering and from oceanic biological activity. Although geophysicists are able to read the magnetic record, this cover often frustrates petrologists attempting to recover samples of basalt by dredging or from submersibles.

Other igneous activity adds material to the oceanic crust during its passage. Large basaltic oceanic plateaus are extruded, perhaps episodically, while other volcanic rocks are added both from individual intra-plate volcanoes and from chains of volcanoes that may be related to hot spots or plumes.

Oceanic crust thus resembles a conveyor belt that is continuously removing basaltic melts and returning them back into the mantle on timescales of up to 200 million years with a mean age of 60 Myr. Elements are extracted from a mantle reservoir at depths between 30 and 100 km, but most are returned back into the mantle as the oceanic crust is subducted, along with some of the sediments that ride on top. Some elements reappear back on the surface as in lavas erupted from intra-plate volcanoes. Others take part in the dehydration or melting of the down-going slab, leading to subduction-zone volcanism that eventually contributes to the growth of the continental crust.

8.2 Structure of the oceanic crust

Like ancient Gaul, the typical oceanic crustal section, 6–7 km thick, is classically composed of three parts. As drilling has reached only shallow depths, much of our information comes from a combination of seismic data and from the examination of locations where oceanic crust has been thrust on to land, the so-called ophiolites. The classic ophiolite sequence from top down contains pillow basalt, dikes and gabbros, either massive or layered, underlain by mantle peridotite and this forms the model for the oceanic crust [6]. But there are many variations in detail both in ophiolite sequences as well as within the oceanic crust. Although the conventional view is that ophiolites represent oceanic crust thrust on to the continental crust, some may be associated with subduction zones. So there are ophiolites and ophiolites as well as local deviations from the notionally "uniform" oceanic crust.

The "standard model" begins with Layer 1 that is about 1 km thick, that has variable P-wave (V_P) seismic velocities up to 2 km/s. It is constituted mostly of siliceous material and carbonates from biological activity along with sediments derived from the weathering and erosion of the continents, with wind-borne dust as an extra component.

Layer 2 is typically about 2.5 km thick, and is composed mostly in the uppermost part (~0.5 km) of basalt with intercalated lenses of sediment. Seismic velocities in the upper parts are <4 km/s, largely because of the great porosity of broken and fractured lava flows that have interacted with seawater. Lower in the section, sheeted dikes that fed the overlying flows are common. Seismic velocities are higher, reaching V_P around 5–6 km/s at depth as the rocks, cooling more slowly, become coarser in grain size and less porous.

Layer 3, about 4 km in thickness, constitutes the main part of the oceanic crust. Although it is probably mainly composed of intrusive gabbros and cumulate rocks, formed in magma chambers below the mid-ocean ridges, it may be underplated or intruded by mantle peridotites. At the base of the oceanic crust, there is an increase in V_P from 6.9 km/s to 7.5 km/s close to the Mohorovicic Discontinuity (that marks the base of the crust), where V_P velocities change to upper-mantle values of 8.1 km/s.

8.3 Mid-ocean ridges

The total volume of basaltic lava erupted at the mid-ocean ridges is about 18 km^3 per year. This is the major expression of volcanism on our planet, accounting for about 90% of the standard estimate of 21 km^3 per year [7]. The igneous activity at the mid-ocean ridges is mostly concentrated in the active or "neovolcanic" zone that varies in width according to the spreading rate of the ridge. The pull-apart of the plates at the mid-ocean ridges facilitates the decompression melting and eruption of lavas. At the ridges, the crust consists of a few kilometers of basalt that have been extruded onto the sea floor that is underlain by a few more kilometers of more coarse-grained, slowly cooled rocks such as gabbro and diabase. As the sea floor spreads away from the ridges, typically at a rate of a few centimeters per year, it cools and thickens while the increase in density causes it to subside. The subduction of the crust back into the mantle on geologically rapid timescales means that the oceanic crust is less than 200 million years old, the oldest being Jurassic. The average age of the oceanic crust is about 60 Myr and the volume of the present terrestrial oceanic crust is 1.7×10^9 km^3 [8]. No fundamental change in the composition of the lavas has been observed over that time.

A mid-ocean ridge famously extends as an underwater mountain range for over 60 000 km around the globe but is much broken up into segments bounded by transform faults and their extensions: fracture zones. These fracture zones extend far across the ocean floor, often exposing lower sections of the crust, although volcanism along them is rare. The ridge segments vary from ten to several hundred kilometers in length. Three classes of ridge, with fast (8–22 cm/year), intermediate (4–8 cm/year) and slow (2–4 cm/year) spreading rates, can be directly related to differences in topography and structure. Recently a new class, with an ultraslow spreading rate of less than 12 mm/year, has been added, together with another intermediate class that has spreading rates between 12 and 20 mm/year. Such ridges with spreading rates below 20 mm/year and low rates of volcanic activity are not trivial in extent, but are typical of about one-third of the total ridge length.

The southern East Pacific Rise is the fastest spreading ridge, while the Gakkel Ridge in the Arctic Ocean has the distinction of being the slowest, with rates between 8 and 13 mm/year. Other ridges with somewhere between slow and ultraslow spreading rates (13–18 mm/year) occur on most of the Arctic Ocean sector as well as on the Southwest Indian Ridge where the rates vary from 14 to 16 mm/year [9].

Not surprisingly, there are differences in the axial ridge morphology that are related to the spreading rate. The fastest spreading ridges are characterized by a broad rise about 4–8 km wide and 200 m high. They usually have a narrow median rift around 100–200 m wide and 5–40 m deep that is the site of the eruptive activity. Those ridges that are spreading at the intermediate spreading rates have a median valley a few kilometers wide and about 1 km deep. The slower spreading ridges, such as the Mid-Atlantic Ridge, have large median valleys, up to 20 km wide and 2 km deep, often floored with hummocky ridges of pillow lavas. This example has become, mostly because of familiarity, the standard image of the mid-ocean ridge topography.

Apart from the layer of overlying sediment, the crustal thickness of basalt produced (about 6.5 km) seems independent of the spreading rate, except for the ridges that are spreading at the ultraslow rate, where it drops to 2 or 3 km or even less. Seismic evidence now shows that slow ridges may have variable thicknesses from "normal" to less than a few kilometers over short distances. Thus it is possible that in places on the very slow-spreading Gakkel Ridge in the Arctic Ocean, mantle rocks may be exposed [10].

However, mantle melting is not just a simple function of the spreading rate. Variations in mantle chemistry (water for example) and temperatures also influence melting. Depending on the degree of partial melting, the melts produced may range from tholeiite to magnesium-rich komatiite, the latter being the product of elevated temperatures and high degrees of melting. Nevertheless the lavas erupted are surprisingly uniform in composition, compared for example to the lunar mare basalts.

8.3.1 Formation processes at mid-ocean ridges

This three-layer model for the structure of the oceanic crust is based on the geophysical interpretation of the seismic data but it is important to realize that drilling has yet to provide a complete section of the crust [11]. The chief problem has been to account for Layer 3 where the change in seismic velocity has been inter-preted to represent the change from dikes to gabbro. But encountering gabbro well within Layer 2 indicates that the boundary between Layer 2 and 3 may have more to do with changes in alteration and porosity than with rock texture or type.

Two models have been advanced to account for the formation and evolution of the oceanic crust at the mid-ocean ridges. One is referred to as the "gabbro glacier" model and postulates that magma rising at the mid-ocean ridge forms as lenses of melt within the crust. The cumulates resulting from crystallization sink as a sort of crystal mush to form the gabbros of Layer 3 [12]. The alternative model, for which the evidence is more persuasive, suggests that Layer 3 results from *in situ* crystal-lization of sills emplaced in the lower crust. This model also receives support from observations of the Oman ophiolite [13].

8.4 Mid-ocean ridge basalts (MORB)

The basalts erupted at the mid-ocean ridges are low-K tholeiites, but they range in composition from picrites with high MgO to ferrobasalts or FeTi basalts. Typical compositions are given in Tables 8.1 and 8.2.

Occasionally crystal fractionation in magma chambers within the ridge, at ridge-transform intersections, or in thickened segments such as Iceland where the ridge intersects a plume, produce rare differentiated products such as icelandites and rhyo-dacites. Mid-ocean ridge basalts (hereafter MORB) are generated by about 10% partial melting of mantle peridotite at depths of about 50 km. It is generally thought that the frequent injection of new MORB-type batches of magma at ridge sites maintains a uniform composition and this concept forms the basis for using N-MORB as the principal component of the oceanic crust. N-type MORB are derived from depleted mantle sources (DMM) [14] as shown by many geochemical and isotopic features, such as their REE patterns that are concave downwards (Fig. 8.1).

A rarer MORB variety, the so-called E-MORB, is moderately enriched in incom-patible elements but its composition varies widely. It occurs in minor amounts on many ridge segments but can form a significant volumetric component around

Table 8.1 *Major element composition (wt%) of the depleted mantle (DMM),*
N-MORB, E-MORB, oceanic island basalt (OIB) and a typical seamount (guyot)

Oxide	DMM[a]	N-MORB[b]	E-MORB[b]	OIB[c]	Seamount[d]
SiO_2	44.9	50.0	51.3	49.0	50.0
TiO_2	0.13	1.11	1.8	2.4	1.3
Al_2O_3	4.3	16.3	15.2	14.5	16.0
FeO	8.1	9.7	9.6	11.8	9.3
MgO	38.2	8.7	7.4	8.0	8.1
CaO	3.5	11.8	10.6	9.6	12.2
Na_2O	0.29	2.5	3.1	2.8	2.7
K_2O	0.07	0.05	0.5	0.83	0.1
Σ	**99.5**	**100.2**	**99.6**	**99.9**	**99.7**

Data sources as follows: [a] Salters, V. J. M. and Stracke, A. (2004) *Geochem. Geophys.*
Geosystems **5**, doi: 10.1029/2003GC000597; [b] Klein, E. M. (2004) *Treatise on Geochemistry*
(eds. H. D. Holland and K. K. Turekian), Elsevier, vol. 3, pp. 433–63; [c] Taylor, S. R. and
McLennan, S. M. (2002) *Encylopedia of Physical Sciences and Technology*, 3rd edn.,
Academic Press, vol. 2, pp. 697–719; [d] Perfit, M. R. *Encyclopedia of Science and*
Technology, Academic Press (in press).

Table 8.2 *Elemental composition of various components of the oceanic crust*

Element	DMM[a]	N-MORB[b]	E-MORB[b]	OIB[c]	Pelagic clay[c]	Deep sea carbonate[c]
Li (ppm)	0.7	–	–	6	57	5
Be (ppb)	25	–	–	–	2600	–
B (ppm)	0.06	–	–	–	230	55
Na (wt%)	0.21	1.85	2.3	2.08	4.0	0.2
Mg (wt%)	23.0	5.2	4.5	4.82	2.1	0.4
Al (wt%)	2.3	8.6	8.0	7.67	8.4	2.0
Si (wt%)	20.9	23.4	24.0	22.9	25.0	3.2
K (ppm)	60	410	4100	6890	5000	3000
Ca (wt%)	2.5	8.4	7.6	6.86	0.93	31.2
Sc (ppm)	16	44	36	30	19	2
Ti (ppm)	0.08	0.67	1.1	1.4	0.46	0.08
V (ppm)	80	280	290	200	120	20
Cr (ppm)	2500	250	–	450	90	11
Mn (ppm)	1050	1150	1250	1200	670	1000
Fe (wt%)	6.3	7.5	7.4	9.17	6.5	0.90
Co (ppm)	105	50	–	50	74	7
Ni (ppm)	1960	120	90	120	230	30
Cu (ppm)	30	70	–	90	250	30
Zn (ppm)	56	80	–	100	200	35
Ga (ppm)	3.2	–	–	19	20	13

Table 8.2 (*cont.*)

Element	DMM[a]	N-MORB[b]	E-MORB[b]	OIB[c]	Pelagic clay[c]	Deep sea carbonate[c]
Ge (ppm)	1.0	–	–	–	2	0.2
As (ppb)	7.4	–	–	–	–	–
Se (ppb)	72	–	–	–	–	–
Rb (ppm)	0.09	0.4	9.0	20	110	10
Sr (ppm)	10	94	180	400	180	2000
Y (ppm)	4.1	25	29	23	40	42
Zr (ppm)	8	60	130	150	150	20
Nb (ppm)	0.2	1.0	11	20	14	–
Mo (ppb)	0.05	–	–	–	27	–
Ru (ppb)	5.7	–	–	–	–	–
Rh (ppb)	1.0	–	–	–	–	–
Pd (ppb)	5.2	–	–	–	–	–
Ag (ppb)	6	–	–	–	–	–
Cd (ppb)	0.01	–	–	–	300	–
In (ppb)	12.2	–	–	–	–	–
Sn (ppm)	1.0	–	–	2.0	3.0	–
Sb (ppb)	2.6	–	–	30	–	–
Te (ppb)	15	–	–	–	–	–
Cs (ppb)	1.3	6	80	600	6000	400
Ba (ppm)	1.2	6.1	120	300	2300	190
La (ppm)	0.23	1.9	11.5	19	42	10
Ce (ppm)	0.77	6.0	26	43	80	–
Pr (ppm)	0.13	0.99	3.4	4.9	10	–
Nd (ppm)	0.71	6.1	17.1	21	41	–
Sm (ppm)	0.27	2.22	4.38	5.4	8.0	–
Eu (ppm)	0.11	0.9	1.54	1.8	1.8	–
Gd (ppm)	0.40	3.5	5.3	5.5	8.3	–
Tb (ppm)	0.08	0.70	0.70	0.9	1.3	–
Dy (ppm)	0.53	4.5	5.2	5.3	7.4	–
Ho (ppm)	0.12	1.1	0.94	1.0	1.5	–
Er (ppm)	0.37	2.6	2.8	2.7	4.1	–
Tm (ppm)	0.06	0.42	0.38	0.3	0.57	–
Yb (ppm)	0.40	2.7	2.7	1.9	3.8	–
Lu (ppm)	0.06	0.40	0.38	0.3	0.55	–
Hf (ppm)	0.2	2.9	2.1	4.0	4.1	–
Ta (ppb)	14	0.1	0.8	–	–	–
W (ppb)	3.5	–	–	0.6	–	–
Re (ppb)	0.16	–	–	–	–	–
Os (ppb)	3.0	–	–	–	–	–
Ir (ppb)	2.9	–	–	–	–	–
Pt (ppb)	6.2	–	–	–	–	–
Au (ppb)	1.0	–	–	–	–	–
Hg (ppb)	10	–	–	–	–	–
Tl (ppb)	0.38	–	–	100	–	–
Pb (ppm)	0.02	0.2	1.0	3.0	30	9

Table 8.2 (cont.)

Element	DMM[a]	N-MORB[b]	E-MORB[b]	OIB[c]	Pelagic clay[c]	Deep sea carbonate[c]
Bi (ppb)	0.4	–	–	–	550	–
Th (ppm)	0.014	0.09	1.1	2.7	13.4	–
U (ppm)	0.005	0.03	0.3	0.7	2.6	–

[a] Values from Salters, V. J. M. and Stracke, A. (2004) *Geochem. Geophys. Geosystems* **5**, doi: 10.1029/2003GC000597, except for K, Rb, Y, Zr, Nb, Ba, REE, Hf, Ta, Th and U from Workman, R. K. and Hart, S. R. (2005) Major and trace element composition of depleted MORB mantle (DMM). *EPSL* **231**, 53–72; [b] Klein, E. M. (2004) *Treatise on Geochemistry* (eds. H. D. Holland and K. K. Turekian), Elsevier, vol. 3, pp. 433–63; [c] Taylor, S. R. and McLennan, S. M. (2002) *Encl. Phys. Sci. Tech.*, Academic Press, vol. 2, pp. 697–719, adapted from McLennan, S. M. and Murray, R. W. (1999) in *Encl. Geochem.* (eds. C. P. Marshall and R. W. Fairbridge), Kluwer Academic Press, pp. 282–92.

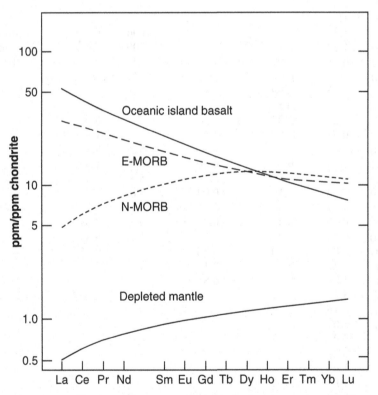

Fig. 8.1 Rare earth element patterns for the depleted mantle and the main components of the oceanic crust. Data from Tables 8.1 and 8.2

regions associated with plume activity such as the Galapagos Islands, Iceland and the Azores [15].

Mid-ocean ridge basalts are much more uniform than the lunar mare basalts as is shown most dramatically by the variation in TiO_2 concentrations (Fig. 8.2).

Although most variations in MORB chemistry are due to variations in melting or crystal fractionation, there are also some regional differences that reflect heterogeneity in the underlying mantle. Thus Indian Ocean MORB has higher $^{238}U/^{204}Pb$ ratios and lower $^{87}Sr/^{86}Sr$ ratios than either Atlantic or Pacific Ocean MORB. Slow-spreading ridges such as the Mid-Atlantic Ridge produce more primitive (high-MgO) MORB. Although MORB is the typical basaltic product of the faster-spreading ridges, more alkaline basalts seem to be erupted at the ridges that are spreading at very slow rates [16]. Variations in water content in the mantle sources may also affect both the amount and degree of melting and the resulting crustal thickness [17].

8.4.1 Interaction with seawater

There is considerable interaction between the ocean and the upper portions of the underlying crust [18]. Studies of ophiolite complexes on land have revealed that

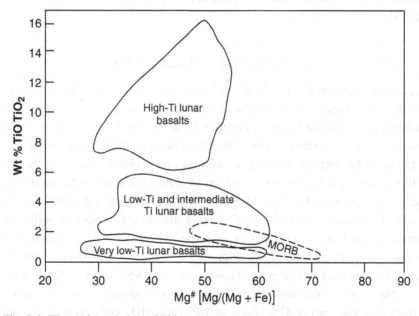

Fig. 8.2 The wide variation of TiO_2 versus Mg# in lunar mare basalts compared to MORB. Data for MORB from [23] and for lunar basalts from Shearer, C. K. *et al.* (2006) Thermal and magmatic evolution of the Moon, in *New Views of the Moon* (eds. B. L. Jolliff *et al.*), Reviews in Mineralogy and Geochemistry 60, pp. 365–518.

most of the process occurs near the ridges between the recently erupted hot crust and seawater. This affects about half the crust at the ridges with water temperatures reaching 400 °C. The hot fluids circulate through the crust and commonly appear to exit the sea floor at temperatures around 300 °C from hydrothermal vents. Sulfides and oxides are precipitated during mixing with cool seawater so that elements such as lead, copper and zinc may be concentrated to such a degree that ore deposits form.

This behavior during such secondary processes leads to the frequently discussed "lead paradox" (see Chapter 11). Seawater circulating through the basalt leaches soluble elements such as the alkalis and uranium. Uranium, oxidized from U^{4+} to U^{6+}, has a complex history as it not only is leached from fresh rock, but is also adsorbed from seawater onto altered basalt. In contrast to uranium, the relatively insoluble rare earth elements are little affected. They are present at very low concentrations in seawater so that equilibration with the REE in the basalt occurs rapidly. Thus little change occurs in the REE patterns or in the Sm/Nd isotopic systematics of MORB. The soluble alkali elements enter seawater, but are readsorbed at lower temperatures [19].

Reaction with seawater alters basaltic glass to palagonite. Furthermore, large quantities of seawater may interact with lava flows from fast-spreading ridges, producing not only pillows, but ubiquitous collapse features from seawater trapped within the pillows, features that typify the upper few hundred meters of the oceanic crust [20].

8.5 Oceanic island basalts (OIB)

In addition to the massive production of lava at the mid-ocean ridges, volcanism within the spreading plates (intra-plate volcanism) produces the many exotic oceanic islands of which the Hawaiian Islands form the most famous example. Their characteristic lavas (oceanic island basalts, hereafter OIB) differ from MORB and appear to be derived from plume sources that originate deeper in the mantle than those erupted at the mid-ocean ridges. Oceanic island basalts are over an order of magnitude less abundant than MORB. The trace-element and isotopic composition of OIB indicates derivation both from a mantle component as well as from ancient subducted recycled oceanic crust, sometimes with minor contributions from sediments or the continental crust. Although often thought to sample "primitive" mantle, OIB share, among other similarities, common Nb/U ratios with MORB. The Nb/U ratio is 30 in the bulk Earth, 10 in the continental crust and island-arc volcanics, but the ratio is 47 in both MORB and OIB. Thus most of the mantle that we sample either through MORB or OIB has been depleted in uranium relative to niobium. The old notion that MORB sampled depleted MORB mantle (DMM) while OIB sampled deeper undepleted mantle is no longer tenable [21].

The most distinctive characteristic of OIB is that they contain high concentrations of the large-ion lithophile elements such as K, Rb, Cs, Ba, LREE and other incompatible elements such as Nb, Ta, Pb, Th and U (Fig. 8.1, Tables 8.1, 8.2). There are several rarer varieties about which much debate persists. HIMU (high μ where $\mu = {}^{238}U/{}^{204}Pb$) has the highest ${}^{238}U/{}^{204}Pb$ ratios and lowest ${}^{87}Sr/{}^{86}Sr$ ratios of any oceanic island basalts. Rocks from St Helena and the Azores provide examples. The source of the geochemical and isotopic peculiarities of this HIMU species of OIB is generally attributed to the incorporation of some old recycled oceanic crust.

Two other species in this zoo are Enriched Mantle 1 (EM1) and Enriched Mantle 2 (EM2) that occupy separate areas on the isotopic diagrams. EM1, from Pitcairn and Tristan da Cunha, seems to have derived its isotopic peculiarities both from recycled oceanic crust and pelagic sediment. EM2 occurs at the Samoa and Societies hot spots and has incorporated a few percent of subducted upper continental crustal material [22]. These two enriched mantle regions, EM1 and EM2, constitute the DUPAL anomaly. The scale of these anomalies and of the recycled crust in the mantle remains small.

8.6 Composition of the oceanic crust

Mid-ocean ridge basalts are relatively uniform in chemical composition compared to the great variety of igneous rocks found on the continents, thus making the oceanic crust much more uniform than the continental crust. Thus despite the difficulty of access, the task is a little simpler than dealing with the heterogeneous continental crust. The composition of the deep oceanic crust has been established through dredging, drilling of the sea floor and direct examination using submersibles [23]. However, the deepest drilling to date has only penetrated to a depth of 1800 m, barely into Layer 3. So despite an enormous amount of effort and data from the original Deep Sea Drilling Project, Ocean Drilling Program and Integrated Ocean Drilling Project, drilling has only penetrated through the upper part of the crust in very few locations [24].

Apart from the production of oceanic crust at mid-ocean ridges, an additional factor that needs to be considered in establishing the composition of the oceanic crust is the contribution from oceanic islands. The total production of intra-plate volcanism, expected to be a mix of tholeiites and alkali basalt, is about 1.5 km^3 per year. This is a highly visible but insignificant component relative to the massive MORB volcanism. While MORB is depleted in large-ion lithophile elements, OIB contributes these elements to the overall composition of the oceanic crust. Other components in the present oceanic crust include pelagic clay and an array of seamounts or guyots (Table 8.2). These form a non-trivial component. They number at least 15 000, have an average height of 2 km and an average diameter of 20 km. They are notably concentrated in the Pacific Ocean.

Yet another underestimated component is that resulting from the presence of basaltic plateaus, such as the massive submerged Ontong Java Plateau in the Western Pacific. These so-called large igneous provinces (LIPs) constitute the most voluminous amounts of basalt erupted from the mantle after that erupted from the mid-ocean ridges. The Ontong Java Plateau is the largest example, covering an area $(1.86 \times 10^6 \text{ km}^2)$ that is about 1% of the surface of the Earth. Its average thickness is 39 km and the lower crust V_p velocities are high, about 7.5 km/s.

Similar examples of these plateaus form the continental flood basalts that occur on land where they are easier to study. Classic examples include the Deccan Traps and the Columbia River basalts. Among their most interesting features is that they were erupted rapidly, typically spanning a few million years at most [25]. In recent times, these massive eruptions were concentrated in the Cretaceous Period. Does this indicate that the eruption of these vast piles is episodic, analogous on a smaller scale to the massive episodic outpouring of basalt that resulted in the resurfacing of Venus?

8.7 Mantle structure

This seems to be an appropriate place in which to comment briefly on the nature of the terrestrial mantle, from which both the oceanic and the continental crusts are ultimately derived. However, we resist the temptation to draw conclusions applicable to other rocky planets from the silicate mantle of the Earth. Even though Venus may possess similar abundances of the major elements, the lack of water and the absence of subduction and recycling will make for a distinctive mantle on our twin planet. Like most other features of the Earth, our mantle is probably unique in detail, such as in the presence of the asthenosphere, critical for the operation of plate tectonics, that is almost certainly missing on Venus.

The terrestrial mantle has well-established discontinuities at a depth of 410 km due to olivine transforming to a denser phase; and at a depth of 660 km, caused by the change to the perovskite and magnesiowüstite phases. This discontinuity is commonly used to separate the upper from the lower mantle. Deeper still, the D″ layer occupies an irregular region about 200 km thick just above the core–mantle boundary that occurs at a depth of 2900 km. This D″ layer is probably due to a pressure-induced phase transition in perovskite that occurs around 2700 km at a temperature of 2500 K and a pressure of 1.25 mbar. But its composition remains enigmatic and it has been suggested to contain a substantial amount of the terrestrial budget of incompatible elements [26]. So it remains as a "loose cannon" in mantle composition models, while our knowledge of the mineral physics of the deep mantle remains limited.

Sampling of the mantle is difficult. We have access to samples such as MORB, OIB and mantle xenoliths from the upper 300 km. But only MORB samples large regions of the mantle, while the xenoliths, in which metasomatic effects are common, may be contaminated either by the enclosing magma or by interaction with the crust. The upper mantle has higher Mg/Si and Ca/Al ratios than either that of the canonical CI initial solar nebular composition or that of most other meteorite classes. Although the high Ca/Al ratio, relative to CI, is usually attributed to melting during magma production, the high Mg/Si seems to be an inherent property.

However, the restricted sampling raises the familiar question of whether the mantle is uniform in chemical composition or layered. This question of mantle homogeneity or heterogeneity has been bedevilled by the familiar difficulty of establishing tests of the various hypotheses. Although commonly divided in the past into a chemically distinct upper and lower mantle, or perhaps composed of a few discrete boxes, such models are now obsolete and the mantle is probably homogeneous in chemical composition on the broadest scale.

The history of this debate is worth recounting. There has been an ongoing controversy between the geochemists and the seismologists concerning the structure of the mantle of the Earth. Geochemists, in particular isotopic specialists, have often subscribed to a two-layered or box mantle. In this model, the oceanic crust (MORB) is derived at the mid-ocean ridges, by partial melting at shallow levels, from an upper mantle source that was previously depleted in the large-ion lithophile elements that now reside in the continental crust. Initial support for this model was given by the apparent reciprocal relation in elemental abundances in these two reservoirs and the approximate geochemical mass balance for incompatible elements between the crust and upper mantle reservoirs. Oceanic island basalts in contrast, were thought to come from an undepleted primitive lower mantle. This notion that OIB were sampling primitive mantle arose from its enrichment in many incompatible elements [27].

The low concentrations of the radioactive heat-producing elements in the depleted mantle source of MORB produce only about 10% of the mantle heat flow. This suggests that there was a deeper mantle source enriched in K, U and Th. This view was reinforced by the undegassed nature of parts of the mantle and the low flux of ^{4}He (about 5% of that produced) at mid-ocean ridges. It was also argued that about 50% of ^{40}Ar produced by the decay of ^{40}K had been retained in the mantle [28] but this was based on a higher terrestrial abundance of K than our estimates (see Table 8.4).

These geochemical considerations led to the concept of a layered mantle. The deeper mantle, below the famous 660 km discontinuity, was thought to be primitive and undifferentiated and OIB was thought to be derived from deep mantle plumes that tapped this region [29].

Seismologists, however, using seismic topography, have been able to trace the passage of descending subducting oceanic crust into the deep mantle, well below the 660 km seismic discontinuity that forms the canonical boundary that had been proposed by the geochemists. This would imply whole-mantle convection (what goes down must come up) and destroys the simple two-layered mantle model.

However, the heat loss–^{4}He paradox remained as strong evidence for a deep undepleted layer, reinforced by the appearance of the stable isotope ^{3}He, thought to be sampling primordial mantle. This ^{4}He dilemma has apparently been resolved by the demonstration that it is the surficial crustal layers, not the 660 km discontinuity, that are hindering the escape of ^{4}He [30]. The removal of these cornerstones of a layered mantle cleared away objections to whole-mantle convection. Overall, the contributions by geochemists to the problem of mantle structure have not been particularly enlightening.

However, much local heterogeneity remains in the mantle. Oceanic crust, derived by partial melting, has been extracted from it for around four thousand million years. But unlike the other planets or the Moon where crustal formation is largely a one-way street, much of the terrestrial oceanic crust has been recycled into the mantle, following extraction of the continental crust. The isotopic and chemical peculiarities of OIB are derived from various recycled crustal materials as discussed above. The mantle is undergoing slow convection at rates between 1 and 5 cm/year but appears to contain ill-defined heterogeneities in composition or blobs that occur in the lower as well as in the upper mantle. Thus the mantle structure may well represent a kind of plum pudding, although the plums are likely rare. Possibly marble cake provides a better analogue [31, 32].

The composition of the depleted mantle (DMM) is given in Tables 8.1 and 8.2. As discussed above, this is taken to represent the present composition of the entire mantle above the D″ layer [33].

8.7.1 Mantle plumes

Mantle plumes have become a well-entrenched model for the source of oceanic island chains and large igneous provinces and this concept has been supported both by laboratory data and computer models [34]. Such hot buoyant upwelling regions beneath the lithosphere have been proposed to be responsible for the large igneous provinces (e.g. Ontong Java Plateau, Deccan Traps, Columbia River basalts that form significant crustal components) and are thought to arise from the base of the mantle. Some of those responsible for oceanic island chains are postulated to arise at the boundary between the upper and lower mantle. The plume hypothesis is consistent with many of the observed geological features of both types of volcanism that are not explained by the plate tectonics paradigm. As Geoff Davies has remarked,

"Plume tectonics is not an alternative to plate tectonics. Mantle plumes plausibly rise from a lower hot thermal boundary layer, whereas plates arise in the upper, cold thermal boundary layer. Plumes and plates are thus complementary, rather than being alternatives" [35].

However, the number of plumes is seriously in dispute. Although numbers of less than 100 are usually cited, these have ranged in the literature from a high of 5200 down to a low of nine [36]. Claims of direct seismic evidence for the existence of up to 32 plumes have been reported but this is based on controversial interpretations and requires confirmation [37].

One of the apparently better-defined plumes is that responsible for the Hawaiian chain. However even this example is more complex than the concept of a simple concentrically zoned plume. Two trends, responsible respectively for Mauna Kea and Mauna Loa have long been recognized from differences in lead, strontium and neodymium isotopes. But more subtle differences exist that include lateral, vertically separate, narrow (50 km wide) compositional streaks [38].

However, the conventional view of the existence of plumes is not universally accepted and it is asserted by several authors that plumes do not exist. In these models, the formation of hot-spot volcanoes, plateau basalts and large igneous provinces is variously attributed to decompression melting, mantle heterogeneity, delamination or even meteorite impact, while island chains might arise through propagating cracks [39].

Overinterpretation of successful models is a common failing in the geological sciences so that there has been the usual tendency to apply the plume model too widely (5200 plumes!). There is much variation among "hot spots" so that it is always possible to find exceptions, apparent or otherwise, to any model. Thus a chain of hot-spot volcanoes near Japan may be due to cracking of the tectonic plate because of flexure rather than resulting from a deep-seated plume [40]. But isotopic evidence that hot spots beneath such volcanoes as Hawaii, Iceland and the Azores indicates that the mantle is both hot and buoyant beneath volcanoes consistent with the plume hypothesis. Unfortunately information on the depth to which this extends is lacking [41]. Thermal and compositional heterogeneity in the upper mantle may account for other examples of "hot-spot" volcanoes.

This debate over plumes is another example of the inherent difficulties of trying to establish general rules to account for geological phenomena. Much of the debate over the reality or otherwise of plumes seems due to the stochastic nature of geological processes. In the spirit of Ockham's razor, the plume model appears to be a reasonable explanation for many intra-plate volcanic chains, but we note that questions about the reality of what is going on in the mantle remain obscure. Possibly some combination of the dynamics of plates and deep-mantle plumes will account for the various hot-spot volcanoes and flood basaltic provinces. But the mantle has been subject to over 3000 Myr of sinking or subducting oceanic crust, the formation of lithospheric keels under cratons and the possibility of foundering of the material from the bases of island arcs, or doubtfully, the lower crust (see discussion on delamination in Chapter 12).

These problems, coupled with the inaccessibility of the mantle to direct observation, call for much caution in trying to unravel the evolution of the mantle of the Earth. Although we have tried to come down on one side or the other of the many controversies discussed in this book, the reality or otherwise of plumes remains to be established unequivocally. Hopefully seismic tomography will provide a definitive answer within the next decade.

8.8 Composition of the Earth

We are now in a position to comment on the composition of the planet itself, important here because of the many constraints it imposes on crustal development. The composition of the Earth would not be such a problem and our task would be much simplified, if all planets were broadly similar in composition, like stars. But the Earth differs significantly in composition from our models of the dust and ice components that were present in the early solar nebula. Indeed, if any planet within the Solar System could be considered anomalous, it is the Earth.

As discussed in Chapter 1, the Earth was assembled from the dry and volatile-element depleted planetesimals that had survived the early violent stages of solar activity. The bulk composition of the Earth depended thus on the sequence of chance events that dictated what planetesimals were accreted, whether they had experienced previous differentiation events and from which regions of the solar nebula they were derived. The bottom line is that the composition of any rocky planet is the end result of stochastic processes and that each one will likely differ in detail from its neighbor.

The similarity in the composition of the carbonaceous chondrites (CI) to that of the Sun for non-gaseous elements has long provided geochemists with the composition of the primitive solar nebula. Early models indeed used the CI abundances to provide the composition of the Earth. But the composition of the Earth is surprising, even compared with such a sample of the original dust component of the nebula (Figs. 1.3, 2.1). The gases and ices are mostly missing from the Earth, swept away in the early turbulence in the nebula before the Earth accreted. The elements that are volatile below 1100 K, such as the alkalis, are also strongly depleted, consistent with the core-accretion or "snow line" model for forming the giant planets (Chapter 1).

Although the composition of the Earth reflects the general volatile-element-depleted nature of the inner Solar System, the point in the present context is that the Earth was assembled from a suite of differentiated planetesimals that formed a random subset of those that occupied the inner solar nebula. Meteorites provide useful but inexact analogues, as there was little input from the asteroid belt at 2 to 4 AU, the present source of most meteorites [42].

8.8.1 Core

We do not consider the composition of the core in detail except to note that its segregation is coincident with the accretion of the Earth. The core consists of 90% iron with minor nickel, cobalt and other trace siderophile elements, together with about 10% of a light element. The question of the light element in the core continues to frustrate workers. Various candidates have included oxygen, silicon and sulfur. In our view, in a subject beset with uncertainty, the light element is probably sulfur. This opinion is based on two observations. Firstly, FeS is a common phase present in iron meteorites. Secondly, the often-stated objection that sulfur is a volatile element and should therefore be depleted in the Earth to the same level as zinc that is of similar volatility, fails to recognize that sulfur is most likely accreted to the Earth in the form of FeS. The presence in Mercury of a partially liquid core, possibly containing FeS, lends credence to this model.

8.8.2 Mantle

The composition of the present (depleted or DMM) mantle was discussed above. Here the composition of the primitive mantle (present mantle + crust = PM) is of more direct interest. Geochemists have long assumed that the refractory element ratios in the Earth are similar to those in the CI chondrites and this assumption is reasonably robust. The higher Ca/Al ratios (1.3 compared to 1.1 for CI) in the mantle of these quintessential refractory elements have usually been attributed to later secondary melting. But as noted above, the upper mantle of the Earth also has a high Mg/Si ratio compared to CI and this has posed a classical dilemma for geochemists. Magnesium and silicon are less refractory than Al, Ca, Ti or the REE and so are not included among the refractory lithophile elements (RLE). Indeed many of the chondritic meteorite groups show differences in Mg/Si ratios [42] indicating that fractionation of these elements has occurred within the inner nebula.

Among the suggestions to account for the high Mg/Si ratio is that the lower mantle has a low Mg/Si ratio, thus balancing that of the high ratio in the upper mantle, perhaps due to olivine fractionation. But this would also change the Ni/Co ratio from its approximately chondritic value in the upper mantle. An alternative idea is that it could be possible, under extremely reducing conditions such as in the enstatite chondrites, to hide the missing Si in the core, so allowing the Earth to retain the Mg/Si ratio of the primitive solar nebula [43].

However, these ideas and others have generally been rejected and the consensus is that the mantle, although heterogeneous in detail, is broadly uniform. So the high Mg/Si ratio observed in the upper-mantle samples is likely representative of the silicate Earth (excluding the D″ layer that remains an unknown, although we regard the possibility that the D″ layer contains a KREEP-rich layer [26] as remote).

There are two main approaches to calculating the problem of the composition of the primitive mantle of the Earth. These may be loosely labeled as cosmochemical (that uses essentially inter-element relationships derived from meteorites, principally CI carbonaceous chondrites) and geological (using the data from mantle xenoliths and basalts erupted from the mantle). In practice, evidence from both sources is employed but with varying emphasis. The usual caveats apply to the use both of meteorites and of mantle xenoliths, that are essentially random samples, while the xenoliths are also subject to metasomatic alteration.

The results of these approaches by Lyubetskaya and Korenaga [44] and Taylor [45] are essentially identical (Table 8.3) but both disagree significantly with the two most widely quoted "standard" estimates of Sun and McDonough [46] and Palme and O'Neill [47]. Major discrepancies between these estimates occur for aluminium, calcium and titanium and the other RLE, while further debate has occurred over the abundance of potassium, rubidium and related elements. Table 8.3 shows these agreements and differences among these various estimates for some critical elements.

Table 8.4 lists our preferred estimates, column A [48], that have been slightly revised with newer data but do not differ significantly from those given by Taylor over 25 years ago [45]. These values agree with the latest detailed evaluation of the problem by Lyubetskaya and Korenaga given in column B [44]. However both these sets of data for the abundances of K, Rb, Al and the other RLE are seriously different from the currently popular estimates (Table 8.3, columns D and E) [46, 47].

Table 8.3 *Five estimates of the abundances of critical elements in the primitive mantle (present mantle + crust) of the Earth*

	A	B	C	D	E
RLE/CI	~2.1	2.23	2.16	2.75	2.80
Al_2O_3 (wt%)	3.30	3.64	3.52	4.43	4.51
K (ppm)	180	180	190	240	260
U (ppb)	18	18	17.3	20.3	22
Th (ppb)	70	64	62.6	79.5	83
Mg/Si	1.14	1.08	1.11	1.04	1.09

A: Taylor [45]; B: This work; C: Lyubetskaya and Korenaga [44]; D: McDonough and Sun [46]; E: Palme and O'Neill [47].

Table 8.4 *Two estimates of the composition of the primitive mantle (present mantle plus crust) of the Earth*

Oxide	A	B
SiO_2	45.6	44.95
TiO_2	0.16	0.158
Al_2O_3	3.64	3.52
FeO	8.0	7.97
MgO	38.45	39.50
CaO	2.89	2.79
Na_2O	0.34	0.298
K_2O	0.02	0.023
Σ	99.2	99.2
Mg#	90.0	89.6

Element	A	B	Element	A	B	Element	A	B
Li (ppm)	1.6	1.6	Rb (ppm)	0.55	0.457	Eu (ppb)	129	123
Be (ppb)	54	54	Sr (ppm)	17	15.8	Gd (ppb)	454	432
B (ppm)	0.26	0.17	Y (ppm)	3.48	3.37	Tb (ppb)	84	80
Na (ppm)	2500	2220	Zr (ppm)	8.6	8.42	Dy (ppb)	566	540
Mg (wt%)	23.2	23.41	Nb (ppm)	0.55	0.46	Ho (ppb)	126	121
Al (wt%)	1.93	1.87	Mo (ppb)	30	30	Er (ppb)	370	340
Si (wt%)	21.4	21.09	Ru (ppb)	4.3	5	Tm (ppb)	57	54
K (ppm)	180	190	Rh (ppb)	1.7	0.9	Yb (ppb)	368	346
Ca (wt%)	2.07	2.00	Pd (ppb)	3.9	3.6	Lu (ppb)	57	54
Sc (ppm)	13	12.7	Ag (ppb)	19	4	Hf (ppm)	0.24	0.23
Ti (ppm)	1021	950	Cd (ppb)	40	50	Ta (ppm)	0.032	0.030
V (ppm)	85	74	In (ppb)	18	10.1	W (ppb)	16	11.9
Cr (ppm)	2540	2645	Sn (ppm)	0.14	0.10	Re (ppb)	0.25	0.32
Mn (ppm)	1000	1020	Sb (ppb)	5	7	Os (ppb)	3.8	3.4
Fe (wt%)	6.22	6.22	Te (ppb)	22	8	Ir (ppb)	3.2	3.2
Co (ppm)	100	105	Cs (ppb)	18	16	Pt (ppb)	8.7	6.6
Ni (ppm)	2000	1985	Ba (ppm)	5.4	5.08	Au (ppb)	1.3	0.88
Cu (ppm)	18	25	La (ppb)	546	508	Tl (ppb)	6	2
Zn (ppm)	50	58	Ce (ppb)	1423	1340	Pb (ppb)	120	144
Ga (ppm)	4	4.2	Pr (ppb)	215	203	Bi (ppb)	10	4
Ge (ppm)	1.2	1.15	Nd (ppb)	1057	994	Th (ppb)	64	62.6
As (ppm)	0.10	0.050	Sm (ppb)	343	324	U (ppb)	18	17.3
Se (ppb)	41	75						

A: Based on Taylor [45] and Taylor and McLennan [48] with the following minor revisions: The refractory lithophile element (RLE = Be, Al, Ca, Sc, Ti, Sr, Y, Zr, Nb, Ba, REE, Hf, Ta, Th and U) abundances have been ratioed to a constant 2.23 × CI (from Table 1.1). Molybdenum data from Sims *et al.* [51] and Li data from Ref. 47. B: See Lyubetskaya and Korenaga [44].

We regard the abundances for these elements in the standard models [46, 47] as another example of overestimates of incompatible elements and prefer our lower values for potassium and the refractory elements. Because of these differences, it is too early to propose a geochemical equivalent of the geophysical Preliminary Reference Earth Model (PREM) although we suspect that it will eventually be close to our estimate. The basic reasons for this comment are that our values in Table 8.4 resolve several dilemmas that beset the standard model. These include the argon budget, the mass balance of the lithophile elements, the concentrations of potassium, uranium and thorium and the global thermal budget [44]. But maybe "this problem is insufficiently well constrained to be solvable at present" [49].

The very low concentrations of siderophile elements in the mantle, such as the chondritic Ni/Co ratios, constitute another controversial problem that we note in passing. A "late veneer" of chondritic material seems to be the only reasonable explanation for the apparently uniform but low (0.008 × CI) concentrations of highly siderophile elements (Os, Ir, Ru, Pt, Pd, Re) in the terrestrial mantle [50].

Bulk planetary compositions influence crustal development mostly through the abundances of the heat-producing elements K, U and Th. Partial melting in planetary mantles produces basalts of various varieties on bodies as similar as the Earth and Venus, or as distinct as the Moon and Mars. But further crustal development can be strikingly dissimilar as shown by the differences among the terrestrial planets that are the main topic of this book. However, even the Earth and Venus, that probably have rather similar abundances of the major elements and, significantly, of the heat-producing elements (Chapter 7), have distinct geological histories.

Mars differs significantly from the Earth in density and possibly in Mg/Si and Al/Si ratios while Mercury is an extreme case that shows that bulk compositions, even of substantial bodies, may be affected by collisions. Perhaps the most striking example of a difference in bulk composition influencing crustal development is the Moon, whose crust of anorthosite may be unique, although the crustal composition of Mercury may be some kind of analog. Most of these variations among the smaller bodies, excluding the Moon, are related to their earlier stage of formation, compared to the later growth of the Earth and Venus (Chapter 1).

Synopsis

The secondary oceanic crust is a terrestrial analog of the lunar maria and of the venusian crust. In contrast to these examples however, plate tectonics on the Earth provides for recycling. Thus the enormous volumes of basaltic crust that are generated over geological time eventually lead to the production of the continental crust. The driving mechanism for the plate movement is mostly slab-pull, with the plates sliding on the weak asthenosphere. This tectonic environment is unique, due to water and the low solubility of water in aluminous orthopyroxene in the upper mantle. On such fine details do the growth of continents depend.

The oceanic crust is composed effectively of mid-ocean ridge basalts (MORB) with minor contributions from OIB, calcareous, siliceous and terrigenous sediment. Eighteen cubic kilometers of lava are erupted per year at the mid-ocean ridges that extend for 60 000 km. Axial ridge morphology varies with the spreading rate that ranges from a few millimeters to 22 cm/yr. Typically 6–7 km thick, the mean age of the crust is 60 Myr and it forms from a combination of lava flows, dikes and intrusions of sills. The principal component, MORB, mostly low-K tholeiite, forms from depleted mantle at around 50 km depth and is comparatively uniform in composition in great contrast to the wide variety of lunar basalts. There is substantial interaction between oceanic crust and seawater particularly at the mid-ocean ridges.

Oceanic island basalts (OIB) are much less abundant and derived from deeper sources, although not from primitive mantle. Some show isotopic variations indicating interaction with recycled oceanic or more rarely continental crust, on a scale difficult to assess.

The present mantle composition is depleted in incompatible elements due to extraction of the continental crust. It is broadly uniform chemically on a large scale, rather than layered or consisting of large isolated boxes. On a smaller scale, it is heterogeneous due to subduction and recycling, possessing a "marble cake" or "plum pudding" structure although the plums are rare enough to present problems for Jack Horner. Probably it has a higher Mg/Si ratio than CI, a consequence of the random processes of planetary accretion from a hierarchy of assorted planetesimals. The reality of mantle plumes, although likely sources for hot spots, remains a current controversy.

The bulk composition of the Earth depends on our understanding of the composition of the primitive silicate mantle. Our estimate, in agreement with the evaluation of Lyubetskaya and Korenaga [44] has significantly lower values for potassium and the refractory lithophile elements (including thorium and uranium) compared to the standard model (Table 8.4).

Notes and references

1. Lyell, C. (1830) *Principles of Geology*, vol. 1, John Murray, p. 81.
2. The mean ocean depth is nearly 4 km (3794 m) while the greatest depth of 11 km is in the Marianas Trough (11 035 m).
3. "Slab" is equivalent to the lithosphere that is composed of the oceanic crust and the uppermost mantle lying above the low-velocity zone or asthenosphere. The plates slide on the asthenosphere, a 150 km thick zone where the presence of hydrous melt lowers the viscosity. The descent of the slabs can be complex, some remaining in the upper mantle while others penetrate the 670 km discontinuity.

4. Slab-pull and suction dominate over ridge-push but the details of subduction processes are highly variable, in common with most other geological processes. See e.g. Conrad, C. P. *et al.* (2004) Great earthquakes and slab pull: Interaction between seismic coupling and plate–slab coupling. *EPSL* **218**, 109–22.

5. Heirtzler, J. R. (1968) Sea floor spreading. *Scientific American*, December 1968, 60–70. Heirtzler, J. R. *et al.* (1968) Marine magnetic anomalies, geomagnetic field reversals, and motions of the ocean floor and continents. *JGR* **73**, 2119–39. We discuss the onset of plate tectonics in Chapter 10 on the Archean crust. See Davies, G. F. (1992) On the emergence of plate tectonics. *Geology* **20**, 963–6.

6. There is a massive amount of literature on ophiolite complexes. A good source is Dilek, Y. and Robinson, P. T. (eds., 2003) *Ophiolites in Earth History*. Geological Society of London Special Publication 218. See also [19].

7. The arc contribution may be too low by a factor between three and five. The volcanic output at subduction zones has been underestimated as many volcanoes in island arcs are submarine. R. J. Arculus, personal communication (Chapter 12).

8. Head, J. W. (1976) Lunar volcanism in space and time. *Rev. Geophys. Space Sci.* **14**, 265–300.

9. Dick, H. J. B. *et al.* (2003) An ultraslow-spreading class of ocean ridge. *Nature* **426**, 405–11.

10. Jokat, W. *et al.* (2003) Geophysical evidence for reduced melt production on the Arctic ultraslow Gakkel mid-ocean ridge. *Nature* **423**, 962–5.

11. Wilson, D. S. *et al.* (2006) Drilling to gabbro in intact oceanic crust. *Science* **312**, 1016–20.

12. Phipps Morgan, J. and Chen, Y. J. (1993) The generation of oceanic crust; magma injection, hydrothermal circulation and crustal flow. *JGR* **98**, 6283–97; Henstock, T. J. *et al.* (1993) The accretion of oceanic crust by episodic sill intrusion. *JGR* **98**, 4143–61.

13. Kelemen, P. B. *et al.* (1997) Geochemistry of gabbro sills in the crust–mantle transition zone of the Oman ophiolite: Implications for the origin of oceanic lower crust. *EPSL* **146**, 475–88; MacLeod, C. J. and Yaouancq, G. (2000) A fossil melt in the Oman ophiolite: Implications for magma chamber processes at fast spreading ridges. *EPSL* **176**, 357–73.

14. Their isotopic data indicate that this mantle depletion extends back for at least one billion years. See Hofmann, A. W. (2004) Sampling mantle heterogeneity through oceanic basalts: Isotopes and trace elements. *Treatise on Geochemistry* (eds. H. D. Holland and K. K. Turekian), Elsevier, vol. 2, pp. 61–101.

15. Yet another variant, labeled T-MORB appears to have resulted from mixing of N- and E-MORB during lava ascent. We apologize for the use of these acronyms (MORB, OIB, HIMU, EM1 etc.) but they are too well established to be uprooted.

16. Michael, P. J. *et al.* (2003) Magmatic and amagmatic seafloor generation at the ultraslow-spreading Gakkel ridge, Arctic Ocean. *Nature* **423**, 956–61.

17. Asimow, W. and Langmuir, C. H. (2003) The importance of water to oceanic melting regimes. *Nature* **421**, 815–20. Magma generation at ridges varies with time; see Bonatti, E. *et al.* (2003) Mantle thermal pulses below the Mid-Atlantic ridge and temporal variations in the formation of oceanic lithosphere. *Nature* **423**, 499–505.

18. Alt, J. C. (2004) Alteration of the upper oceanic crust: Mineralogy, chemistry and processes, in *Hydrogeology of the Oceanic Lithosphere* (eds. E. Davis and H. Elderfield), Cambridge University Press, pp. 456–88. Much of the cooling of the lower crust takes place through interaction with circulating seawater within a few kilometers of the ridge axis. Maclennan, J. *et al.* (2005) Cooling of the lower oceanic crust. *Geology* **33**, 357–60.

19. Staudigel, H. (2004) Hydrothermal alteration processes in the oceanic crust. *Treatise on Geochemistry* (eds. H. D. Holland and K. K.Turekian), Elsevier, vol. 3, pp. 511–35.

20. Perfit, M. R. *et al.* (2003) Interaction of sea water and lava during submarine eruptions at mid-ocean ridges. *Nature* **426**, 62–5.

21. The fact that so many OIB have low $^{87}Sr/^{86}Sr$ and high $^{143}Nd/^{144}Nd$ also supports their derivation from sources that were once depleted. The DUPAL anomaly is an extensive region in the Indian Ocean mantle that has undergone a long evolution with high Th/U and Rb/Sr and low U/Pb and Sm/Nd ratios compared to Atlantic–Pacific Ocean mantle. Zindler, A. and Hart, S. (1986) Chemical geodynamics. *Ann. Rev. Earth Planet. Sci.* **14**, 493–521; Sun, S.- S. and McDonough, W. F. (1989) Chemical and isotopic systematics of oceanic basalts: Implications for mantle composition and processes, in *Magmatism in the Ocean Basins* (eds. A. D. Saunders *et al.*), Geological Society of London Special Publication 42, pp. 313–45; Hofmann, A. W. (2004) Sampling mantle heterogeneity though oceanic basalt: isotopes and trace elements. *Treatise on Geochemistry* (eds. H. D. Holland and K. K. Turekian), Elsevier, vol. 2, pp. 61–101.

22. Stracke, A. *et al.* (2005) FOZO, HIMU and the rest of the mantle zoo. *Geochem. Geophys. Geosystems* **6**, doi: 10.1029/2004GC00082; Hofmann, A. W. (2004) Sampling mantle heterogeneity through oceanic basalts: isotopes and trace elements. *Treatise on Geochemistry* (eds. H. D. Holland and K. K. Turekian), vol. 2, pp. 66–101; Chauvel, C. *et al.* (1992) HIMU–EM; the French Polynesian connection. *EPSL* **110**, 99–119. Sobolev *et al.* attempt to quantify the amount of recycled oceanic crust, see Sobolev A. V. *et al.* (2007) The amount of recycled crust in sources of mantle-derived melts. *Science* **316**, 412–17. Although the EM2 composition is due to the presence of continental-derived sediments associated with subducted oceanic crust, it is sometimes claimed to be delaminated lower continental crust, see Escrig, S. *et al.* (2004) Osmium isotopic constraints on the nature of the DUPAL anomaly from Indian mid-ocean-ridge basalts. *Nature* **431**, 59–63. However the elemental ratios have upper, not lower crustal signatures. Most of these problems have been resolved by the study of Jackson, M. G. *et al.* (2007) The return of subducted continental crust in Samoan lavas. *Nature* **448**, 684–7. They have been able to demonstrate definitively that the EM2 signatures are derived from ancient recycled upper continental crust.

23. Data sources include *Basaltic Volcanism on the Terrestrial Planets*, Pergamon, sect. 1.2.5 Ocean floor basaltic volcanism (pp. 132–60) and section 1.2.6 Oceanic intraplate volcanism (pp. 161–92). Data are available from the websites for DSDP (Deep Sea Drilling Project), ODP (Ocean Drilling Program) and IODP (Integrated Ocean Drilling Project).

24. Deeper drilling is urgently awaited, with the long-term possibility of a complete section through the crust into the upper mantle. This prospect, already proposed half a century ago, may answer many of the questions about the oceanic crust. Bascom, W. (1961) *A Hole in the Bottom of the Sea*, Doubleday. Drilling at the faster spreading sites, where much of the crust is generated, is also badly needed. Wilson, D. S. *et al.* (2003) Superfast spreading rate crust. *Ocean Drill. Program Initial Rep.* **206** (College Station, TX); CD ROM.

25. Coffin, M. F. and Eldholm, L. T. (1994) Large igneous provinces: Crustal structure, dimensions and external consequences. *Rev. Geophys.* **32**, 1–36. There is considerable debate over the origin of these large igneous provinces. Sun, S. S. and McDonough, W. F. (1989) Chemical and isotopic systematics of oceanic basalts: Implications for mantle composition and processes, in *Magmatism in the Ocean Basins* (eds. A. D. Saunders *et al.*), Geological Society of London Special Publication 42, pp. 313–45; Saunders, A. D. (2005) Large igneous provinces: Origin and environmental

consequences. *Elements* **1**, 259–63; Wessel, P. (2001) Global distribution of seamounts inferred from gridded Geosat/ERS-1 altimetry. *JGR* **106**, 19,431–41.

26. Murakami, M. *et al.* (2004) Post-perovskite phase transition in $MgSiO_3$. *Science* **304**, 855–8; Hernlund, J. W. *et al.* (2005) A doubling of the post-perovskite phase boundary and structure of the Earth's lowermost mantle. *Nature* **434**, 882–6. See also Williams, Q. and Revenaugh, J. (2005) Ancient subduction, mantle eclogite and the 300 km seismic discontinuity. *Geology* **33**, 1–4. See Boyet, M. and Carlson, R. W. (2005) [142]Nd evidence for early (> 4.53 Ga) global differentiation of the silicate Earth. *Science* **309**, 576–81 for the concept that the D″ layer may be enriched in incompatible elements and be the terrestrial equivalent of lunar KREEP (see discussion in Chapter 9).

27. As noted, both MORB and OIB contain the high Nb/U ratios (47) and low Pb/U ratios that rule out derivation from primitive mantle where Nb/U = 30. This results in OIB, like MORB, plotting to the right of the mantle isochron on the Pb–Pb evolution diagram.

28. Allègre, C. J. *et al.* (1994) The argon constraints on mantle structure. *GRL* **23**, 3555–7. This view was in turn strongly supported by the discrepancy between the high heat flow and the low flux of ^4He. Radioactive decay of ^{238}U, ^{235}U and ^{232}Th that produces much of the heat, also produces alpha particles, the nuclei of ^4He. Accordingly the model proposed that ^4He was trapped beneath the 660 km discontinuity, while the heat escaped by conduction. e.g. Oxburgh, E. R. and O'Nions, R. K. (1987) Helium loss, tectonics and the terrestrial heat budget. *Science* **237**, 1583–8.

29. Hofmann, A. W. (1997) Mantle geochemistry: The message from oceanic volcanism. *Nature* **385**, 219–29. Helffrich, G. R. and Wood, B. J. (2001) The Earth's mantle. *Nature* **412**, 501–7. One model suggests that there is a deep layer about 4% denser than the overlying mantle, that begins at a depth of around 1600 km. Kellogg, L. H. *et al.* (1999) Compositional stratification of the deep mantle. *Science* **283**, 1881–4. Some equate this deep region with the D″ layer just above the core–mantle boundary. e.g. Tolstikhin, I. N. and Hofmann, A. W. (2002) Geochemical importance of the core–mantle transition zone. AGU Fall Meeting, San Francisco, CA, December 8–12, 2002, Abst. U62A-U6. See Hofmann, A. W. (2004) Sampling mantle heterogeneity through oceanic basalts: Isotopes and trace elements. *Treatise on Geochemistry* (eds. H. D. Holland and K. K. Turekian), Elsevier, vol. 2, pp. 61–101.

30. Castro, M. C. *et al.* (2005) 2-D numerical simulations of ground water flow, heat transfer and ^4He transport – implications for the terrestrial He budget and the mantle helium heat imbalance. *EPSL* **237**, 893–910. The high $^3He/^4He$ ratios also may not be primordial but may merely be residual from older episodes of partial melting in the mantle. Xie, S. and Tackley, P. J. (2004) Evolution of helium and argon isotopes in a convecting mantle. *PEPI* **146**, 417–39.

31. Lyubetskaya, T. and Korenaga, J. (2007) Chemical composition of the Earth's primitive mantle and its variance: 2 Implications for global geodynamics. *JGR* **112**, doi: 10.1029/2005JB004224.

32. van Keken, P. E. *et al.* (2002) Mantle mixing: The generation, preservation and destruction of chemical heterogeneity. *Ann. Rev. Earth Planet. Sci.* **30**, 493–525; Trampert, J. *et al.* (2004) Probabilistic tomography maps chemical heterogeneities throughout the lower mantle. *Science* **306**, 853–6; Stevenson, D. J. (2003) Styles of mantle convection and their influence on planetary evolution. *C. R. Geosciences* **335**, 99–111. See also van der Hilst, R. D. *et al.* (eds., 2005) *Earth's Deep Mantle: Structure, Composition and Evolution.* AGU Monograph 160; and Schubert, G. *et al.* (2001) *Mantle Convection in the Earth and Planets*, Cambridge University Press; Campbell, I. H. and Griffiths, R. W. (1993) The evolution of the mantle's chemical-structure. *Lithos* **30**, 389–99; Campbell, I. H. and Griffiths, R. W. (1992) The changing nature of

mantle hotspots through time – Implications for the chemical evolution of the mantle. *J. Geol.* **100**, 497–523.

33. Two estimates for the composition of the depleted (present) mantle have been given by Salters, V. J. M. and Stracke, A. (2004) Composition of the depleted mantle. *Geochem. Geophys. Geosystems* **5**, doi: 10.1029/2003GC000597 and Workman, R. K. and Hart, S. R. (2005) Major and trace element composition of depleted MORB mantle (DMM). *EPSL* **231**, 53–72. Values are similar for the major elements but the Workman and Hart data are significantly lower for K, Rb, Y, Zr, Nb, Ba, REE, Hf, Ta, Th and U and are adopted here as they appear to be based on fewer assumptions.

34. Ritsema, J. and Allen, R. M. (2003) The elusive mantle plume. *EPSL* **207**, 1–12; Campbell, I. H. (2005) Large igneous provinces and the mantle plume hypothesis. *Elements* **1**, 265–9; Hill, R. I. *et al.* (1992) Mantle plumes and continental tectonics. *Science* **256**, 186–93.

35. Davies, G. F. (1992) On the emergence of plate tectonics. *Geology* **20**, 965. See also Stein, M. and Hofmann, A. W. (2002) Mantle plumes and episodic crustal growth. *Nature* **372**, 63–8.

36. Malamud, B. D. and Turcotte, D. L. (1999) How many plumes are there? *EPSL* **174**, 113–24; Courtillot, V. *et al.* (2003) Three distinct types of hotspots in the earth's mantle. *EPSL* **205**, 295–308; Abbott, D. H. and Isley, A. E. (2002) The intensity, occurrence and duration of superplume events and eras over geological time. *J. Geodynamics* **34**, 265–307.

37. Montelli, R. *et al.* (2004) Finite-frequency tomography reveals a variety of plumes in the mantle. *Science* **303**, 338–43.

38. Abouchami, W. *et al.* (2005) Lead isotopes reveal bilateral asymmetry and vertical continuity in the Hawaiian mantle plume. *Nature* **434**, 851–6.

39. Anti-plume views may be found in Foulger, G. R. *et al.* (2006) *Plates, Plumes and Paradigms* (eds. G. R. Foulger *et al.*), GSA Special Paper 388. Anderson, D. L. (2001) The thermal state of the upper mantle: No role for mantle plumes. *GRL* **27**, 3623–6; Anderson, D. L. (2001) Top-down tectonics? *Science* **293**, 2016–18; Foulger, G. R. (2002) Hotspots, plumes or plate tectonics? *Astron. Geophysics* **43**, 6.19–6.23; Jones, A. P. (2005) Meteorite impacts as triggers to large igneous provinces. *Elements* **1**, 277–81; Abbott, D. and Isley, A. E. (2002) Extraterrestrial influences on mantle plume activity. *EPSL* **205**, 53–62; Elkins-Tanton, L. T. (2005) Continental magmatism caused by lithospheric delamination, in *Plates, Plumes and Paradigms* (eds. G. R. Foulger *et al.*), GSA Special Paper 388, pp. 449–62. But see Ivanov, B. A. and Melosh, H. J. (2003) Impacts do not initiate volcanic eruptions. *Geology* **31**, 869–72.

40. Hirano, N. *et al.* (2006) Volcanism in response to plate flexure. *Science* **313**, 1426–8.

41. Bourdon, B. *et al.* (2006) Insights into the dynamics of mantle plumes from uranium-series geochemistry. *Nature* **444**, 713–17.

42. Drake, M. J. and Righter, K. (2002) Determining the composition of the Earth. *Nature* **416**, 39–44; Righter, K. *et al.* (2006) Compositional relationships between meteorites and terrestrial planets, in *Meteorites and the Early Solar System II* (eds. D. S. Lauretta and H. Y. McSween), University of Arizona Press, pp. 803–28.

43. Allègre, C. J. *et al.* (1995) The chemical composition of the Earth. *EPSL* **134**, 515–56. But in that case, it would be necessary to hide lots of other elements more readily reduced than silicon in the core, something that has not happened. The suggestion that silicon is the light element, based on slight differences in silicon isotopes between the Earth and meteorites remains to be substantiated; Georg, R. B. *et al.* (2007) Silicon in the Earth's core. *Nature* **447**, 1102–6. One is reminded of the many attempts to account for the low K/U ratio in the Earth (compared to CI) by hiding the potassium in the core.

Thus it has sometimes been suggested that potassium might enter the core under high pressures. e.g. van Westrenen, W. *et al.* (2002) Potassium in planetary cores? An experimental study of potassium partitioning between metal and silicate liquids. *LPSC* **XXXIII**, abst. 1524. However the low K/U ratios that are endemic throughout the inner Solar System occur in bodies that vary in size from the parent bodies of some meteorites to that of Mars and the Earth, so that arguments based on pressure or temperature in large bodies are invalid. Furthermore the Earth is depleted not only in potassium but a whole suite of elements whose only common characteristic is their relative volatility. Most of these elements do not have geochemical affinities that would cause them to enter metal, rather than silicate phases. Indeed potassium has a low solubility in metal, so that the K content of the core may be even less than 1 ppm. See Chabot, N. L. and Drake, M. J. (1999) Potassium solubility in metal. *EPSL* **172**, 323–35.

44. Lyubetskaya, T. and Korenaga, J. (2007) Chemical composition of the Earth's primitive mantle and its variance: 1. Method and results. *JGR* **112**, doi: 10.1029/2005JB004224. 2. Implications for global geodynamics. *JGR* **112**, doi: 10.1029/2005JB004224. We recommend these papers for their detailed discussion of the overall problem of mantle composition and of the difficulties with the "standard models". Thus the abundances of the refractory elements (RLE) in the "standard model" of Palme and O'Neill [47] turn out to be very sensitive to slight changes in the adopted value for Mg#.

45. Taylor, S. R. (1980) Refractory and moderately volatile element abundances in the Earth, Moon and meteorites. *Proc. Lunar Planet. Sci. Conf.* **11**, 333–48.

46. McDonough, W. and Sun, S. S. (1995) The composition of the Earth. *Chem. Geol.* **120**. 223–53.

47. Palme, H. and O'Neill, H. St. C. (2004) in *Treatise on Geochemistry* (eds. H. D. Holland and K. K. Turekian), Elsevier, vol. 2, pp. 1–38.

48. See Ref. 45 and Taylor, S. R. and McLennan, S. M. (1985) *The Continental Crust: Its Composition and Evolution*, Blackwell, Table 11.3 and Taylor, S. R. (2001) *Solar System Evolution*, 2nd edn., Cambridge University Press, Table 12.6. The refractory lithophile elements (RLE) are Be, Al, Ca, Sc, Ti, Sr, Y, Zr, Nb, Ba, REE, Hf, Ta, Th and U; Fe, Ni and Co are siderophile elements.

49. Stevenson, D. J. (1998) The world between crust and core. *Science* **281**, 1462.

50. Day, J. M. D. *et al.* (2007) Highly siderophile element constraints on accretion and differentiation of the Earth–Moon system. *Science* **315**, 217–19.

51. Sims, K. W. W. *et al.* (1990) Chemical fractionation during formation of the Earth's core and continental crust, in *Origin of the Earth* (eds. H. E. Newsom and J. H. Jones), Oxford University Press, pp. 291–317.

9

The Hadean crust of the Earth

I find no traces of a beginning

(James Hutton) [1]

In the following four chapters, we deal with the development of the continental crust on the Earth. The history of this planet, except for the past 200 Myr, is contained almost entirely in the continental crust that is subject to so many factors (erosion, tectonic activity, differentiation, metamorphism, volcanism, break-up and re-accretion among others) that it is surprising how good the record is. We begin with that dark period from which no rocks have survived. This however has not prevented, but rather encouraged speculation about the nature of the crust in that remote epoch. This, the so-named Hadean Eon, extends for several hundred million years, from the formation of the Earth to the first known occurrence of a preserved rock record, a period of time comparable to the extent of the Phanerozoic [2].

9.1 The Hadean crust and mantle

What indeed was the nature of the crust of the Hadean Earth? Extensive searches have failed to reveal rocks older than somewhere between 3850 and 4030 Myr. The only earlier remnants that have survived to record the existence of Hadean surface rocks are some relict detrital zircon crystals up to 4100 Myr in age, with a handful as old as 4363 Myr, that are found in younger sedimentary rocks in the Jack Hills in Western Australia [3]. These ancient zircon grains contain low abundances of scandium (that is concentrated in basic rocks) and other trace-element signatures suggestive of derivation from felsic igneous rocks [4]. Although this implies the presence of silica-rich rocks, it leaves unanswered the significant question of how extensive such rocks might have been. So what happened in that obscure period that followed the accretion of the Earth and before the survival of the oldest rocks?

9.2 A terrestrial magma ocean

In the beginning, melting of the Earth during accretion from planetesimals was likely and the energies involved during the Moon-forming impact render it inevitable. This massive event, the last major impact during the accretion of the Earth, would have over-ridden any previous smaller-scale melting episodes caused by earlier planetesimals as they plunged into the growing Earth. However, it is not clear whether re-equilibration of metal and silicate occurred during these events. Models suggest that the core of the Moon-forming impactor accreted directly to the core [5, 6] so it is a moot point whether that metal re-equilibrated with the mantle. The subsequent history of the terrestrial mantle is a matter of intense geochemical debate, fuelled by a lack of much evidence for early differentiation.

Is there indeed any evidence of the element fractionation that might be expected from the crystallization of such a mass of magma? None of the expected variations from CI ratios are evident. However a wide range of terrestrial rocks have an $^{142}Nd/^{144}Nd$ ratio that is 20 ppm higher than that in many chondritic meteorites (^{146}Nd decays to ^{142}Nd with a half-life of 103 Myr) [7]. This observation has been interpreted to indicate that the mantle experienced a very early global differentiation while ^{142}Nd was "alive". Such an event is proposed to have resulted in the production of two reservoirs from a global magma ocean; one enriched in light REE (low Sm/Nd) and the other depleted in LREE (high Sm/Nd) and in other incompatible elements. Such an enriched layer would be analogous to KREEP on the Moon (Chapters 2 and 3). It has been estimated that it would contain perhaps 30–50% of the Earth's budget of the incompatible elements, notably the heat-producing elements potassium, uranium and thorium. It is further postulated that this layer sank to the base of the mantle where it has remained inaccessible throughout geological time.

However, the stable isotopes ^{137}Ba and ^{138}Ba likewise show a 50 ppm difference between chondrites and the Earth [8]. Thus the ^{142}Nd data more likely reflect incomplete mixing of nucleosynthetic products (mainly s and r process) in the primordial solar nebula rather than indicating that a major early terrestrial differentiation event occurred.

If extensive differentiation did not happen, what then was the crystallization history of this stupendous mass of molten rock? The crystallization of large bodies of molten rock, such as in the Skaergaard or Stillwater layered intrusions, might lead one to expect the segregation in the mantle of zones of differing mineralogy, whose elemental and isotopic fingerprints would inform us of this early crystallization in the mantle. Such was the course of events on the Moon.

However, if the mantle in the early Earth was molten, the amount of melt was so large that conventional terrestrial analogies, derived from layered intrusions several orders of magnitude smaller, or from the Moon that is two orders of magnitude smaller than the Earth, may simply be inadequate to describe the solidification of such a mass of material. The absence of evidence on the Earth, such as we have from the Moon, indicates that the mantle crystallized without producing separate zones of minerals, as various theoretical studies have suggested [9].

9.3 The early crust

Although it is probable that the Earth initially was melted and that a magma ocean existed, no anorthositic crust, analogous to the lunar highland crust, appears to have formed. A variety of observations support this conclusion. For example, the Earth is less well endowed with calcium and aluminium than the Moon, so that plagioclase would not be an early crystallizing phase from a terrestrial magma ocean (Chapter 2) and in any event, is not stable below about 40 km. But if plagioclase had precipitated, it would have sunk in the wet terrestrial magma whereas it floated on the bone-dry lunar magma ocean. Further, no Eu anomalies that would result from such an anorthositic crust appear in the oldest clastic sediments [10] nor in the REE patterns of the Hadean zircons. Equally, there is no sign of significant evolution in $^{87}Sr/^{86}Sr$ ratios during the Hadean, yet this would be expected if a plagioclase-rich crust, capable of fractionating Rb from Sr, had remained isolated from the Hadean mantle.

Another suggestion is that the earliest crust was composed of the Mg-rich lava, komatiite [11]. Such high-Mg melts might be produced under the early high-temperature conditions in the Hadean Earth, as heat production, particularly from ^{235}U and ^{40}K, was about four times that of the present. Crusts so formed would be too refractory to melt during subduction, but they could have been recycled into the deep mantle without the tell-tale signature of LREE and incompatible-element enrichment subsequently reappearing in surface rocks. But the zircons were derived from felsic rocks. These were much more likely to be derived from felsic differentiates of basalts rather than from such refractory Mg-rich komatiitic rocks.

It has been further suggested by analogy with Mars, Venus and Vesta and because of the high heat production in the Hadean, that perhaps the early crust was basaltic. If so, small areas of felsic rocks perhaps formed from remelting of this basaltic crust as it sank back into the mantle; outcrops of these siliceous rocks might be the source of the resistant zircons that appear in the younger sediments. Other alternatives include the formation of small amounts of differentiated rocks from the ubiquitous basalt or differentiation within pools of impact melt, analogous to the Sudbury Igneous Complex.

However, the basic problem is that such a primordial crust, if it ever existed, seems to have vanished without providing a signature that can be recognized from currently available chemical or isotopic data. For example, no Archean or later rocks have been derived from such a primitive crust (which might resemble an alkali basalt in composition and be enriched in LREE, with low Sm/Nd ratios) [12]. But before continuing with these speculations, it is necessary to discuss the effect on the Earth of the late cataclysmic bombardment, the popular misconception of early widespread granitic crusts in the Hadean and examine the zircon evidence in more detail.

9.3.1 The bombardment record

It is often thought that the growth of early primary crusts proceeded in a turbulent environment. But the sweep-up of large planetesimals was probably completed shortly following the final assembly of the Earth and Venus. The long interval between 4.4 and 4.0 Gyr may have been punctuated only by an occasional, albeit large impact and so be quieter than sometimes supposed. In any event, as discussed in Chapter 2, a cataclysmic bombardment appears to have struck the inner Solar System later around 4.1–3.85 Gyr [13].

How long did this episode last? The ages of the Imbrium collision and the slightly younger Orientale Basin formation place the terminal stages of these events on the Moon at about 3850 Myr. Lunar accretion was essentially completed by about 4460 Myr, that appears to represent the age of the crystallization of the highland crust. However, between about 4100 and 3850 Myr, some 80 basins (> 300 km diameter) and over 10 000 craters with diameters in the range 30–300 km formed on the Moon. In the same interval, over 200 multiring basins (> 1000 km in diameter) probably formed on the Earth [14].

On the Moon, the principal effect was the production of the great basins and craters with extensive brecciation of the samples from the lunar highlands. Any earlier record was obliterated on Venus by resurfacing with basalt less than 1 Gyr ago, but it has been preserved on Mars in the southern old heavily cratered terrain (although with much degradation of the craters). The cratered basement in the northern hemisphere of Mars was covered by younger lavas while a cratering record, somewhat resembling that on the Moon, has been preserved on the battered face of Mercury.

This bombardment is often thought to explain the absence of identifiable rock units on the Earth older than 3850 Myr. But extensive searches for traces of this bombardment in Early Archean rocks at Isua, Greenland, have yet to find unequivocal evidence of this catastrophe. Neither shocked minerals nor a geochemical signature such as excess iridium, that is characteristic of meteoritic impacts, have been identified [15]. However, enrichments in the tungsten isotope ^{182}W have been reported in metasedimentary rocks 3.7 to 3.8 Gyr in age from Isua, Greenland [16]. The meteoritic tungsten in these samples is considered to be a precipitate rather than a detrital component. If true, it would mean that the tungsten in seawater was derived from meteorites and so perhaps be evidence of a widespread, rather than local event [10].

However, on the Earth the record is less likely to be preserved for two reasons. Firstly terrestrial eroding and transporting processes working on the debris resulting from several hundred impacts on the scale that produced Mare Imbrium and Mare Orientale on the Moon likely removed the evidence (although not mantle keels, if present). Possibly the explanation is that the oldest surviving rocks on the Earth postdate the cataclysm that ended at 3850 Myr, so that the record simply is missing, but the effect of the cataclysmic bombardment on the evolution of the crust of the Earth remains an open question.

9.4 The early continental crustal myth

Like the heads of the Hydra, the notion of an early granitic continental crust continues to reappear. A vast edifice has been erected on the basis of the few surviving zircon grains, although in the opinion of one commentator, "the earliest surviving materials (e.g. detrital zircons) impose only anecdotal constraints" [17]. The most extreme views portray an Earth 4.4 Gyr ago that included oceans and large areas of continental crust that would look familiar to *Homo sapiens* [18]. But one swallow does not make a summer, a few zircon grains do not make a continent and no good evidence exists for that enduring geological myth of an extensive primordial crust of "sial" or granite. Such models seem to have originated through false analogies between the production of a silicic residuum during crystallization of basaltic magmas and conditions in an early molten Earth, coupled with a lack of appreciation of the difficulties of producing granite (see Chapters 11, 12).

Several observations inform us that early "granitic" crusts were very limited in extent. There was no land vegetation in the Hadean or indeed through the Archean or Proterozoic. Any rock exposed above sea level would be barren, with the consequent formation and wind transport of dust. Due to dust blown from deserts, present-day mid-ocean sediments indeed show the fingerprint of Eu depletion in the upper continental granitic crust. But such a signature of Eu depletion is missing from virtually all Archean sedimentary rocks except for those found adjacent to local cratonic areas or in sediments dating from the latest Archean.

Zircon, a highly durable mineral, survives through many cycles of weathering, erosion and deposition of sediments. The mineral also appears to be suited for large-scale wind transport as witnessed by the enrichment of zirconium in loess deposits and the presence of substantial exotic zircon in ancient soils. Thus if there were extensive regions of early granitic or felsic crust, then a significant population of ancient zircons derived from them by erosion should have survived and be recycled into younger Archean sedimentary rocks.

However, zircon populations, dated by the [176]Lu–[176]Hf technique, in Early Archean quartzites, are mostly the same age as the terrain in which they are found. There is a scarcity or absence of zircons of Hadean age [19]. This conclusion is reinforced by later studies that show, despite extensive searches, that few zircons have survived from times before 3900 Gyr ago. This evidence disposes of the notion that there were large areas of granitic crust older than 4.0 Gyr [20] (Fig. 9.1). The conclusion is that old sialic or granitic continental crust was rare.

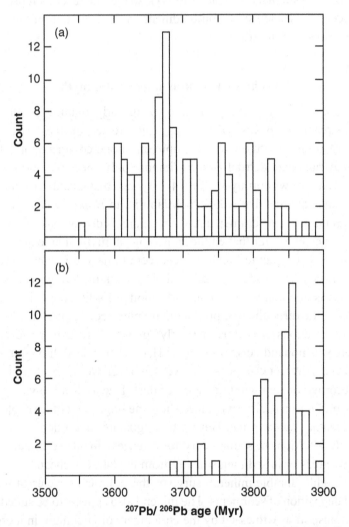

Fig. 9.1 Ages of detrital zircons in younger sedimentary rocks: (a) for strata between 3500 and 3600 Myr ($n = 117$) and (b) for strata between 3700 and 3800 Myr ($n = 54$). Adapted from Nutman, A. P. (2001) On the scarcity of > 3900 Ma detrital zircons in > 3500 Ma metasediments. *Precamb. Res.* **105**, 108, Fig. 4.

Nevertheless, claims continue to be made that the early Earth had extensive areas of continental crust about 4.3 Gyr ago and that subduction processes were operating, based both on the ∂^{18}O data and the high REE contents of zircons. These high ∂^{18}O values of the zircons have been taken to be indicative of the presence of water at that remote epoch, a cool early Earth and conditions similar enough to that of the present regime that *Homo sapiens* would find agreeable; whether life itself was present in that remote epoch remains uncertain [18, 20].

However a detailed examination of the zircons reveals that most have undergone U–Pb disturbance under high-temperature conditions and later reactions with fluids. The oldest zircon unequivocally of magmatic origin has a ∂^{18}O value of 4.8 ± 0.4 ‰, consistent with formation from an igneous rock derived from the mantle [21]. The high abundances of the LREE (La–Sm) in such zircons have also been taken as evidence that they originated from continental-crustal-type rocks [22, 23]. But such high LREE concentrations are commonly observed in zircons and are attributable to a variety of causes, rather than necessarily being derived from a magma inferred to have been produced in a wet subduction zone. Such a conclusion would imply that conditions during the Hadean resembled those in the modern Earth, complete with plate tectonics.

However zircons from the lunar highlands also display similar "overabundances" of REE (albeit with a greater depletion in europium, due to the extraction of europium into the highland crust) (Fig. 9.2). They also display ∂^{18}O values that overlap with those of "supracrustal" zircons on the Earth [23]. But the lunar minerals have neither been near water nor a subduction zone and so this interpretation of the zircon data by Cavosie *et al.* [23] is non-unique [24]. Thus great caution is warranted in extrapolating to the distant past from present-day conditions on the surface of this planet.

9.5 Isotopic constraints: ^{142}Nd and ^{176}Lu

The possibility that significant constraints on the history of the early Earth can be gained from a study of both short and long-lived isotopes has resulted in a huge amount of work, mostly beset by contradictory claims. Samarium-147 decays to ^{143}Nd with a half-life of 10.6×10^{10} years and so this system can document extraction of the LREE from the mantle. However the record in early rocks is bedevilled by open-system behavior during episodes of later metamorphism and evidence for a widespread early crust that depleted the LREE in the mantle has not been generally accepted. This has focused attention on the ^{146}Sm–^{142}Nd isotopic system that has a half-life of 103 Myr and accordingly could record mantle fractionation events occurring before about 4.3 Gyr ago.

However, the measurement of possible enrichment of ^{142}Nd close to the currently attainable limits of precision has proven both difficult and contentious, with some

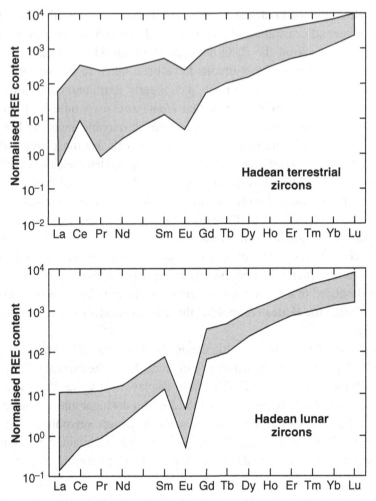

Fig. 9.2 The REE abundances in Hadean zircons show the same degree of enrichment over chondritic abundances as zircons from the lunar highlands except for the major depletion in Eu in the latter. Adapted from Whitehouse, M. J. and Kamber, B. (2002) On the overabundance of light rare earth elements in terrestrial zircons and its implication for Earth's earliest magmatic differentiation. *EPSL* **204**, 344, Fig. 8.

samples from Isua apparently showing an anomaly while others do not. There is no sign of ^{142}Nd anomalies in the present upper mantle, suggesting that if there was major mantle–crust fractionation in the Hadean, the mantle has subsequently been remixed. Probably the ^{142}Nd data reflect the existence of local Sm/Nd fractionation in the mantle [25].

The ^{176}Lu–^{176}Hf isotopic system that has a half-life of about 37 Gyr also has great potential for studies of early zircons [26], for which independent, highly precise U–Pb ages can also be obtained. While hafnium is concentrated in zircon, commonly at 2% levels, lutetium is so low in abundance that the ^{176}Hf/^{177}Hf ratio barely changes and so a fossil record is preserved of their initial hafnium isotopic ratio. However another problem besets this useful system. The decay constant is not well known, with values that range from 1.865×10^{-11}/year [27] up to 1.983×10^{-11}/year. Here we adopt the lower value (1.865×10^{-11}/year) that is consistent with geochemical measurements [28].

Extraction of crust from the mantle will preferentially incorporate hafnium, relative to the HREE element, lutetium, so that the ^{176}Hf/^{177}Hf ratio will increase at a much faster rate in the mantle compared to that in the crust. The present mantle has a high ^{176}Hf/^{177}Hf ratio that is consistent with the extraction and isolation of significant amounts of continental crust from the mantle throughout geological time [29].

In some remnant zircons (ranging in age from 3.96 to 4.35 Gyr) found in younger sedimentary rocks (deposited about 3000 Myr) at Jack Hills, Western Australia, heterogeneities have been claimed in the ^{176}Hf/^{177}Hf ratios. These have been interpreted to indicate extensive differentiation of the mantle in that remote epoch [30]. However none of the younger zircons, with ages extending to as old as 3500 Myr appear to be derived from a mantle with anomalously high ^{176}Hf/^{177}Hf ratios, so that they provide, once again, no evidence for much in the way of continental crust prior to 3900 Myr. But they give some information on when these processes started. The Jack Hills zircons were derived from material that was extracted from the mantle about 4360 Myr ago.

What was the source of the zircons that survived into the younger sedimentary rocks? Possibly thicker portions of a basaltic crust melted at their base to produce occasional islands of felsic igneous rock. Alternatively these zircon crystals, might have originated in felsic differentiates from pools of melt generated during large impacts. A later example is the Sudbury structure, about 250 km in diameter, that formed 1850 Myr ago in the continental crust and produced a pool of impact melt 3 km deep and 90 km in diameter. Sixty percent of the resulting Sudbury Igneous Complex is composed of felsic rocks [31]. In the Hadean, the crust was probably more mafic than at 1850 Myr (see Chapter 11) but production of felsic differentiates during large impacts remains a distinct likelihood. But drawing analogues between later times with the Hadean remains a hazardous enterprise [24].

9.5.1 Mantle keels in the Hadean?

Continental crust is typically underlain by low-density regions (or keels) in the mantle (see Section 10.5.1). If extensive areas of continental crust had existed in the Hadean, one might predict that such keels could have survived even if the overlying crust had vanished by various recycling processes. However there is no sign of them, although such keels underlie most Archean stable continental blocks (or cratons).

The age of formation of the peridotites that dominate such lithospheric keels has proven difficult to establish by conventional dating techniques such as the Rb–Sr, Sm–Nd, U–Pb, Th–Pb and Lu–Hf isotopic systems. However the ^{187}Re–^{187}Os system with a half-life of 4.23×10^{10} years, seems more immune to later metasomatic activity and provides the most reliable dating for lithospheric keels. The oldest evidence for development of a depleted sub-continental lithosphere occurs in the Zimbabwe Craton at 3.8 Gyr, the same age as the oldest detrital zircon from that craton [32]. The consensus is that there is no evidence of Hadean ages in the Re–Os studies of peridotites from the extensive Kaapvaal craton suites [33]. So the mantle keels that are related to the formation of evolved crust, are Archean rather than Hadean in age and provide no signature for the former existence of Hadean continental crust.

9.6 A model for the Hadean

Here we summarize our interpretation of the enigmatic and contradictory evidence that is available for this remote epoch and present a hazy view of the crust of the Earth in Hadean times. Following the accretion of the Earth, it was mostly melted, either from the energy of the incoming planetesimals or from the Moon-forming impact or both. Mantle convection was vigorous, due both to initial melting and to the enhanced radioactive heating from the presence of the relatively short-lived isotopes ^{40}K and ^{235}U.

Mantle temperatures during the Hadean were probably significantly higher due to a combination of thermal inputs from impacts of large bodies, and a much higher radiogenic heat production. Over half the heat produced by the decay of ^{235}U to ^{207}Pb in the Earth was released during Hadean time alone, so adding up to a few hundred degrees to the Earth's internal temperature. Major impact events after the accretion of the Earth and before the late heavy bombardment were probably infrequent.

Some constraints are appearing regarding the nature of a primordial crust. Neither anorthositic nor ultrabasic (komatiitic) crusts appear likely. Following the solidification of the mantle, isotopic evidence and the high Hadean heat production suggest that a basaltic crust formed within 100–200 Myr of T_{zero}. This crust appears to have persisted as a "stagnant lid" for a few hundred million years [34]. How extensive this crust was remains uncertain.

Evidence for the widespread notion of early granitic crusts is mostly lacking. The data from the handful of zircons that have survived point to the previous existence of small amounts of felsic rocks. The oldest zircons yet recorded are those from 3000 Myr sediments from the Jack Hills of Western Australia. They preserve $^{176}Hf/^{177}Hf$ ratios indicating formation ages around 4360 Myr ago. Other zircons (e.g. from the Acasta Gneiss, Itsaq Gneisses and from the Pilbara region of Western Australia) were apparently derived from precursors about 4100 Myr ago. None of this amounts to evidence for vast areas of continental crust. Rather the scarcity of such detrital zircons in Archean sedimentary rocks calls for sources from isolated outcrops of siliceous rocks, that provided the tiny remnant of resistant zircons as the sole reminder of their existence [35].

Felsic rocks might be generated in various ways: by localized foundering of eclogite beneath islands of thick basaltic crust (Iceland analogues), followed by mantle remelting producing surficial patches of felsic rocks or by the production of localized patches of differentiated rocks. Another plausible alternative is that some felsic differentiates were produced during crystallization of pools of basaltic impact melts. The record from the surviving zircons is clear, in our judgement, that there is no sign of any extensive areas of felsic, granitic or sialic crust older than about 3.6–3.9 Gyr in contrast to the suggestions that vast regions of continental crust existed in these remote epochs [36].

What happened to the primitive basaltic crust? A popular view is that it was destroyed by a late heavy bombardment, analogous to the lunar "cataclysm" but that event remains undecipherable on the Earth. Other possibilities include either a massive mantle overturn (Venus analogue?) [37] or that it was recycled into the mantle, a process that began perhaps about 3750 Myr ago [38]. Such a process could lead to the production of felsic (TTG) rocks in the Early Archean and to the first major preservation of the geological record, as well as mantle keels. These Archean felsic rocks (the so-called TTG Suite; see Chapter 10) have U/Pb ratios that are too high to have been derived from the mantle but are consistent with derivation by melting of a basaltic precursor that formed about 4.3 Gyr ago. The production of such a basaltic crust early in Earth history can be predicted to extract LREE from the mantle, so that some evidence of ^{142}Nd excesses may be expected, consistent with the enigmatic record of their detection. But radiogenic isotope data "cannot constrain the fraction of the total mantle that was depleted by crust formation" [34, p. 85].

Synopsis

The Hadean Eon, highly controversial in all aspects, covers the first several hundred million years that are missing from the rock record. The only survivors are a handful of zircon grains derived from felsic rocks, that are found in isolated younger sedimentary rocks. The Earth was melted during accretion and formed a magma ocean. However, its solidification does not seem to have resulted in discernible elemental fractionation. Excess ^{142}Nd, compared to chondrites, in Early Archean rocks, has been interpreted as recording early differentiation of the Hadean magma ocean, that resulted in a KREEP-rich layer now forming the D″ layer. However this isotopic difference seems more likely to have resulted from incomplete mixing of nucleosynthetic products in the nebula.

The nature of the early crust remains enigmatic. It was not anorthositic as on the Moon nor granitic but was likely basaltic, from which small amounts of felsic rocks differentiated, that emerged as islands from an early ocean. Isotopic evidence suggests that this crust seems to have persisted for some hundreds of Myr [34, 38].

Although the Hadean is often viewed as a Dantean version of hell, with a continuing bombardment, impacts following initial accretion were probably sporadic, until the late heavy bombardment. This traumatic event lasted, based on the lunar dates, from 4 to 3.85 Gyr but decisive terrestrial evidence has yet to be found.

There is no evidence for any early widespread sialic continental crust. The interpretation of the Sm–Nd and Lu–Hf isotopic systems in zircons remains schismatic, but is consistent with an origin from felsic rocks derived from basaltic precursors. In contrast to younger periods, zircons derived from pre-3.9 Gyr rocks are rare, consistent with derivation from scattered outcrops rather than from substantial cratons.

If one travelled back into the Hadean, one might view from a time capsule a broad ocean with rare islands that included felsic rocks of which the zircons are the only surviving remnants. A submersible would encounter basaltic pillow lavas. Only intermittently would a major impact disrupt the tranquil scene.

Notes and references

1. Hutton, James (1788) *The Theory of the Earth*, T. Cadell.
2. The term Hadean Eon is due to Preston Cloud (see Chapter 2, *Oasis in Space*). An excellent review is given by Kramers, J. (2007) Hierarchical Earth accretion and the Hadean Eon. *J. Geol. Soc. London* **164**, 3–17. The Hadean extends from the end of the accretion of the Earth around 4500–4450 Myr ago to the formation of the oldest recognized rocks that have ages somewhere between 4030 and 3850 Myr. The Nectarian (3850–?4200 Myr) and Pre-Nectarian Systems (?4200–?4460–?4527 Myr) on the Moon and the Noachian (?4500–3750 Myr) on Mars cover this missing period of time in the terrestrial rock record. Bowring, S. and Williams, I. P. (1999) Priscoan (4.00–4.03 Gyr) orthogneisses from northwestern Canada. *Contrib. Mineral. Pet.* **134**, 3–16, but see Whitehouse, M. J. *et al.* (2001) Priscoan (4.00–4.03 Gyr) orthogneisses from northwestern Canada – a discussion. *Contrib. Mineral. Pet.* **141**, 248–50.
3. The oldest reliable age for the Hadean zircons is 4363 ± 20 Myr. Nemchin, A. A. *et al.* (2006) Re-evaluation of the origin and evolution of > 4.2 Gyr zircons from the Jack Hills metasedimentary rocks. *EPSL* **244**, 218–33. Cavosie, A. J. *et al.* (2004) Internal zoning and U–Th–Pb chemistry of Jack Hills detrital zircons: A mineral record of early Archean to Mesoproterozoic (4348–1576 Ma) magmatism. *Precambrian Res.* **135**, 251–79.

4. Taylor, S. R. and McLennan, S. M. (1985) *The Continental Crust: Its Composition and Evolution*, Blackwell, p. 296, note 31; Maas, R. *et al.* (1992) The Earth's oldest known crust. *GCA* **56**, 1281–300. Although trace element abundances in zircons can be used in principle to infer the composition of the magma in which the minerals grew, the wide variation (up to six orders of magnitude for the REE) in zircon-melt distribution coefficients renders further conclusions suspect. See Hanchar, J. M. and van Westrenen, W. (2007) Rare earth element behavior in zircon-melt systems. *Elements* **3**, 37–42.

5. Stevenson, D. J. (1987) Origin of the Moon – the collision hypothesis. *Ann. Rev. Earth Planet Sci.* **15**, 271–315.

6. Canup, R. and Asphaug, E. (2001) Origin of the Moon in a giant impact near the end of the Earth's formation. *Nature* **412**, 708–12; Canup, R. M. (2004) Dynamics of lunar formation. *Ann. Rev. Astron. Astrophys.* **42**, 441–75.

7. Boyet, M. and Carlson, R. W. (2005) [142]Nd evidence for early (> 4.53 Gyr) global differentiation of the silicate Earth. *Science* **309**, 576–81. The nature of the D″ layer remains enigmatic.

8. Ranen, M. C. and Jacobsen, S. (2006) Barium isotopes in chondritic meteorites: Implications for planetary reservoir models. *Science* **314**, 809–12; see also Andreasen, R. and Sharma, M. (2006) Solar nebula heterogeneity in p-process samarium and neodymium isotopes. *Science* **314**, 806–9.

9. Tonks, W. B. and Melosh, H. J. (1990) The physics of crystal settling and suspension in a turbulent magma ocean, in *The Origin of the Earth* (eds. H. Newsom and J. H. Jones), Oxford University Press, pp. 151–74; Rubie, D. C. *et al.* (2003) Mechanisms of metal–silicate equilibration in the terrestrial magma ocean. *EPSL* **205**, 239–55; Solomatov, V. S. (2000) Fluid dynamics of a terrestrial magma ocean, in *Origin of the Earth and Moon* (eds. R. Canup and K. Righter), University of Arizona Press, pp. 323–38.

10. Bolhar, R. *et al.* (2005) Chemical characterization of Earth's most ancient clastic metasediments from the Isua greenstone belt, southern West Greenland. *GCA* **69**, 1555–73.

11. Chase, C. G. and Patchett, P. J. (1988) Stored mafic/ultramafic crust and early Archean mantle depletion. *EPSL* **91**, 66–72.

12. Galer, S. J. G. and Goldstein, S. L. (1991) Early mantle differentiation and its thermal consequence. *GCA* **55**, 227–39. The debate over the nature of the earliest, now vanished crust, continues without respite. A good summary of the current position can be found in Caro, G. *et al.* (2006) High-precision [142]Nd/[144]Nd measurements in terrestrial rocks: Constraints from the early differentiation of the Earth's mantle. *GCA* **70**, 164–91. See also Kamber, B. *et al.* (2003) Inheritance of early Archean Pb-isotope variability from long-lived Hadean protocrust. *Contrib. Mineral. Petrol.* **145**, 25–46. For examples of the controversies that surround these rocks, see Moorbath, S. *et al.* (1997) Extreme Nd-isotopic heterogeneity in the early Archean – fact or fiction? *Chem. Geol.* **135**, 213–31; Bennett, V. C. and Nutman, A. P. (1998) Extreme Nd-isotopic heterogeneity in the early Archean – fact or fiction? – Comment. *Chem. Geol.* **148**, 213–17; Kamber, B. *et al.* (1998) Extreme Nd-isotopic heterogeneity in the early Archean – fact or fiction? – Reply. *Chem. Geol.* **148**, 219–24; Myers, J. S. (2001) Protoliths of the 3.8–3.7 Gyr Isua greenstone belt, West Greenland. *Precambrian Res.* **105**, 129–41; Friend, C. R. L. *et al.* (2002) Protoliths of the 3.8–3.7 Gyr Isua greenstone belt – comment. *Precambrian Res.* **117**, 145–9; Myers, J. S. (2002) Protoliths of the 3.8–3.7 Gyr Isua greenstone belt, West Greenland – reply. *Precambrian Res.* **117**, 151–6. See also Rollinson, H. (2002) The metamorphic history of the Isua greenstone belt, West Greenland, in Geological Society of London Special Publication 199, *The Early Earth: Physical, Chemical and Biological Development* (eds. C. M. R. Fowler *et al.*), pp. 329–50.

13. Although the debate continues whether this was due to a spike or "cataclysm" in the cratering flux, or to a declining flux of impacting objects, we find the evidence persuasive for the "cataclysm" on the Moon and throughout the inner Solar System. See Chapter 2; Kring, D. A. and Cohen, B. A. (2002) Cataclysmic bombardment throughout the inner solar system. *JGR* **107**, doi: 1029/2001JE001529.

14. Grieve, R. A. F. (1998) Extraterrestrial impacts on Earth: the evidence and the consequences, in *Meteorites: Flux with Time and Impact Effects* (2006, eds. M. M. Grady *et al.*), Geological Society of London Special Publication 140, pp. 105–31; Lunine, J. I. (2006) Physical conditions on the early Earth. *Phil. Trans. Royal Soc.* **B 361**, 1721–31; Cockell, C. K. (2006) The origin and emergence of life under impact bombardment. *Phil. Trans. Royal Soc.* **B 361**, 1845–56; Kasting, J. F. and Howard, M. T. (2006) Atmospheric composition and climate on the early Earth. *Phil. Trans. Royal Soc.* **B 361**, 1731–42.

15. Ryder, G. *et al.* (2000) Heavy bombardment of the Earth at 3.85 Gyr: The search for petrographic and geochemical evidence, in *Origin of the Earth and Moon* (eds. R. Canup and K. Righter), University of Arizona Press, pp. 475–92. Here the authors discuss the extensive search for petrographic and geochemical evidence of a late heavy bombardment around 3850 Myr in the Early Archean rocks at Isua, Greenland and also in the younger Archean sediments in South Africa. See also the comprehensive review by Koeberl, C. (2006) The record of impact processes on the early Earth: A review of the first 2.5 billion years. *Processes on the Early Earth* (eds. W. U. Reimold and R. L. Gibson), GSA Special Paper 405, pp. 1–22.

16. Hafnium-182 decays to ^{182}W with a half-life of 9 Myr so that the ^{182}W enrichment in Isua rocks implies a contribution from a meteoritic source. Schoenberg, R. *et al.* (2002) Tungsten isotopic evidence from 3.8 Gyr metamorphosed sediments for early meteorite bombardment of the Earth. *Nature* **418**, 403–5.

17. Stevenson, D. J. (1998) Physical and chemical constraints for the early Earth and Moon. *Origin of the Earth and Moon.* LPI contribution 957, pp. 42–3.

18. Wilde, S. A. *et al.* (2001) Evidence from detrital zircons for the existence of continental crust and oceans on the Earth 4.4 Gyr ago. *Nature* **409**, 175–7; Mojzsis, S. J. *et al.* (2001) Oxygen-isotope evidence from ancient zircons for liquid water at the Earth's surface 4300 Ma ago. *Nature* **409**, 178–81; Watson, E. B. and Harrison, T. M. (2005) Zircon thermometer reveals minimum melting conditions on earliest Earth. *Science* **308**, 841–4. But the interpretation of the environment in which these zircons formed remains controversial. See Williams, I. S. (2007) Old diamonds and the upper crust. *Nature* **448**, 880–1 and Menneken, M. *et al.* (2007) Hadean diamonds in zircon from Jack Hills, Western Australia. *Nature* **448**, 917–20.

19. Stephenson, R. K. and Patchett, P. J. (1990) Implications for the evolution of the continental crust from Hf isotopic systematics of Archean detrital zircons. *GCA* **54**, 1683–97. Hadean zircons are indeed rare. "Only two rocks, both from Western Australia, have been found with an 'abundance' of the grains (which means about 1 part per million by weight)". Williams, I. S. (2007) Old diamonds and the upper crust. *Nature* **448**, p. 881.

20. e.g. Nutman, A. P. (2001) On the scarcity of > 3900 Ma detrital zircons in > 3500 Ma metasediments. *Precambrian Res.* **105**, 93–114. Suggestions that acidic early oceans might have destroyed zircon are refuted by the presence of marble in the oldest sedimentary sequences. Lunine, J. I. (2006) Physical conditions on the early Earth. *Phil. Trans. Royal Soc.* **B 361**, 1721–31. Quartz is another highly resistant mineral that survives cycles of erosion and deposition. A good example comes from the Archean Pilbara succession of Western Australia that began about 3515 Myr ago. The first

significant appearance of detrital quartz derived from granitic rocks does not appear until 500 Myr later, in the Lalla Rookh Sandstone, a member of the De Grey Group, dated at 2950 Myr ago, consistent with the major increase in continental growth and the development of granitic cratons in the Late Archean. Lindsay, J. A. *et al.* (2005) The problem of deep carbon: An Archean paradox. *Precambrian Res.* **143**, 1–22; Van Kranendonk, M. J. *et al.* (2000) Geology and tectonic evolution of the Archean North Pilbara terrain, Pilbara Craton, Western Australia. *Econ. Geol.* **97**, 695–32.

21. Peck, W. H. *et al.* (2001) Oxygen isotope ratios and rare earth elements in 3.1 to 4.4 Gyr zircons: Ion microprobe evidence for high $\partial^{18}O$ continental crust and oceans in the early Archean. *GCA* **65**, 4215–29.

22. Valley, J. W. *et al.* (2002) A cool early Earth. *Geology* **30**, 351–4.

23. Cavosie, A. J. *et al.* (2005) Magmatic $\partial^{18}O$ in 4400–3900 Ma detrital zircons: A record of the alteration and recycling of crust in the early Archean. *EPSL* **235**, 663–81.

24. Whitehouse, M. J. and Kamber, B. (2002) On the overabundance of light rare earth elements in terrestrial zircons and its implication for Earth's earliest magmatic differentiation. *EPSL* **204**, 333–46. Nemchin, A. A. *et al.* (2006) Heavy isotope composition of oxygen in zircon from soil sample 14163: Lunar perspective of an early ocean on Earth. *LPSC* **37**, Abst. 1593.

25. Caro, G. *et al.* (2003) $^{146}Nd–^{142}Nd$ evidence from Isua metamorphosed sediments for early differentiation of the Earth's mantle. *Nature* **423**, 428–31; Boyet, M. *et al.* (2002) ^{142}Nd anomaly confirmed at Isua. *GCA* **66**, A99; Papanastassiou, D. A. *et al.* (2003) No ^{142}Nd excess in early Archean Isua gneiss IE 715–28. *LPSC* **XXXIV**, Abst. 1851; Boyet, M. *et al.* (2003) ^{142}Nd evidence for early Earth differentiation. *EPSL* **214**, 427–42; Caro, G. *et al.* (2005) High precision $^{142}Nd/^{144}Nd$ measurements in terrestrial rocks: Constraints on the early differentiation of the Earth's mantle. *GCA* **70**, 164–91.

26. Zircon is inevitably enhanced, often to percent levels, in Hf^{4+}, that has the same ionic radius as Zr^{4+} (camouflaged in Goldschmidtian terminology). The trivalent REE ion lutetium, of which the 176 isotope decays to ^{176}Hf, is depleted in zircon. The consequence of the resulting low Lu/Hf ratio is that the initial ^{176}Hf value (customarily given as the $^{176}Hf/^{177}Hf$ ratio) will be little changed over time.

27. Scherer, E. *et al.* (2001) Calibration of the lutetium–hafnium clock. *Science* **293**, 683–7.

28. Bizzaro, M. *et al.* (2003) Early history of the Earth's crust–mantle system inferred from hafnium isotopes in chondrites. *Nature* **421**, 931–3; Kramers, J. (2007) Hierarchical Earth accretion and the Hadean Eon. *J. Geol. Soc. London* **164**, 3–17; Kramers, J. (2001) The smile of the Cheshire Cat. *Science* **293**, 619–20. The value of the decay constant is crucial to the interpretation of the lutetium–hafnium clock. However this uncertainty of 6% in the decay constant at present makes the use of the Lu–Hf system uncertain. See Patchett, P. J. (2004) Lu–Hf and Sm–Nd isotopic systematics in chondrites and their constraints on the Lu–Hf properties of the Earth. *EPSL* **222**, 29–41. The value of 1.865×10^{-11}/year adopted here was obtained from terrestrial rocks and is consistent with measurements from phosphate minerals in meteorites. Amelin, Y. (2005) Meteorite phosphates show constant ^{176}Lu decay rate since 4557 million years ago. *Science* **310**, 839–41.

29. Blichert-Toft, J. and Albarède, F. (1997) The Lu–Hf geochemistry of chondrites and the evolution of the mantle–crust system. *EPSL* **148**, 243–58.

30. Wilde, S. A. *et al.* (2001) Evidence from detrital zircons for the existence of continental crust and oceans on the Earth 4.4 Gyr ago. *Nature* **409**, 175–7.

31. Grieve, R. A. F. *et al.* (1991) The Sudbury Structure: Controversial or misunderstood? *JGR* **96**, 22,753–64; Stöffler, D. *et al.* (1994) The formation of the Sudbury Structure, Canada. *Large Meteorite Impacts and Planetary Evolution* (eds. B. O. Dressler *et al.*),

GSA Special Paper 293, pp. 303–18; Norman, M. D. (1994) Sudbury Igneous Complex: Impact melt or endogenous magma? Implications for lunar crustal evolution. *Large Meteorite Impacts and Planetary Evolution* (eds. B. O. Dressler *et al.*), GSA Special Paper 293, pp. 331–41; Grieve, R. A. F. *et al.* (2000) Vredefort, Sudbury and Chicxulub: Three of a kind. *Ann. Rev. Earth Planet. Sci.* **28**, 305–38; Therriault, A. M. *et al.* (2002) The Sudbury Igneous Complex: A differentiated impact melt sheet. *Econ. Geol.* **97**, 1521–40; Grieve, R. A. F. *et al.* (2006) Large-scale impacts and the evolution of the Earth's crust: The early years *Processes on the Early Earth* (eds. W. U. Reimold and R. L. Gibson), GSA Special Paper 405, pp. 23–32. However, production of impact melt sheets may be inhibited by target properties, such as high water content. See Artemieva, N. (2007) Possible reasons of shock melt deficiency in the Bosumtwi drill cores. *MPS* **42**, 883–94.

32. Nägler, T. F. *et al.* (1997) Growth of subcontinental lithospheric mantle beneath Zimbabwe started at or before 3.8 Ga: Re–Os studies of chromites. *Geology* **25**, 983–6.

33. Walker, R. J. *et al.* (1989) Os, Sr, Nd and Pb isotopic systematics of southern African peridotite xenoliths: Implications for the chemical evolution of sub-continental mantle. *GCA* **53**, 1583–95.

34. Kamber, B. S. (2007) The enigma of the terrestrial protocrust: Evidence of its former existence and the importance of its complete disappearance, in *The Earth's Oldest Rocks* (eds. M. van Kranendonk *et al.*), Elsevier, pp. 75–89.

35. Nutman, A. P. (2001) On the scarcity of >3900 Ma detrital zircons in >3500 Ma metasediments. *Precambrian Res.* **105**, 93–114.

36. e.g. Bowring, S. A. and Housh, T. (1995) The Earth's early evolution. *Science* **269**, 1535–40; Bowring, S. A. *et al.* (2001) Evidence from detrital zircons for the existence of continental crust and oceans on the Earth 4.4 Gyr ago. *Nature* **409**, 175–8.

37. Kramers, J. D. (2007) Hierarchical Earth accretion and the Hadean Eon. *J. Geol. Soc. London* **164**, 3–17.

38. Kamber, B. S. *et al.* (2005) Volcanic resurfacing and the early terrestrial crust: Zircon U–Pb and REE constraints from the Isua greenstone belt, southern West Greenland. *EPSL* **240**, 276–90; Kamber, B. *et al.* (2003) Inheritance of early Archean Pb-isotope variability from long-lived Hadean protocrust. *Contrib. Mineral. Petrol.* **145**, 25–46; Whitehouse, M. J. *et al.* (1999) Age significance of U–Th–Pb zircon data from early Archean rocks of west Greenland – a reassessment based on combined ion–microprobe and imaging studies. *Chem. Geol.* **160**, 201–24.

10

The Archean crust of the Earth

That dark backward and abysm of time

(William Shakespeare) [1]

10.1 The Archean

More so than most of the past, the Archean is truly another country, with a geological record that is distinct from that of more recent epochs. The Archean covers a crucial 1500 Myr of Earth history, nearly three times the length of the entire Phanerozoic, from the earliest recorded rocks at its beginning to the growth of 60–70% of the continental crust by its close [2].

Much confusion has arisen through the imprecise use of the term Archean, or even Precambrian, in referring to the "Archean crust". Thus the "Precambrian" includes two totally distinct periods of Earth history that are separated by the great transition between the Archean and Proterozoic. Although the Archean has been formally divided into the following eras: Eoarchean (3800?–3600 Myr), Paleoarchean (3600–3200 Myr), Mesoarchean (3200–2800 Myr) and Neoarchean (2800–2500 Myr) [3] it will be interesting to see if this classification is widely adopted. However, we are less concerned here with the details of the tectonic evolution of the Archean terrains to which this scheme might be applicable, so that we use the somewhat broader and commonly employed subdivision of that epoch into Early (3.9–3.5 Gyr), Middle (3.5–3.0 Gyr) and Late Archean (3.0–2.5 Gyr).

Yet even within the Archean, there is a vast difference between the scattered remnants that remain of the earliest crust, preserved at locations such as Isua in Greenland and the massive cratons in Canada, Australia, Africa and elsewhere, that developed in the Late Archean over a billion years later. Rocks older than 3.5 Gyr occupy only a tiny area of the globe and growth of the continental crust was slow during that time. Massive crustal growth occurred in the Middle and particularly in the Late Archean. In the 500 Myr from 3.5 to 3.0 Gyr ago, about 20% of the continental crust formed while 500 Myr later by the close of the Archean, 60–70% was in place [4] (Fig. 10.1).

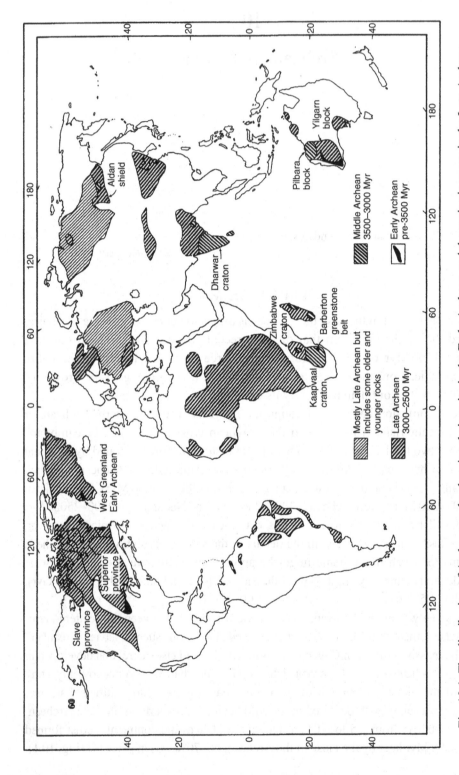

Fig. 10.1 The distribution of Archean rocks showing their rarity in the Early Archean and the major increase in the Late Archean. Map courtesy Professor Don Lowe, Stanford University.

So crustal growth reached a peak in the Late Archean, slowing down again during the Proterozoic. It was during the period of the Late Archean, that extends between 3.0 and 2.5 Gyr, that major mantle differentiation occurred making this a critical period in the evolution of the Earth. Indeed this interval of time was distinguished by Preston Cloud [5] as the Proterozoic Revolution, "transitional" between the Archean and Proterozoic, although we resist the temptation here to distinguish this important period in Earth history with yet another name. We comment further on the importance of the Archean–Proterozoic transition in the next chapter.

There is considerable variability among Archean regions that has hindered workers in these difficult terrains from arriving at some overall consensus. Thus there are significant differences among the preserved remnants of Late Archean cratons such as the Superior, Slave and Kaapvaal cratons that indicate that they could not have been derived from a Late Archean supercontinent [6], although that is a commonly held view. This observation reinforces one of the messages in this book. Seeking general laws or overall crustal compositions from the geology of local regions is hazardous as stochastic processes dominate much of the tectonic evolution of the crust.

Despite decades of intensive geological study of the Archean terrains, driven in the main by their extensive mineral resources, many problems remain intractable. This is perhaps best illustrated by the proliferation of review volumes and the great length of the reference lists in papers dealing with Archean geology. Bill Menard has commented that as a general index of a dormant or moribund science "an increasing fraction is literature on literature and bibliographies grow longer and citations grow older" [7]. Arguments rather than data have tended to predominate. Thus the Kaapvaal Craton on Southern Africa has been studied by "a team of over 100 scientists, students and technicians…However, many fundamental problems remained unanswered. Among others, these included (1) 'Why and how was the Archaean different …?' and (2) 'Is cratonization…separate from crust formation?'" [8].

Such massive approaches to the problem yield much valuable detail, but rarely seem to answer the broader questions. We are reminded of the intensive study of lunar highland breccias or of meteoritic chondrules. There is a fine line between recalling that the devil is in the details and that too much detail numbs the mind. Nevertheless, some resolution of the broader problems posed by the Archean record is beginning to emerge through the fog.

10.1.1 The earliest Archean rocks

The Acasta gneiss from the Slave Craton in the Canadian shield is often cited as the "oldest rock" but its true age remains in doubt [9]. This example of ancient crust may indeed be somewhere between 3.94 and 4.03 Gyr but perhaps it is a younger

rock (3.65 Gyr) that contains older (4 Gyr) inherited zircons of uncertain provenance [10].

The oldest well-dated continental rocks occur in West Greenland with ages that
range from 3.65 to 3.82 Gyr. The Isua greenstone belt contains a sequence of
metamorphosed volcanics and sedimentary rocks, indicative of the presence of liquid
water as an eroding agent at that time. Metasediments from the Isua greenstone belt
that are dated at about 3800 Myr have been firmly established as the oldest documented clastic sediments in the geological record [11]. To the north, these are
bounded by the slightly younger Amitsoq gneisses (3.65–3.70 Gyr) that were originally granitic rocks of the tonalite–trondhjemite–granodiorite suite (hereafter TTG).
South of the Isua belt are granitic gneisses with ages up to 3.82 Gyr but that contain
enclaves of sedimentary and volcanic rocks, older to an uncertain degree [12].

10.1.2 Akilia island, southwest Greenland

The reason for discussing this small island on the southwest coast of Greenland is
that it provides a good example of the problems and controversies encountered in
dealing with these oldest surviving rocks. It contains banded quartz–pyroxene rocks
that have been claimed to have originally been a sedimentary banded iron formation,
older than 3.85 Gyr. Apatite crystals from this sequence have also been claimed to
contain graphite with low $^{13}C/^{12}C$ ratios. This isotopically light carbon was heralded
as a signal that life was present at that remote epoch [13].

This potential and widely publicized evidence for life at this distant time has been
subject to intensive investigations. However much uncertainty surrounds this enigmatic outcrop and the claims of evidence for life at > 3.8 Gyr. Although interpreted
as sedimentary [13], the rocks do not resemble well-preserved ancient banded iron
formations that occur nearby at Isua [14]. The age is also debatable and the rocks
may be at least 100 Myr younger than the ages cited. To geochemists, the presence
of several thousand ppm Cr in these rocks suggests that they were probably
originally ultrabasic igneous rocks rather than sediments, in which case the carbon
isotopic measurements would be irrelevant to the question of early life [15].

However, the outcome of this controversial episode has become of less interest as
other workers have not been able to find graphite and hence evidence of early life in
the original samples [16]. So Akilia remains as a cautionary tale of the difficulties in
dealing with these ancient rocks.

10.2 The Archean upper crust

What is the composition of the Archean crust and does it differ from that of more
recent epochs? Our approach to this problem, as with our studies of the continental

crust generally, has been to derive the composition of the Archean crust from the record that is preserved in the sedimentary rocks rather than to deal with the complex and fragmentary tectonic history that varies widely with locality. Sampling of the bulk crustal composition by the patient long-term natural processes of erosion and sedimentation is, in our experience, an infinitely more efficient process than any alternative. For the interested reader, we give the details of the use of sedimentary rocks to establish crustal composition of the Earth in the appendices in Chapter 11 [17].

The geochemical evidence so derived from the Archean sedimentary sequences clearly reveals that the formation of the Archean crust was distinct from that of later times [18]. This is shown most clearly by the differences in the REE patterns. In contrast to the great uniformity of the REE patterns in Post-Archean sedimentary rocks (Chapter 11), those in Archean sediments show great variability. For example very steep LREE/HREE patterns that are typical of the TTG suite, contrast with flat REE patterns that are typical of those in basalts, in closely contiguous sedimentary rocks (Fig. 10.2).

These steep REE patterns occur in sediments that were derived from the erosion of the TTG igneous suites and their volcanic equivalents (e.g. dacites). In contrast, the flat REE patterns are derived from erosion of basaltic rocks. Although the Archean crust is sometimes thought to be composed solely of the TTG suite, these two contrasting REE patterns and mixtures of them, in Archean sedimentary

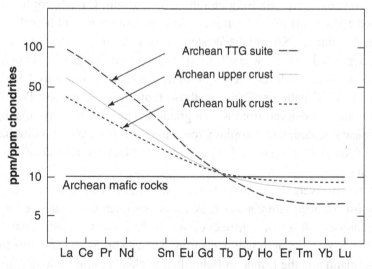

Fig. 10.2 Rare earth element patterns for the Archean upper and bulk crusts and for Archean mafic and TTG rocks, the principal components of the "bimodal" suite making up the Archean crust. Note the absence of Eu anomalies. Data from Table 10.1.

rocks, clearly indicate the bimodal nature of the bulk crust. Significantly, the REE patterns in Archean sedimentary rocks rarely show any depletion in europium. These are crucial differences between the composition of the upper crust in Archean and that of later times that reveal significant changes in the evolution of the continental crust [19]. The composition of the Archean upper crust as recorded in the compositions of Archean sediments seems to have been composed of similar amounts of basalt-komatiite and TTG intrusions and extrusives with a few very localized cratonic regions.

The average REE pattern of the Archean crust (Fig. 10.2) resembles that of modern andesites. But this similarity is deceptive as the crust is a mixture of two distinct rock types. This forms a good example of the sort of problems that beset geology. Later we comment on the comparison between modern adakites and Archean tonalites. Nature has many traps for the unwary and similarity is no proof of the operation of identical processes.

10.2.1 Archean high-grade terrains

The regions that were the sources of the Archean sediments were mostly terrains dominated by the "bimodal" suite of basalts and the sodium-rich rocks of the TTG suite that were the principal building blocks of the Archean continental crust. In contrast to the widespread greenstone belts typical of much of the Middle Archean, small areas of high-grade metamorphic terrains occur [20].

Those exposed at Kapuskasing and the Quetico Belt, Canada are highly metamorphosed equivalents of Archean greenstone-belt sediments. Others that occur in Wyoming–Montana, USA and the Western Gneiss Terrain, Australia are sediments that were deposited on a stable shelf environment on small regions of cratonic crust. These sediments are distinguished from other Archean sedimentary REE patterns in displaying REE patterns similar to those typical of the present Post-Archean upper crust and are derived from K-rich granites. The REE patterns, distinguished by a significant depletion in europium, resemble the Post-Archean Australian Shale (hereafter PAAS) REE patterns [21] (Chapter 11) that are typical of the present upper crust.

How extensive were these Archean examples of evolved granitic crust? As is the case for most Archean sedimentary rocks, these sediments were derived from very localized sources. Thus many distinct classes of REE patterns, steep, flat, "andesitic" and PAAS types occur in sediments from Archean high-grade terrains such as in India, Greenland and the Limpopo Belt, often in close proximity to one another.

That the high-grade terrains were of very local extent is also shown by their close association with greenstone-belt sediments. Although these rocks are often within a few kilometers of one another, the greenstone-belt sediments never display REE signatures with the depletion in europium that is characteristic of the K-rich granites. This is surprising as bare rock exposures must have been common due to the absence of land-based vegetation and wind-borne dust would be expected to have distributed the chemical signatures of these granites widely. The absence of such europium-depleted REE signatures in the great majority of Archean sediments is accordingly strong evidence for the limited area of outcrop of such granitic terrains. Based on this evidence from the composition of Archean sedimentary rocks, the high-grade granitic terrains of the Early and Middle Archean, that resemble modern continental crust likely occupied less than 10 percent of the exposed Archean crust of that age [22].

10.3 The Archean bulk crust

In summary, there is little evidence in the Archean for the presence of massive upper crustal granites that is consistent with the lack of extensive differentiation within the Archean crust. In contrast, intra-crustal partial melting was widespread in the Post-Archean crust and led to the present highly stratified upper and lower crust.

For this reason, the concentration of heat-producing elements in the upper crust in the Archean was much less than in later times. However, the less dense felsic members of the TTG suite are more dominant in the upper crust than the denser basaltic components (Fig. 10.2). Accordingly, the Archean crust was likely modestly stratified, but mainly in the ratio of basalt to TTG rocks, rather than by the more profound intra-crustal differentiation processes characteristic of the Post-Archean crust.

Thus the Archean upper crust that has a potassium content about 1.5% (1.81% K_2O) compared to 1% for the bulk Archean crust derived from heat-flow data (Table 10.1) – is much less enriched than the Post-Archean upper crust where the potassium concentration is 2.8% (compared to 1.1% (1.32% K_2O) for the bulk crust – see Table 11.4). The difference in the upper crusts between the Archean and the Post Archean lies primarily in the amount of intra-crustal differentiation that dominated the latter crust [23].

The Archean crust makes up about 60% of the present crust. It is the contribution of nickel, cobalt and chromium from the Archean basaltic component that leads to the present crustal abundances of these elements that are higher than those currently being supplied by andesites.

10.3.1 The Archean oceanic crust

Due to the lack of preservation of Archean or indeed of pre-Jurassic oceanic crust, little can be said about the Archean oceanic crust. But because of the enhanced heat

Table 10.1 *Chemical composition of Archean mafic and felsic end-member composition, average Archean shale and the Archean upper and bulk crusts*

	Archean mafic	Archean felsic	Shale	Archean upper crust	Archean bulk crust
Oxide					
SiO_2	50.6	69.7	60.4	60.1	56.9
TiO_2	1.2	0.5	0.8	0.83	1.00
Al_2O_3	14.9	15.8	17.1	15.3	15.2
FeO	12.9	3.1	9.5	8.0	9.60
MnO	0.2	0.15	0.1	0.14	0.15
MgO	8.1	1.4	4.3	4.7	5.90
CaO	9.5	2.9	3.2	6.20	7.30
Na_2O	2.3	4.7	2.1	3.05	3.00
K_2O	0.3	1.7	2.3	1.81	1.20
Σ	**100.0**	**100.0**	**99.8**	**100.1**	**100.3**
Element					
Ba	130	650	575	390	300
Rb	10	65	60	50	28
Sr	175	300	180	240	215
La	3.7	37	20	20	15
Ce	9.6	75	42	42	31
Pr	1.4	8.5	4.9	4.9	3.7
Nd	7.1	33	20	20	16
Sm	2.3	5.7	4.0	4.0	3.4
Eu	0.87	1.6	1.2	1.2	1.1
Gd	3.0	3.7	3.4	3.4	3.2
Tb	0.58	0.6	0.57	0.57	0.59
Dy	3.8	3.1	3.4	3.4	3.6
Ho	0.85	0.6	0.74	0.74	0.77
Er	2.5	1.7	2.1	2.1	2.2
Tm	0.36	0.22	0.30	0.30	0.32
Yb	2.5	1.5	2.0	2.0	2.2
Lu	0.38	0.22	0.31	0.31	0.33
La_N/Yb_N	1.0	16.7	6.8	6.8	4.6
Y	21	14	18	18	19
Th	0.8	6.8	6.3	5.7	3.8
U	0.2	1.8	1.6	1.5	1.0
Zr	50	200	120	125	100
Hf	2	4	3.5	3	3
Cr	330	30	205 (675)*	180	230
V	345	45	135	195	245
Sc	40	5	20	14	30
Ni	185	20	100 (425)*	105	130
Co	40	10	40	25	30

* The Cr and Ni values in parentheses are for Early Archean shales. Revised from Taylor and McLennan (1985) Table 7.8 [18].

flow the oceanic crust was probably thicker, hotter and more buoyant in the Archean than in later epochs [24]. In the likely absence of plate tectonics, perhaps plumes were the major source. Extensive eruptions of flood basalts occurred in the Archean [25]. However the common notion that the ubiquitous Archean greenstone belts represent oceanic crust is negated by the observation that they were deposited on felsic (TTG) crust and are an integral part of the Archean continental crust [26].

Present-day occurrences of ophiolite complexes have been extensively investigated as they provide onshore examples of oceanic crust that can be studied in great detail [27]. Some workers claim that there are Archean examples of ophiolites that shed light on the nature of the Archean oceanic crust [28]. But this claim is controversial, like so much else in the Archean geological record and other geologists [29] dispute these observations, flatly stating that there are no known ophiolites in the Archean.

A possible example of Archean oceanic crust has been reported from a claimed ophiolite sequence in China and this occurrence, the Dong Wanzi complex, has been used in another attempt to resolve the question of whether plate tectonics occurred in the Archean [30]. As an illustration of the problems that beset Archean field work, the geological evidence itself that this is an occurrence of ophiolites has been seriously disputed [31].

However, this is also a good example of the confusion arising from the use of the term Archean. These rocks, formed in the latest Archean, are dated at 2505 Myr and so are barely "Archean" and are even "Proterozoic" within the error limits. Modern crust-forming processes were well underway in the Late Archean in the "Archean–Proterozoic transition zone". So this example, even if the disputed interpretation of the field and geological evidence is correct, in fact is not very relevant to the overall question. By 2505 Myr ago, nearly everyone is agreed that plate tectonics in the modern sense was operating.

10.3.2 Heat flow in the Archean

The Earth cooled dramatically throughout the 1500 Myr timespan of the Archean (Fig. 10.3). The presence of komatiites in the Archean, although limited in extent, is often cited as evidence of a hotter Archean mantle. However, basalts rather than komatiites dominate in Archean terrains and the mantle was probably only about 150 K hotter than present mantle [32].

There is a steep offset in the heat-flow data at the Archean–Proterozoic boundary, with a decrease for Archean terrains compared to that observed in Proterozoic and Phanerozoic crust, even when the tectonic heat contribution is removed for the younger crust. As erosional levels in Archean terrains are not significantly deeper, the lower heat flow is not due to removal of an upper crust enriched in potassium,

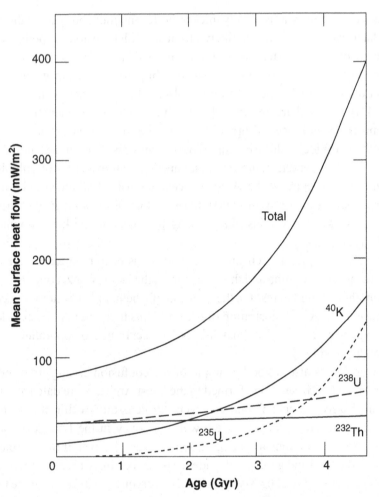

Fig. 10.3 The decrease in heat flow with time for the major radioactive isotopes.

uranium and thorium. It is due instead to the increased thickness of the subcrustal lithosphere beneath Archean cratons with perhaps a small contribution from the slightly different bulk crustal compositions between the Archean and Post Archean [33].

10.4 The formation of the TTG suite

These Na-rich granites that are dominant in the Archean contrast with the K-rich granites that characterize the upper crust in Post-Archean times. The REE patterns of the Archean TTG suites are steep, with high La/Yb ratios and typically lack depletions or enrichments in europium (Fig. 10.2). These steep patterns are a consequence of the formation of the TTG suite by partial melting in the presence of garnet, that has a reciprocal REE pattern and that remains as a residual phase in the source. As garnet is stable in mafic–ultramafic systems below about 40 km, the TTG suite thus formed at mantle depths.

However, the process forming the TTG suite, that has often been ascribed to melting of a young hot subducting slab [34] is the subject of continuing controversy. Quite apart from the debate on whether plate tectonics was operating in the Early and Middle Archean, such a model is different from the origin of most Post-Archean arc rocks, in which the slab dehydrates rather than melts, providing the flux to melt the overlying mantle. However, modern analogues of the "Archean-type" TTG suite (e.g. boninites and adakites) are indeed produced (e.g. in the southern Andes) by melting of the young hydrated subducting slab [35]. This example has been interpreted as indicative of similar subduction of hot young crust in the Archean, hence conventional plate tectonics and for a multitude of small plates, that we regard as an unlikely scenario.

10.4.1 Adakites and the TTG suite

Although "both 'adakite' and 'TTG' … individually encompass such a complex range of compositions [that] clearly complicates any attempt to draw petrogenetic analogies between the two groups" [36], adakites indeed show significant differences from the Archean TTG suite. Thus modern adakites have relatively high $Mg/(Mg + Fe)$, Ni and Cr abundances and relatively low SiO_2 concentrations that show that they have undergone interaction with mantle peridotite [37].

These geochemical signatures of reaction with mantle peridotite are lacking in the Archean TTG suite. The lower Ni, Cr concentrations and lower $Mg/(Mg + Fe)$ ratios of the Archean TTG rocks are consistent with its derivation from melting of hydrated basalt, without contributions of nickel and chromium from the mantle. Thus the TTG suite that appears as early as 3.8 Gyr ago and dominates crustal growth throughout the Archean, does appear to have resulted from the partial remelting of basaltic crust.

However, as the Archean oceanic crust was likely hotter and more buoyant, modern-style subduction and plate tectonics seems unlikely to have occurred before the Late Archean. If subduction is an unlikely mechanism, how might the TTG suite be generated [38]? One possibility is that cooling at the base of this thick basaltic crust could have allowed the transition to denser amphibolite or eclogite and so promote subsidence in a different style to that of modern plate tectonics [39]. Foundering of eclogite in this context is a reasonable phenomenon, allows both for sinking, "dripping" and remelting to produce the TTG suite by melting at depth and in turn allows for the upward flow of fertile peridotite to produce more basalt at the surface [40] (Fig. 10.4).

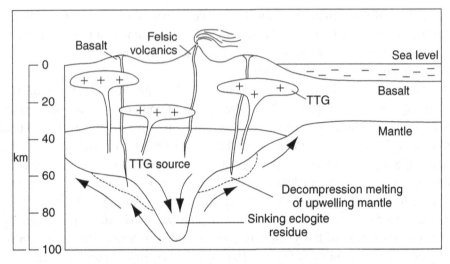

Fig. 10.4 Sketch of the formation of the TTG suite through melting of dense amphibolite or eclogite beneath a thickened (Iceland analog) region of Archean basaltic crust. The thickening may be due to the presence of a mantle plume or compressional tectonics. Melting of upwelling mantle around sinking eclogite blobs provides fresh basalt as a source for greenstone belts and the subsequent formation of TTG suite intrusives and volcanics. Based partly on Smithies, R. H. et al. (2003) Formation of Earth's early Archean continental crust. *Precambrian Res.* **127**, 98, Fig. 5.

Experimental models indicate that melting of foundering slabs of hydrated metabasalts, either amphibolites or eclogites can produce the TTG suite at pressures around 20 kbar, a region where garnet, but not plagioclase is stable [41]. Thus as the thick buoyant Early Archean oceanic crust predicted by geophysical considerations of a hotter Earth make subduction an unlikely mechanism, it is more likely that sinking and melting of denser amphibolites or eclogites produces the characteristic Archean TTG suite [42]. The significance of this is that the TTG suite can be produced without involving modern-style plate tectonics [43].

10.4.2 The evidence from sanukitoids

Minor occurrences of sanukitoids, that are Mg-rich diorites, occur in the Late Archean TTG suites, particularly in the southwestern Superior Province of Canada [44]. None are known before the Late Archean. The Mg/(Mg+Fe) ratios in these suites of TTG rocks are uniformly higher than those generated by melting of basaltic compositions under a wide variety of experimental conditions and compositions. This would imply some interaction with mantle peridotite during the production of these magmas and this particular species of TTG suite rocks may be derived as mantle melts that have been enriched in LIL elements shortly (< 100–200 Myr) before their intrusion.

There are some similarities in the compositions of high-Mg sanukitoid suites and modern boninites (high-Mg andesites and adakites [35]) in island-arc environments. Generation of such rocks appears to require low pressure/high temperature conditions. In the Archean, such conditions might have been achieved by mantle melting associated with dehydration and/or partial melting of subducted young and hot oceanic lithosphere. This would imply that the appearance of sanukitoids in the Late Archean is an example of rocks derived by melting of a mantle wedge and so imply the operation of subduction processes. We thus interpret the sanukitoid suite as providing some of the earliest evidence of the operation of the modern plate-tectonic regime.

10.5 Plate tectonics in the Archean?

Like the notion of a primitive world-encircling sialic (granitic) crust, the question whether a plate-tectonic regime operated early in the Archean reappears continually in the geological literature and remains the subject of schismatic controversy as has been apparent from the preceding discussion.

There is a wide variety of opinion. Some workers place the beginning of the modern tectonic regime as late as 2.0 Gyr ago in line with the notion that high temperatures in the early Earth resulted in a hot thick buoyant Archean oceanic crust. Subduction was thus inhibited in this model until the oceanic crust became cool enough to lose buoyancy and to subside back into the mantle [45].

At the other extreme, it is suggested that subduction was responsible for the generation of siliceous igneous rocks from which the 4.2 Gyr zircons were derived [46]. Others appeal to a thin Mg-rich komatiite crust that could allow for conventional plate tectonics to operate. But komatiites, although common, are swamped by the abundance of basalts in Archean terrains of all ages, while Archean crusts are expected to be thicker in any event due to a hotter mantle [47]. Other workers place the onset of modern-style plate tectonics at around 4 Gyr, essentially at the beginning of the growth of the preserved continental crust [48].

Field evidence has been interpreted in many ways. Thus on one hand "There is ... mounting support for Archean continental evolution via plate-tectonic processes" [49] but alternatively, "few [cratons] are big enough to preserve complete tectonic systems" [50, p. 152] and there is a consensus that the Archean greenstone belts are not preserved oceanic crust [51]. "Few Archaean granite–greenstone terrains show a fully convincing suite of plate tectonic elements ... and a uniformitarian application of plate tectonics to Archaean time seems naïve" [50, p. 171].

The whole question is made even more complex as the development of modern plate tectonics likely evolved over a significant period of time and geological evidence for its operation may be preserved in the rock record at different times in different regions, possibly extending back before 3000 Myr ago in South Africa or Greenland. Thus the evidence from one region may appear to contradict that from another region of broadly similar age.

It is indeed difficult in Archean terrains to assemble the appropriate field evidence for the former presence of arcs, back-arc basins and the like, that might point with less ambiguity to the operation of typical modern-style plate tectonics. However if the tectonic processes were different in the Archean and large cratons were not produced at least until the Late Archean, then such geological evidence for the operation of modern-style plate tectonics will of necessity be lacking.

There is reasonable evidence to support the concept that modern-style plate tectonics began in the Late Archean, around 3000 Myr ago, or in some locations perhaps as early as 3300 Myr ago in the late Middle Archean [52]. Tectonic processes in the earlier Archean seem to have involved the production of thick basaltic crust, foundering of denser hydrated amphibolite (and eclogite) that partially melted at around 10–30 kbar pressure, to form the voluminous TTG suite. This process results in the upward movement of fresh fertile peridotite, that forms additional new basaltic crust and the cycle is repeated (Fig. 10.4).

This model accords with our geochemical observations on Archean sedimentary rocks, that indicate that bulk composition of the Archean continental crust was close to a 50/50 mixture of basalt and the TTG suite. Occasional small amounts of intra-crustal melting produced small regions of K-rich granite with REE patterns similar to those that dominate the Post-Archean upper crust [53]. In the simplest form of the model that we develop in this book, continents were thus formed during the Archean from sinking basaltic material that produced the TTG suite and from subduction-zone andesites in the Post Archean.

In summary, our assessment is that tectonic styles were different in the Archean, involving growth of thick buoyant basaltic crust. Basal sections transformed to denser phases that sank and melted to produce the TTG suite. The current familiar plate-tectonic regime with its andesitic volcanoes produced by slab interaction with the overlying mantle wedge was only established around the time of the Archean–Proterozoic transition when the Earth had cooled sufficiently so that oceanic crust could lose buoyancy and be subducted in a coherent slab back into the mantle.

This fundamental change in the tectonic regime of the Earth is related to the declining heat production and the development of large stable cratons that impose long linear obstructions to the spreading basalts of the oceanic floor. This controls the geometry of modern plates and results in the present-day subduction of mostly cool oceanic crust. The point here is that this crust was hotter and produced the TTG suite rather than the present dominance of andesites.

10.5.1 The subcrustal lithosphere and mantle keels

An interesting problem, closely related to the development of the crust, is the nature of the subcrustal lithosphere. The basic observation is that fast seismic P-wave velocities extend to depths of around 220 km indicating the existence of deep keels, refractory in composition, with high Mg/Fe ratios beneath continental regions [54]. These iron-poor regions are also less dense than the surrounding mantle. This imparts stability to the keels, which thus "float" under the continents and consequently are thought to be major factors in preserving the stability of cratons. For this reason, they require some discussion in this book [55].

The general interpretation from the seismic data is that these deep roots represent refractory lower-density material, consistent with their being the residue remaining from the extraction of an Fe-rich melt, leaving behind an Mg-enriched zone of lower density [56]. Heat flow is low, presumably due to the depletion in heat-producing elements. This view has received considerable support from the evidence from inclusions in diamonds that low temperatures have persisted beneath shields for up to 3000 Myr [57]. The keel ages, best constrained by the Re–Os isotopic system, are typically Late Archean and indicate that such regions can remain isolated from the rest of the mantle for periods exceeding two billion years. Thus keels formed in the Archean but not in the Hadean, consistent with the lack of other evidence for continental-type crust in that epoch [58].

Although the low Fe/Mg ratio and low Ca and Al contents are generally attributed to the extraction of a partial melt, this "sub-continental lithospheric mantle" (SCLM) is enriched in incompatible elements such as Ba, Th, U, Ta, Nb, La, Ce, Nd and depleted in HREE, Ti, Sc, V, Al and Ca relative to average abundances in the mantle. This element pattern indicative of both enrichment and depletion indicates that multistage processes must have occurred, with an initial extraction of a partial melt, followed by at least one stage of a secondary enrichment, often referred to as a "metasomatic" event.

Many models have been proposed to account for this enigmatic composition. The first constraint is the serious sampling problem. The geochemical evidence comes from data from xenoliths. But because of secondary alteration from the enclosing lavas, considerable caution is warranted in interpreting these data. It is also unclear how representative the xenoliths are and whether they represent refractory survivors of multiple extractions of partial melts.

A viable location both for enriching portions of the mantle in incompatible elements, and for the extraction of partial melts would seem to occur above subduction zones, either by dehydration or partial melting of the down-going slab. Thus the formation of the keels would be closely connected with the processes of the formation of the continental crust from the mantle in this model. The Late Archean Re–Os ages of many keels are consistent with other evidence for massive episodic continental growth at that time.

Among other important effects, the keels apparently deflect mantle heat, possibly the cause of apparent low mantle heat flows under Archean cratons. The keels presumably also form barriers to underplating the crust by rising mantle plumes [59]. Such buoyant low-density keels would also place difficulties in the way of the popular model of delamination of dense material from the base of the continental crust back into the mantle [60].

10.6 Meteorite impacts in the Archean

The "late heavy bombardment" or "cataclysm" that is addressed in Chapter 2 on the Moon lasted from about 4.0 to 3.85 Gyr. One of the last events was the formation of the Orientale Basin at 3850 Myr. No younger basin-forming events occurred on the Moon so predating all but the earliest Archean events recorded on the Earth. Although often considered to be responsible for the absence of old crust on the Earth, no definitive evidence of this bombardment has been found in the most ancient rocks at Isua, despite extensive searches [61] except for some enigmatic evidence from tungsten isotopic anomalies [62]. No traces of impact-induced shock features have been reported in any of the surviving zircon crystals [63]. But perhaps these rocks are too young to record the late heavy bombardment. An alternative speculation is that a major basaltic resurfacing covered the evidence [64].

The first apparent evidence of meteorite impacts in the Archean is the existence of several extensive beds of impact-produced spherules in Australia and South Africa [65]. These beds have ages between 2.5 and 3.47 Gyr and were deposited in deep water. The spherules are distinct from volcanic lapilli, are typically about 1 mm in diameter and are interpreted to have condensed in the atmosphere from vapor following large impacts. The original glass has been replaced by secondary minerals. These Archean examples seem to have been originally basaltic in composition in contrast to Phanerozoic spherules suggesting that "the target materials were … more mafic early in Earth history" [66] consistent with our model for continental growth. Most contain high concentrations of platinum-group elements, probably indicative of an origin from meteoritic impact although these are not in meteoritic ratios, perhaps due to alteration. The presence of ^{53}Cr excesses derived from the decay of ^{53}Mn (half-life 3.7 Myr) is the best evidence that they were derived from an impacting meteorite or asteroid [67]. But no parent craters are preserved [68], nor indeed expected in such ancient terrains. Meanwhile the interesting question of why these spherule beds are restricted to the Archean remains an enigma.

Among other misconceptions about the Archean, a confusing literature has emerged concerning the role of impacts of asteroids, comets and meteorites in the Archean and their supposed effects both on tectonics and the generation of melts from the mantle. These ideas continue to flourish, despite repeated efforts to point out the problems associated with them [69] and claims continue to appear linking impacts with tectonic and magmatic events [70]. These models seem to have originated by the mistaken notion that the basalts filling the lunar maria were a consequence of the basin-forming impacts. However there are many large lunar basins that contain little mare basalt. A good example is the largest documented basin on the Moon, the South Pole–Aitken Basin, that is 2500 km in diameter, but that contains only trivial amounts of mare basalt (Fig. 2.5).

Put most simply, the impact of a 20 km diameter body on the Earth at 15 km/s will form a crater about 300 km diameter with up to 20 000 km^3 of melt. However the crater collapses and "even an impact this large is insufficient to raise mantle material above the peridotite solidus due to decompression only" [69]. Thus while it is possible that a pool of molten mantle material might be created, there is no effective pressure release to initiate continued mantle melting on the scale observed in flood basalt provinces. In fact most impact melt that is observed at large impact-produced structures is thoroughly crustal.

The prime example is the 1850 Myr old Sudbury structure (initially 200 km diameter) in Canada, where even the nickel-rich ore body is derived from within the crust [71]. In this example, as at Vredefort, impact-initiated mantle volcanism has not occurred. Evidence of large impacts during the Archean (and the Proterozoic) continues to accumulate but these events do not compare with the lunar basin-forming impacts formed by the late heavy bombardment before 3.8 Gyr [63, 66]. The cratering flux on the Earth appears to have doubled in the past 100 Myr, likely due to a collision that produced the Baptistina Family of asteroids while some of the debris drifted into Earth-crossing orbits. The most dramatic consequences were that large fragments from this collision were probably responsible both for the lunar crater Tycho and for the K–T boundary impact, that set the scene for the rise of primates on the Earth [72].

Synopsis

The enigmatic Archean Eon covers a crucial 1500 Myr of Earth history, from the first traces of continental rocks to the formation of massive cratons in the Late Archean when upwards of 50–70% of the crust was in place and modern-style plate tectonics were operating. The Late Archean (3.0–2.5 Gyr), that might better be termed the Archean–Proterozoic transition, reflects a fundamental change in crustal evolution, linked to the long-term cooling of the Earth. Controversy surrounds the oldest rocks that are somewhere between 3.6 and 4 Gyr in age.

The REE patterns of Archean sedimentary rocks reveal a complex crust dominated by basalts and the sodium-rich tonalite, trondhjemite, granodiorite (TTG) suite and their volcanic equivalents in contrast to the K-rich granitic rocks of the Post-Archean upper crust. Intra-crustal differentiation, so common later, was rare until the Late Archean. The consequence was that the composition of the Archean upper crust differed little from that of the bulk crust, in great contrast to the layered crust of Post-Archean times. Nothing significant remains of the Archean oceanic crust, but due to the higher heat production, a buoyant crust, thicker than that at present was likely, so making subduction difficult.

In our interpretation, the TTG granites formed (with the REE patterns indicating that garnet was a residual phase) due to melting of basalt, transforming to eclogite at mantle depths, that foundered beneath the thick basaltic crust. We judge the alternate model of formation by subduction of young hot basaltic crust as plausible but less likely as modern adakites formed by this process show significant differences from the Archean TTG suite.

We conclude that the modern plate-tectonic regime became operational around 3.0 Gyr when subduction of the cooler oceanic crust became possible. This change was related to the long-term reduction in heat production and the formation of thinner oceanic crust. Sub-crustal lithospheric keels also seem to have formed concurrently with the growth of the continental crust, no Hadean-aged examples being known.

No definitive evidence of the late heavy bombardment, so apparent on the Moon, has yet appeared on Earth. Later impacts, that formed beds of spherules, were on a smaller scale. Although often postulated to have widespread dramatic magmatic and tectonic consequences, the scale of even the largest impacts is too small to produce other than local melt pools, one of the best documented being the younger Proterozoic Sudbury Igneous Complex that is 3 km thick.

Notes and references

1. Shakespeare, W. (1610–11) *The Tempest*, Act I, Scene II.
2. In this and in the following chapters dealing with the Earth, when we use the terms "crust" or "crustal growth" we are referring to the continental crust.
3. Bleeker, W. (2004) Towards a 'natural' timescale for the Precambrian – a proposal. *Lethaia* **37**, 219–22.
4. Taylor, S. R. and McLennan, S. M. (1995) The geochemical evolution of the continental crust. *Rev. Geophys.* **33**, 241–65; Bleeker, W. (2002) Archaean tectonics: A review, with illustrations from the Slave Craton, in *The Early Earth: Physical, Chemical and Biological Development* (eds. C. M. R. Fowler *et al.*), Geological Society of London Special Publication 199, p. 1.

5. Cloud, P. (1988) *Oasis in Space*, Norton, ch. 7.
6. Bleeker, W. (2003) The Late Archean record: A puzzle in *c.* 35 pieces. *Lithos* **71**, 99–134. Other workers also note that "In some ways, the Superior Province appears different from other Archaean cratons world-wide ... All this suggests that not all Archaean cratons were created in the same way". Kendall, J.-M. *et al.* (2002) Seismic heterogeneity and isotropy in the Western Superior Province, in *The Early Earth: Physical, Chemical and Biological Development* (eds. C. M. R. Fowler *et al.*), Geological Society of London Special Paper 199, p. 41.
7. Menard, W. H. (1971) *Science Growth and Change*, Harvard University Press, p. 144. Thus the book by Benn, K. *et al.* (2006) *Archean Geodynamics and Environments*. AGU Geophysics Monograph 64, contains over 1600 references, a characteristic of volumes about the Archean Eon, while *The Earth's Oldest Rocks* (eds. M. Kranendonk *et al.*, 2007), Elsevier, cites about 3000 papers.
8. James, D. E. and Fouch, M. J. (2002) Formation and evolution of Archean terrains: Insights from southern Africa, in *The Early Earth: Physical, Chemical and Biological Development* (eds. C. M. R. Fowler *et al.*), Geological Society of London Special Publication 199, pp. 1–26. The publication of massive review volumes seems to be a characteristic resulting from the study of Archean geology. Examples can be seen by comparing *Archean Geology* (eds. J. E. Glover and D. I. Groves, 1981), Geological Society of Australia, that deals with many of the same problems as *The Early Earth: Physical, Chemical and Biological Development* (eds. C. M. R. Fowler *et al.*, 2002), Geological Society of London Special Publication 199, p. 41 published two decades later. Another good example is the large treatise *Greenstone Belts* (eds. M. De Wit and L. D. Ashwal, 1997), Oxford University Press, which contains the following apologia in the foreword by Kevin Burke: "because there are nearly 100 authors there are naturally a lot of conflicting statements ... [and] ... the book ... invites some attempts at synthesis that its editors have understandably avoided" (p. vii). But the flood continues unabated. Another volume deals with *The Precambrian Earth: Themes and Events* (eds. P. G. Eriksson *et al.*, 2004), Elsevier. Yet another major edited volume on the Archean has appeared *Archean Geodynamics and Environments* (eds. K. Benn *et al.*, 2006), American Geophysics Union, while the Elsevier Series *Developments in Precambrian Geology* has now reached volume 15. These books are supplemented by the specialist journal *Precambrian Geology*. This mass of literature on the Archean has meant that many significant papers cannot be referenced. We apologize to those authors. A representative set of more recent works includes the following: Barley, M. E. and Loader, S. E. (eds., 1998) The tectonic and metallogenic evolution of the Pilbara terrain. *Precambrian Res.* **88**, 1–267; Van Kranendonk, M. J. *et al.* (2000) Geology and tectonic evolution of the Archean North Pilbara terrain, Pilbara Craton, Western Australia. *Econ. Geol.* **97**, 695–732; Huston, D. L. *et al.* (eds., 2002) Early to Middle Archean mineral deposits of the North Pilbara Terrain, Western Australia. *Econ. Geol.* **97**, 691–895; Jones, A. G. *et al.* (eds., 2003) A tale of two cratons: The Slave–Kaapvaal Workshop. *Lithos* **71**, 99–596 and Bleeker, W. (2003) The late Archean record: A puzzle in ca. 35 pieces. *Lithos* **71**, 99–134. These papers provide many examples of the difficulties in interpreting the geological record in Archean terrains.
9. Bowring, S. and Williams, I. P. (1999) Priscoan (4.00–4.03 Gyr) orthogneisses from northwestern Canada. *Contrib. Mineral. Petrol.* **134**, 3–16, but see Whitehouse, M. J. *et al.* (2001) Priscoan (4.00–4.03 Gyr) orthogneisses from northwestern Canada – a discussion. *Contrib. Mineral. Petrol.* **141**, 248–50. This study highlights the probability of subsequent metamorphic alteration of the Sm–Nd isotopic system.

10. Iizuka, T. *et al.* (2007) Geology and zircon geochronology of the Acasta gneiss complex, northwestern Canada: New constraints on its tectonothermal history. *Precambian Res.* **153**, 179–208; Iizuka, T. *et al.* (2007) The Early Archean Acasta gneiss complex, in *The Earth's Oldest Rocks* (eds. M. Van Kranendonk *et al.*), Elsevier, pp. 127–47. Although zircons as old as 4.03 Gyr have been reported, it remains unclear whether these dates come from inherited zircons that occur in younger rocks. See Moorbath, S. (2005) Oldest rocks, earliest life, heaviest impacts and the Hadean–Archean transition. *Applied Geochem.* **20**, 819–24.

11. Bolhar, R. *et al.* (2005) Chemical characterization of earth's most ancient metasediments from the Isua greenstone belt, southern West Greenland. *GCA* **69**, 1555–73.

12. Nutman, A. *et al.* (2004) Inventory and assessment of Palaeoarchean gneiss terrains and detrital zircons in southern West Greenland. *Precambrian Res.* **135**, 281–314; Kamber, B. *et al.* (2005) Volcanic resurfacing and the early terrestrial crust: Zircon, U/Pb and REE constraints from the Isua greenstone belt, southern West Greenland. *EPSL* **240**, 276–90; Moorbath, S. (2005) Oldest rocks, earliest life, heaviest impacts and the Hadean–Archean transition. *Applied Geochem.* **20**, 819–24; Kamber, B. *et al.* (2001) The oldest rocks on the Earth, in *The Age of the Earth* (eds. C. L. E. Lewis and S. J. Knell), Geological Society of London Special Publication 190, pp. 177–203.

13. Mojzsis, S. J. *et al.* (1996) Evidence for life on Earth before 3800 million years ago. *Nature* **384**, 55–9.

14. Stephen Moorbath (2006) personal communication.

15. Kamber, B. and Moorbath, S. (1998) Initial Pb of the Amitsoq gneiss revisited. *Chem. Geol.* **150**, 19–41; Fedo, C. M. and Whitehouse, M. (2002) Origin and significance of Archean quartzose rock at Akilia, Greenland: Comment. *Science* **298**, 917a; Mojzsis, S. J. and Harrison, T. M. (2002) Origin and significance of Archean quartzose rock at Akilia, Greenland: Comment. *Science* **298**, 917a.

16. Allen Nutman (2006) personal communication. For summaries of this lamentable episode, see Moorbath, S. (2005) Dating earliest life. *Nature* **434**, 155 and Moorbath, S. (2005) Oldest rocks, earliest life, heaviest impacts and the Hadean–Archean transition. *Appl. Geochem.* **20**, 819–24. See also Lepland, A. *et al.* (2005) Questioning the evidence for Earth's earliest life – Akilia revisited. *Geology* **33**, 77–9; Eiler, J. M. (2007) The oldest fossil or just another rock. *Science* **317**, 1046–7; and Whitehouse, M. J. and Fedo, C. M. (2007) Searching for Earth's earliest life in southern west Greenland – history, current status and future prospects, in *The Earth's Oldest Rocks* (eds. M. Van Kranendonk *et al.*), Elsevier, pp. 841–53.

17. See also McLennan, S. M. (2001) Relationship between the trace element composition of sedimentary rocks and upper continental crust. *Geochem. Geophys. Geosystems* **2**, doi: 10.1029/2000GC000109.

18. See details in Taylor, S. R. and McLennan, S. M. (1985) *The Continental Crust: Its Composition and Evolution*, Blackwell; Taylor, S. R. and McLennan, S. M. (1995) The geochemical evolution of the continental crust. *Rev. Geophys.* **33**, 241–65; McLennan, S. M. *et al.* (2006) Composition, differentiation and evolution of continental crust: Constraints from sedimentary rocks and heat flow, in *Evolution and Differentiation of the Continental Crust* (eds. M. Brown and T. Rushmer), Cambridge University Press, pp. 92–134.

19. Taylor, S. R. and Hallberg, J. A. (1977) Rare earth elements in the Marda calc-alkaline suite: An Archean geochemical analogue of Andean-type volcanism. *GCA* **41**, 1125–9. The differences between the REE patterns of Archean and Post-Archean sediments has led various authors to argue that this contrast between the Archean and Post-Archean continental crust is the result of differing tectonic settings rather than revealing any

significant difference between processes in the Archean and later times. Gibbs, A. K. *et al.* (1986) The Archean–Proterozoic transition: Evidence from the geochemistry of metasedimentary rocks of Guyana and Wyoming. *GCA* **50**, 2125–41; Condie, K. C. (1992) Chemical composition and evolution of the upper continental crust: Contrasting results from surface samples and shales. *Chem. Geol.* **104**, 1–37. But Archean turbidites differ in their mineralogy from their modern analogs. They are not derived from island-arc andesites but from the mixing of the bimodal suite of basaltic and felsic igneous rocks (the TTG suite) and their volcanic equivalents that are dominant in the Archean. Thus this analogy between conditions in the Archean and modern arc environments is an illusion, as andesites are exceedingly rare in the Archean. See McLennan, S. M. (1984) Petrological characteristics of Archean graywackes. *J. Sediment. Petrol.* **54**, 889–98.

20. Examples occur in Greenland, India, Montana and Wyoming, the Canadian Shield, the Western Gneiss Terrain, Australia and in the Limpopo Belt, South Africa. These high-grade belts are preferentially preserved as the greenstone-belt environments are more readily destroyed by erosion. Veizer, J. and Jansen, S. L. (1985) Basement and sedimentary recycling 2, Time dimension to global tectonics. *J. Geol.* **93**, 625–43.

21. The depletion in europium, with Eu/Eu* typically 0.65, is a consequence of the formation of granites by melting in the deep crust (see Chapters 11 and 12). Plagioclase, stable within the crust, is a residual phase and retains divalent Eu, under reducing conditions. The remainder of the REE are trivalent and, as incompatible elements, are enriched in the granitic melts that dominate the upper crust in Post-Archean time.

22. Taylor, S. R. *et al.* (1986) Rare earth element patterns in Archean high-grade metasediments and their tectonic significance. *GCA* **50**, 2267–79.

23. An extended discussion on the Archean bulk crustal composition is given in Taylor, S. R. and McLennan, S. M. (1985) *The Continental Crust: Its Composition and Evolution*, Blackwell, Section 7.10.2. The potassium contents, revised since that work, are 1.0% for the Archean bulk crust and 1.1% for the Post-Archean bulk crust (Chapter 11).

24. Bickle, M. J. (1986) Implications for melting and stabilization of the lithosphere and heat loss in the Archaean. *EPSL* **80**, 314–24; Davies, G. F. (1992) On the emergence of plate tectonics. *Geology* **20**, 963–6.

25. Arndt, N. (1999) Why was flood volcanism on submerged continental platforms so common in the Precambrian? *Precambrian Res.* **97**, 155–64.

26. Hamilton, W. B. (1998) Archean magmatism and deformation were not products of plate tectonics. *Precambrian Res.* **91**, 143–79; Bickle, M. J. *et al.* (1994) Archean greenstone belts are not oceanic crust. *J. Geol.* **102**, 121–38.

27. See Dilek, Y. and Robinson, P. T. (eds., 2003) *Ophiolites in Earth History.* Geological Society of London Special Publication 218.

28. Kusky, T. M. (1990) Evidence for Archean ocean opening and closing in the southern Slave Province. *Tectonics* **9**, 1533–63; Kusky, T. M. and Kidd, W. S. F. (1992) Remnants of an Archean oceanic plateau, Belingwe greenstone belt. *Geology* **20**, 43–6.

29. Hamilton, W. B. (1998) Archean magmatism and deformation were not products of plate tectonics. *Precambrian Res.* **91**, 143–79.

30. Karson, J. A. (2001) Oceanic crust when Earth was young. *Science* **292**, 1076–9; Kusky, T. M. *et al.* (2001) The Archean DongWanzi ophiolite complex, North China craton: 2.505-billion year old oceanic crust and mantle. *Science* **292**, 1142–5.

31. Zhai, M. *et al.* (2002) Is the DongWanzi complex an Archean ophiolite? *Science* **295**, p. 923.

32. Durrheim, R. J. and Mooney, W. D. (1994) Evolution of the Precambrian lithosphere: Seismological and geochemical constraints. *JGR* **99**, 15,359–74.

33. Watson, J. (1976) Vertical movements in Proterozoic structural provinces. *Phil. Trans. Royal Soc.* **A280**, 629–40. But see Nyblade, A. A. and Pollack, H. N. (1993) A global analysis of heat flow from Precambrian terrains. *JGR* **98**, 12,207–18; Martin, H. and Moyen, J.-F. (2002) Secular changes in tonalite-trondhjemite-granodiorite composition as markers of the progressive cooling of the Earth. *Geology* **30**, 319–22.

34. See for example Drummond, M. S. and Defant, M. J. (1990) A model for trondhjemite-tonalite-dacite genesis and crustal growth by slab melting: Archean to modern comparisons. *JGR* **95**, 21,503–21; Rapp, R. P. *et al.* (1991) Partial melting of amphibolite/eclogite and the origin of Archean trondhjemites and tonalites. *Precambrian Res.* **51**, 1–25; Martin, H. (1994) The Archean grey gneisses and the genesis of continental crust, in *Archean Crustal Evolution* (ed. K. C. Condie), Elsevier, pp. 205–59; Evans, O. C. and Hanson, G. N. (1997) Post-kinematic Archean tonalites, trondhjemites and granodiorites of the S. W. Superior Province: Derivation through direct mantle melting, in *Greenstone Belts* (eds. L. Ashwal and M. De Wit), Oxford University Press, pp. 280–95.

35. Adakites are rocks derived by slab melting, with steep REE patterns signifying that garnet was a residual phase. Boninites are high-Mg# andesites, with U-shaped REE patterns, indicating a complex history of melting of a depleted source followed by enrichment of LREE and other incompatible elements from hydrated mantle above subduction zones. Arndt, N. (2003) Komatiites, kimberlites and boninites. *JGR* **108**, doi: 1029/2002JB002157. Their rarity makes them unlikely candidates as major contributors to continental growth. But see Kelemen, P. B. (1995) Genesis of high Mg# andesites and the continental crust. *Contrib. Mineral. Petrol.* **120**, 1–19.

36. Martin, H. *et al.* (2005) An overview of adakite, tonalite–trondhjemite–granodiorite (TTG) and sanukitoid: relationships and some implications for crustal evolution. *Lithos* **79**, p. 5.

37. Smithies, R. H. (2000) The Archean tonalite–trondhjemite–granodiorite (TTG) series is not an analogue of Cenozoic adakite. *EPSL* **182**, 115–25.

38. Foley, S. F. *et al.* (2003) Evolution of the Archean crust by delamination and shallow subduction. *Nature* **421**, 249–52.

39. Bédard, J. H. (2006) A catalytic delamination-driven model for coupled genesis of Archean crust and sub-continental mantle. *GCA* **70**, 1188–214.

40. Smithies, R. H. *et al.* (2003) Formation of the Earth's early Archean continental crust. *EPSL* **127**, 89–101.

41. Foley *et al.* (2002) suggest that the geochemistry of the TTG suite is consistent with melting of garnet-bearing amphibolitic rather than eclogitic precursors. This is based on the observation that the TTG suite has low Nb/Ta ratios relative to MORB. Higher Nb/Ta ratios are expected from the melting of eclogites, where these elements are concentrated in rutile. However, garnet is required to be present in the residue from partial melting to account for the steep LREE-enriched patterns of the TTG suite, as well as for the lack of Eu anomalies. Foley, S. F. *et al.* (2002) Growth of early continental crust controlled by melting of amphibolite in subduction zones. *Nature* **417**, 837–40. See also van Thienen, P. *et al.* (2004) On the formation of continental silicic melts in thermo-chemical mantle convection models: Implications for early Earth and Venus. *Tectonophysics* **394**, 111–24.

42. Rapp, R. P. *et al.* (1991) Partial melting of amphibolite/eclogite and the origin of Archean trondhjemites and tonalites. *Precambrian Res.* **51**, 1–25.

43. Kamber, B. S. *et al.* (2002) Fluid-mobile trace element constraints on the role of slab melting and implications for Archean crustal growth models. *Contrib. Mineral.*

Petrol. **144**, 38–56; Davies, G. F. (1992) On the emergence of plate tectonics. *Geology* **20**, 963–6.

44. e.g. Evans, O. C. and Hanson, G. N. (1997) Post-kinematic Archean tonalites, trondhjemites and granodiorites of the S. W. Superior Province: Derivation through direct mantle melting, in *Greenstone Belts* (eds. L. Ashwal and M. De Wit), Oxford University Press; 280–95 Rollinson, H. and Martin, H. (2005) Geodynamic controls on adakite, TTG and sanukitoid genesis: Implications for models of crust formation. *Lithos* **79**, ix–xii.

45. Hamilton, W. B. (1998) Archean magmatism and deformation were not products of plate tectonics. *Precambrian Res.* **91**, 143–79.

46. Harrison, T. M. *et al.* (2005) Heterogeneous Hadean hafnium: Evidence of continental crust at 4.4 to 4.6 Gyr. *Science* **310**, 1947–50. But see Kramers, J. D. (2007) Hierarchical Earth accretion and the Hadean Eon. *J. Geol. Soc. London* **164**, 3–17.

47. Arndt, N. (1983) Role of a thin, komatiite-rich oceanic crust in the Archean plate-tectonic process. *Geology* **11**, 372–5; Nisbet, E. G. and Fowler, C. M. R. (1983) Model for Archean plate tectonics. *Geology* **11**, 376–9; Abbott, D. H. and Isley, A. E. (2002) The intensity, occurrence and duration of superplume events and eras over geological time. *J. Geodynamics* **34**, 265–307.

48. De Wit, M. J. (1998) On Archean granites, greenstones, cratons and tectonics: Does the evidence demand a verdict? *Precambrian Res.* **91**, 181–266. See also Furnes, H. *et al.* (2007) A vestige of the Earth's oldest ophiolite. *Science* **315**, 1704–7 who "contend that the ISB (Isua Supracrustal Belt) preserves vestiges of Earth's earliest ophiolite and oceanic crust. This implies that … Phanerozoic-like plate tectonics were operating 3.8 billion years ago" (p. 1706). This follows an earlier claim from the same area by Kohima, T. *et al.* (1999) Plate tectonics at 3.8–3.7 Ga: Field evidence from the Isua accretionary complex, southern West Greenland. *J. Geol.* **107**, 515–54. Extraordinary claims require extraordinary evidence, difficult to obtain from this enigmatic terrain.

49. Kendall, J. -M. *et al.* (2002) Seismic heterogeneity and isotropy in the Western Superior Province, in *The Early Earth: Physical, Chemical and Biological Development* (eds. C. M. R. Fowler *et al.*), Geological Society of London Special Publication 199, p. 41.

50. Bleeker, W. (2002) Archaean tectonics: A review, with illustrations from the Slave craton, in *The Early Earth: Physical, Chemical and Biological Development* (eds. C. M. R. Fowler *et al.*), Geological Society of London Special Publication 199.

51. Ojakangas, R. W. *et al.* (2001) The Mesoproterozoic midcontinent rift system, Lake Superior region, USA. *Sediment. Geol.* **141–142**, 421–42. See also Bickle, M. J. *et al.* (1995) Archean greenstone belts are not oceanic crust. *J. Geol.* **102**, 121–38.

52. Davies, G. F. (1992) On the emergence of plate tectonics. *Geology* **20**, 963–6; A potential late Middle Archean example is reported by Garde, A. A. (2007) A mid-Archean island arc complex in the eastern Akia terrane, Godthäbsfjord, southern West Greenland. *J. Geol. Soc. London* **164**, 565–79. This complex is dated from zircons at 3070 ± 1 Myr and so perhaps formed at the end of the Middle Archean. However, as is common in Archean terrains, "the poor state of preservation prevents the establishment of a proper volcanic stratigraphy…and hinders detailed comparison with well-preserved Archean or modern arc environments elsewhere" (p. 565). Another example of about the same age is reported by Smithies, R. H. *et al.* (2005) Modern-style subduction processes in the Mesoarchean: Geochemical evidence from the 3.12 Ga Whundo intra-oceanic arc. *EPSL* **231**, 221–37.

53. Foley, S. F. *et al.* (2003) Evolution of the Archean crust by delamination and shallow subduction. *Nature* **421**, 249–52.

54. Jordan, T. H. (1988) Structure and formation of the continental tectosphere. *J. Petrol.* (*Lithosphere Vol.*) **11**, 37.
55. De Smet, J. H. *et al.* (2000) Early formation and long-term stability of continents resulting from decompression in a convecting mantle. *Tectonophysics* **322**, 19–33; Forte, A. M. and Perry, H. K. C. (2000) Geodynamic evidence for a chemically depleted continental tectosphere. *Science* **290**, 1940–4; Rudnick, R. L. *et al.* (1998) Thermal structure, thickness and composition of continental lithosphere. *Chem. Geol.* **145**, 395–411.
56. There are the usual discrepancies in the interpretation of the seismic data. Thus in one version, high-velocity mantle roots extend to 250–300 km beneath Kaapvaal Craton, but others note that the depth of the high velocity zone below southern Africa does not extend below 160 km. Priestley, K. and McKenzie, D. (2002) The structure of the upper mantle beneath southern Africa, in *The Early Earth: Physical, Chemical and Biological Development* (eds. C. M. R. Fowler *et al.*), Geological Society of London Special Publication 199, pp. 45–64. That is in rough agreement with the 200–225 km depths of the continental roots beneath the Australian shield. Debayle, E. and Kennett, B. L. N. (2000) The Australian continental upper mantle: Structure and deformation inferred from surface waves. *JGR* **105**, 25,423–50. The sub-cratonic lithosphere has two main mineral components: forsterite (Mg olivine) and Mg-rich orthopyroxene, both the residual products of high degrees of partial melting. In this interpretation, the keels are the residues from partial melting events, at pressures around 60–70 kbar. Such pressures seem too high for melting at mid-ocean ridges or in mantle wedges at present. Herzberg, C. T. (1999) Phase equilibrium constraints on the formation of cratonic mantle, in *Mantle Petrology* (ed. F. R. Boyd), Geochemical Society Special Publication 6, pp. 241–57. Walter, M. J. (1999) Melting residue of fertile peridotite and the origin of cratonic lithosphere, in *Mantle Petrology* (ed. F. R. Boyd), Geochemical Society Special Publication 6, pp. 225–39. James, D. E. and Fouch, M. J. (2002) Craton development in Southern Africa, in *The Early Earth: Physical, Chemical and Biological Development* (eds. C. M. R. Fowler *et al.*), Geological Society of London Special Publication 199, p. 41.
57. Boyd, F. R. and Gurney, J. J. (1986) Diamonds and the African lithosphere. *Science* **232**, 472–7.
58. There is no evidence of Hadean ages in any of the more than 100 samples of peridotite xenoliths so far dated. Walter, M. J. (1999) Melting residue of fertile peridotite and the origin of cratonic lithosphere, in *Mantle Petrology* (ed. F. R. Boyd), Geochemical Society Special Publication 6, pp. 225–39. Data from the Sm–Nd, Rb–Sr, Lu–Hf and Th–U–Pb isotopic systems are readily affected by contamination from the enclosing lavas and by the secondary enrichment events. Only the Re–Os system seems capable of giving reliable ages for the partial melting events. Rhenium is a moderately incompatible element, that is partitioned preferentially into the melt phase during partial melting. The daughter product of ^{187}Re, ^{187}Os is compatible and remains with the peridotite residue. There the Re/Os ratio is so low and the ^{187}Re half-life so long that the osmium isotopic ratios are not readily influenced by later metasomatic episodes.
59. James, D. E. and Fouch, M. J. (2002) Craton development in Southern Africa, in *The Early Earth: Physical, Chemical and Biological Development* (eds. C. M. R. Fowler *et al.*), Geological Society of London Special Publication 199, p. 21.
60. Further discussion of these inaccessible regions is beyond the scope of this book. The interested reader who wishes to delve further into the controversies surrounding them is referred to the following works: Arndt, N. T. *et al.* (2002) Strange partners: Formation and survival of continental crust and lithospheric mantle, in *The Early Earth: Physical,*

Chemical and Biological Development (eds. C. M. R. Fowler *et al.*), Geological Society of London Special Publication 199, pp. 91–103. See Silver, P. G. (1996) Seismic anisotropy beneath the continents; probing the depths of geology. *Ann. Rev. Earth Space Sci.* **24**, 385 for a discussion of these relations. Moser, D. E. *et al.* (2001) Birth of the Kaapvaal tectosphere 3.08 billion years ago. *Science* **291**, 465–8. Pearson, D. G. *et al.* (2002) The development of lithospheric keels beneath the earliest continents, in *The Early Earth: Physical, Chemical and Biological Development* (eds. C. M. R. Fowler *et al.*), Geological Society of London Special Publication 199, pp. 65–90.

61. Frei, R. and Rosing, M. T. (2005) Searches for traces of the late heavy bombardment on Earth – results from high precision chromium isotopes. *EPSL* **236**, 28–40; Grieve, R. A. F. *et al.* (2006) Large-scale impacts and the evolution of the Earth's crust: The early years, in *Processes on Early Earth* (eds. W. U. Reimold and L. Gibson), Geological Society of America Special Paper 405, pp. 23–32.

62. Debatable evidence of a meteoritic contribution appears from the enrichment of ^{182}W in Isua rocks. Schoenberg, R. *et al.* (2002) Tungsten isotopic evidence from 3.8 Gyr metamorphosed sediments for early meteorite bombardment of the Earth. *Nature* **418**, 403–5.

63. Koeberl, C. (2006) The record of impact processes on the early Earth: A review of the first 2.5 billion years, in *Processes on Early Earth* (eds. W. U. Reimold and L. Gibson), Geological Society of America Special Paper 405, pp. 1–22.

64. Kamber, B. S. *et al.* (2005) Volcanic resurfacing and the early terrestrial crust: Zircon U–Pb and REE constraints from the Isua greenstone belt, southern West Greenland. *EPSL* **240**, 276–90. Such resurfacing is analogous to the later event on Venus (Chapter 7).

65. Simonson, B. N. and Glass, B. P. (2004) Spherule layers: Records of ancient impacts. *Ann. Rev. Earth Planet. Sci.* **32**, 329–61. These beds of spherules occur in the 3.2–3.5 Gyr Fig Tree Group (Barberton, South Africa) and Warrawoona Group (Western Australia). Although probably the consequence of impacts, there is little sign of shocked minerals, breccias or other impact-derived material. One shocked grain of quartz has been reported by Rasmussen, B. and Koeberl, C. (2004) Iridium anomalies and shocked quartz in a Late Archean spherule layer from the Pilbara craton: New evidence for a major asteroid impact at 2.63 Gyr. *Geology* **32**, 1029–32. Other evidence suggestive of an extra-terrestrial impact, in the presence of ^{53}Cr anomalies, has been found in at least one bed (S4) dated at 3.22 Gyr in the Fig Tree sequence in South Africa. The ^{53}Cr values, produced by the radioactive decay of ^{53}Mn, are distinct from terrestrial values. These are identical to those in CI meteorites and also to those in the Cretaceous–Tertiary boundary clay and constitute evidence of the impact of a massive, perhaps 20 km diameter projectile. See Shukolyukov, A. *et al.* (2000) Early Archean spherule beds: Confirmation of impact origin. *Meteor. Planet. Sci.* **35**, A146. For further discussions on impacts and their effects, see Cockell, C. K. (2006) The origin and emergence of life under impact bombardment. *Phil. Trans. Royal Soc.* **B361**, 1845–56; Kasting, J. F. and Howard, M. T. (2006) Atmospheric composition and climate on the early Earth. *Phil. Trans. Royal Soc.* **B361**, 1731–42; and Lunine, J. I. (2006) Physical conditions on the early Earth. *Phil. Trans. Royal Soc.* **B361**, 1721–31.

66. Simonson, B. N. and Glass, B. P. (2004) Spherule layers: Records of ancient impacts. *Ann. Rev. Earth Planet. Sci.* **32**, p. 351.

67. Shukolyukov, A. *et al.* (2000) The oldest impact deposits on Earth – first confirmation of an extraterrestrial component, in *Impacts on the Early Earth* (eds. I. Gilmour and C. Koeberl), Springer, pp. 99–115.

68. The oldest impact structure on the Earth is the Vredefort structure, South Africa dated at 2023 ± 4 Myr followed by the impact at Sudbury, Ontario at 1850 ± 3 Myr (Ref. 63).

69. Ivanov, B. A. and Melosh, H. J. (2003) Impacts do not initiate volcanic eruptions. *Geology* **31**, 869–72.

70. The original suggestion appears to be due to Green, D. H. (1972) Archean greenstone belts may include terrestrial equivalents of lunar maria? *EPSL* **15**, 263–70. This notion has been taken up by many subsequent workers. More recent examples include Alt, D. *et al.* (1988) Terrestrial maria: The origins of large basaltic plateaux, hotspot tracks and spreading ridges. *J. Geol.* **96**, 647–62; Glikson, A. Y. (1999) Oceanic mega-impacts and crustal evolution. *Geology* **27**, 387–90; Jones, A. P. *et al.* (2002) Impact induced melting and the development of large igneous provinces. *EPSL* **202**, 551–61; Abbott, D. H. and Isley, A. E. (2002) Extraterrestrial influences on mantle plume activity. *EPSL* **205**, 53–62; Elkins-Tanton, L. T. (2005) Continental magmatism caused by lithospheric delamination, in *Plates, Plumes and Paradigms* (eds. G. R. Foulger *et al.*), GSA Special Paper 388, pp. 449–62; Elkins-Tanton, L. T. *et al.* (2004) Magmatic effects of the lunar late heavy bombardment. *EPSL* **222**, 17.

71. Grieve, R. A. F. *et al.* (1991) The Sudbury structure: Controversial or misunderstood? *JGR* **96**, 22,753–64.

72. Bottke, W. F. *et al.* (2007) An asteroid breakup 160 million years ago was the probable source of the K-T impactor. *Nature* **449**, 48–53.

11

The Post-Archean continental crust

> Possibly many may think that the deposition and consolidation of fine-grained mud must be a very simple matter and the results of little interest. However....it is soon found to be so complex a question...that one might feel inclined to abandon the enquiry, were it not that so much of the history of our rocks appears to be written in this language.
>
> *(Henry C. Sorby) [1]*

The continental crust of the Earth is so familiar to us that perhaps we underestimate its significance as a platform for human existence while at the same time overestimate its significance for understanding planetary crusts. Without such a haven above sea level, the later stages of evolution would have taken a very different course. If oceanic islands had formed the only dry land, birds rather than mammals might have become dominant as they did in Mauritius and New Zealand. However, the buoyant extensive continental crust has provided a useful platform for the land-based stages of evolution. After the extinction of the dinosaurs and much else 65 Myr ago, the way was cleared on the continental massifs for mammalian evolution to flourish. This led ultimately to the emergence of primates and to the appearance of many species of the genus Homo, ultimately enabling this account.

11.1 The Archean–Proterozoic transition

Following the Archean, which had lasted for 1500 Myr, the Proterozoic Eon continued for an even longer period (2000 Myr). The lumping together of these two disparate eons into the Precambrian, although understandable from a paleontological perspective, is an unsatisfactory union from the viewpoint of crustal development. This is because at the transition between the Archean and Proterozoic, the processes of crustal formation, both the volume and differentiation within the crust underwent significant changes. This fundamental distinction between the Archean and the

Proterozoic was initially recognized by Sir William Logan in 1845 in Canada when he drew attention to the "great unconformity" between the Huronian sedimentary sequence and the underlying "granitic" basement in Ontario [2].

This unconformity marks the profound "boundary" between the Archean and Proterozoic eons and reflects fundamental changes in the way in which the continental crust evolved. However, the processes that were responsible were non-synchronous, extending over several hundred million years in different areas to the despair of workers attempting to delineate precise geological boundaries. Thus although the formal boundary is set at 2500 Myr, except for catastrophic events such as occurred at the K-T boundary, nature rarely makes such sudden changes. So the switch from Archean-style crustal development to our familiar plate-tectonic regime extended from around 3000 Myr ago or a little earlier to well into the Proterozoic.

It was during the Proterozoic, that encompasses half of the recorded history of the Earth, that the continental crust achieved its present familiar form. Compared with the Early- or Middle-Archean terrains, a geologist interested in continents who travelled back in time for a couple of billion years would still have found a recognizable landscape. Large cratonic masses were established by that time, leading to the development of shallow shelves that were intermittently flooded, providing multiple habitats for life to flourish. These cratonic masses were split up and reassembled in a jumble of terranes whose wanderings have been laboriously tracked by geological and paleomagnetic studies [3].

The Earth has cooled substantially over the past 4.5 billion years, due both to whole-Earth cooling and to the slow decay of the radiogenic elements, particularly that of the shorter-lived ^{235}U and ^{40}K (Fig. 10.3). Consequently, there can be little doubt that the thermal regime of continental-crust formation and crustal differentiation has changed over Earth history. As the Earth cooled, the oceanic crust became "negatively buoyant" and was able to sink back into the mantle. The consequence was that the tectonic regime of the Early and Middle Archean was replaced by the development of modern-style subduction zones. The present system of plate tectonics was established in the Late Archean and seems to have been fully operational since that time.

The rapid increase in crustal volume during the Late Archean was due to little-understood episodic processes in the mantle. In contrast, the causes of the change in upper-crustal composition are known to be due to the major emplacement of K-rich granitic rocks in the upper crust, a process that began in earnest in the Late Archean. Massive melting deep within the crust produced granites, depleted in europium, that eventually dominated the upper crust. Sometimes called "cratonization", this process "stabilized" the crust and transferred the heat-producers and many other elements upwards from the lower crust.

11.2 Changes in crustal composition during the Archean–Proterozoic transition

During the Archean, the upper crust was a mixture of the "bimodal suite" of basalts (with subordinate komatiites) and the sodium-rich tonalite–trondhjemite–granodiorite (TTG) suite, in contrast to the dominance of K-rich granites since the Late Archean. Sedimentary rocks record the signature of this significant difference in the composition of the upper continental crust between the Archean and Post Archean, as for example reflected in an increase in K_2O/Na_2O ratios. There were many other changes [4].

There was thus a secular change in the rare earth element distribution of sedimentary rocks marked by an abrupt change at the Archean–Proterozoic transition. Post-Archean sedimentary rocks are characterized by very uniform REE patterns and a uniform depletion in europium (average $Eu/Eu^* = 0.65 \pm 0.05$) whereas those in Archean sedimentary rocks are both much more variable and rarely show any depletion in europium [5]. Post-Archean sedimentary rocks are characterized by relatively flat HREE distributions ($Gd_N/Yb_N = 1.0$–2.0) whereas Archean sedimentary rocks are characterized by variable HREE patterns many with the steep HREE patterns characteristic of the Archean TTG suite (Fig. 10.2). However the bulk composition of the Archean and Post-Archean upper crusts have similar HREE ratios because the steep patterns of the Archean TTG igneous suite are balanced by the approximately equal abundance of basaltic rocks with flat patterns. The consequence is that the overall Archean REE pattern mimics that of modern andesites.

The relatively low $^{87}Sr/^{86}Sr$ and high $^{143}Nd/^{144}Nd$ ratios of Archean seawater, reflected in unaltered marine carbonates and banded iron formations, were due to the much higher contribution of mantle-derived components to the oceans during the Archean. The upper crust was enriched of ^{87}Rb in the Late Archean and Early Proterozoic, a consequence of the intrusion of K-rich granites into the upper crust. Subsequent erosion of the upper crust and transfer to the ocean led to an increase in the oceanic $^{87}Sr/^{86}Sr$ ratio. This signature became incorporated in marine carbonate sediments, leading to an enhanced $^{87}Sr/^{86}Sr$ ratio in such sediments from the Proterozoic onwards [6].

Among other changes, there is a dramatic increase in the Th/Sc ratio at the Archean–Proterozoic boundary. This element pair form an especially sensitive and reliable measure of the ratio of acidic to basic chemical compositions [7] (Fig. 11.1).

There is also an abrupt change in Th/U ratios and a marked discontinuity in lead isotopic characteristics in sediments with neodymium-model ages older than Early Proterozoic. These are derived from the Archean upper crust indicating that it had much lower U/Pb and Th/Pb ratios compared to those of the Post-Archean upper continental crust [8]. In addition a marked shift in the oxygen isotope signature

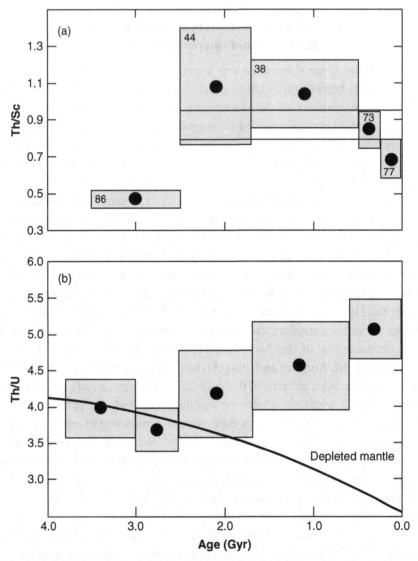

Fig. 11.1 (a) The variation in Th/Sc ratios with age for 300 samples of sedimentary rocks versus stratigraphic age showing a major break at the Archean–Proterozoic boundary. Data from McLennan and Hemming (1992) [7]. The apparent "reversal" during the Mesozoic and Tertiary is probably a transient feature (also shown by Sm/Nd and K_2O/Na_2O) that is always present in the young tectonically active settings (effectively tracking first-cycle greywacke versus cratonic shale). (b) The variation in Th/U ratios in fine-grained sedimentary rocks with age. The curved line marked "depleted mantle" represents the evolution of the Th/U ratio in the upper mantle from Collerson and Kamber (1999) [8].

occurs between the Archean and Proterozoic eons. Archean sediments had $\partial^{18}O$ values around $7 \pm 0.5‰$ but following the Archean–Proterozoic transition, the values began to rise to 10‰ [9].

The extensive uranium deposits that occur in basal Proterozoic sediments are a consequence of intra-crustal melting leading to the enrichment of the upper crust in incompatible elements of which uranium is a quintessential example. The widespread proliferation of stromatolites in the Proterozoic was a consequence of the development of stable shelves around the continents, until the emergence of metazoan grazing species caused the stromatolites to decline in the Phanerozoic. Banded iron-formations likewise proliferated in the Late Archean and Early Proterozoic partly due to the development of stable shelves. It was at this time that the first supercontinent formed [10].

In summary many geological events correlate with a major change during the Archean–Proterozoic transition and thus are not due to long-term recycling in the sedimentary record [11]. By about 2000 Myr ago, this major period of crustal growth was over and continental volumes have continued to grow at a slower rate. Continents and oceans have retained similar relative volumes. This is the famous "freeboard constraint", that then becomes a constraint applicable to models of continental growth.

11.3 The Post-Archean upper crust

What is the composition of the present (Post-Archean) upper crust and how might it be determined? One approach has involved widespread sampling programs of surface rocks such as those carried out over the Canadian shield and across the USSR. These estimates generally indicate that the major-element composition of the Post-Archean upper crust is mostly similar to that of granodiorite [12].

Because of the extreme variability in the exposed upper crust, obvious to the most casual observer, such programs have to be carried out over continent-wide areas. Indeed as the number of regional sampling programs has grown it is becoming clear that there are some variations [13]. This can be seen in Table 11.1 where several of the regional averages are compared to our upper-crustal estimates for selected trace elements.

Several causes of this variability are apparent: the intrinsic uncertainty of such approaches, analytical difficulties for individual elements, real regional differences in composition or some combination of these. A number of other estimates have been made using a variety of approaches, including igneous rock balances, sedimentary compositions and "hybrid" models where different elements are derived from a variety of sources [14]. Clearly in trying to estimate the composition of either the upper or the bulk continental crust, reliance on local regions will introduce errors. A much wider sampling strategy is called for.

Table 11.1 *Estimates for various trace elements from different cratonic regions compared with several compositions for the upper continental crust. See Note 44 for references*

Element (ppm)	Canadian Shield[a]	Canadian Shield[b]	East China[c]	Scotland[d]	New Mexico[e]	Colorado[f]	Upper crust[g]	Upper crust[h]	Upper crust[i]	Upper crust[j]	Upper crust[k]	Upper crust[l]
Sc	7.0	12	15	–	–	16	13.3	–	11	–	14	13.6
Ti	3120	3180	3900	2400	2700	4200	3300	–	3000	4555	3840	4100
V	53	59	98	–	–	–	86	–	60	–	97	107
Cr	35	76	80	<50	19	82	104	–	35	–	92	83
Co	12	–	17	35	8	17	18	–	10	–	17.3	17
Ni	19	19	38	25	13	43	56	–	20	–	47	44
Rb	110	–	82	85	187	72	83	–	112	–	84	(112)
Zr	237	190	188	135	180	148	160	–	190	–	193	(190)
Nb	26	–	12	4	–	9.1	9.8	–	25	13.7	12	12
Cs	–	–	3.55	–	–	–	–	5.8	3.7	7.3	4.9	4.6
Ba	1070	730	678	795	590	749	633	668	550	–	624	(550)
La	32.3	71	34.8	55	43	27	28.4	–	30	–	31	(30)
Hf	5.8	–	5.12	–	–	4.4	4.3	–	5.8	–	5.3	(5.8)
Ta	5.7	–	0.74	–	–	0.67	0.79	1.5	2.2	0.96	0.9	1.0
Pb	17	18	18	–	–	14	17	–	20	–	17	17
Th	10.3	10.8	8.95	–	13	6.2	8.6	–	10.7	–	10.5	(10.7)

Data sources: [a] Average Canadian shield (Shaw *et al.*, 1986); [b] Average Canadian shield (Fahrig and Eade, 1968; Eade and Fahrig, 1971, 1973); [c] Average Central East China calculated on carbonate-free basis (Gao *et al.*, 1998); [d] Average of crystalline basement of northwest highlands of Scotland (Bowes, 1972); [e] Average Precambrian surface terrane, New Mexico (Condie and Brookins, 1980); [f] Average of Colorado Plateau upper crust derived from equal proportions of northwest and southeast sections (Condie and Selverstone, 1999); [g] Average upper continental crust from "Map Model" (Condie, 1993); [h] Average upper continental crust (Wedepohl, 1995) where his values differ from those of Shaw *et al.* (1986); [i] Average upper crust (Taylor and McLennan, 1985, 1995); [j] Upper continental crust derived from marine sedimentary record (Plank and Langmuir, 1998); [k] Average upper continental crust derived from a combination of approaches (Rudnick and Gao, 2003); [l] Average upper continental crust from the sedimentary record (McLennan, 2001) and Table 11.4. Elements that are unchanged from Taylor and McLennan (1985) are in parentheses.

In contrast, our approach has been that only the geological processes of erosion and sedimentation have efficiently sampled the heterogeneous upper crust [15]. Thus an alternative procedure to massive and expensive sampling programs to obtain the composition of the upper crust, is to use sedimentary rocks and sediments such as loess. Nature has mimicked the procedures of geochemists but on a scale many orders of magnitude larger and so provides a natural sampling of the crust. Of course the usual caveats apply to such procedures (see appendices).

The usefulness of this approach can be seen by comparing the data obtained from 23 Post-Archean Australian shales that range in age from Proterozoic to Triassic with the similar results obtained from the 40 780 samples collected in the USSR study. This demonstrates the advantages of employing natural sampling processes that integrate the geological complexities (Fig. 11.2).

The most dramatic observation is that the REE abundances in Post-Archean clastic sedimentary rocks are remarkably uniform in great contrast to the REE patterns in sediments of Archean age. This uniformity extends both within and between continents so that composite shale samples from Europe (European shale composite = ES), North America (North American shale composite = NASC) and

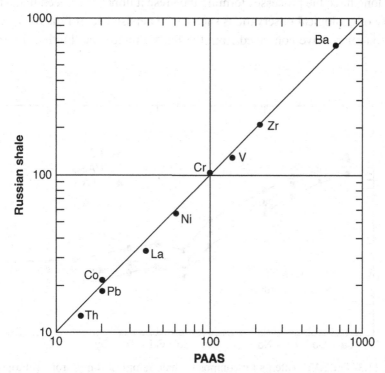

Fig. 11.2 A comparison of estimates of the composition (in ppm) of the upper crust from 23 Post-Archean Australian Shales (PAAS) [15] with data from 40 780 Russian shales (data from Ronov *et al.* (1992) [12]).

Australia (Post-Archean Australian shale = PAAS) have REE patterns that are effectively identical (Fig. 11.3, Table 11.2).

The abundances of other insoluble elements (e.g. thorium and scandium) are likewise a useful measure of upper-crustal abundances and of intra-crustal differentiation as noted earlier [16].

McLennan [17] has re-evaluated the upper-crustal trace element abundances from sedimentary data for other trace elements that are transferred essentially quantitatively from the upper crust into clastic sediments, including Ti, Zr, Hf, Nb, Ta, Rb, Cs, Pb, Cr, V, Ni and Co. The abundances were obtained by using the ratios of the REE (notably La) to these various trace elements to obtain clastic sedimentary rock averages. Using elemental ratios that remain relatively constant (e.g. $K/U = 10^4$, $K/Rb = 250$, $Th/U = 3.8$, $Rb/Sr = 0.3$) it is possible to extrapolate to calculate upper-crustal abundances for many other elements. This composition is valid for the upper 10 km or so of the continental crust exposed to weathering and erosion but such a composition cannot be representative of the entire crust as heat-flow data inform us that the lower crust must be depleted in potassium, uranium and thorium (Tables 11.3, 11.4).

How long have the processes forming the present upper crust been in operation? A variety of geological observations such as the persistence of Proterozoic cratons suggest that they have continued from the Early Proterozoic. Further information

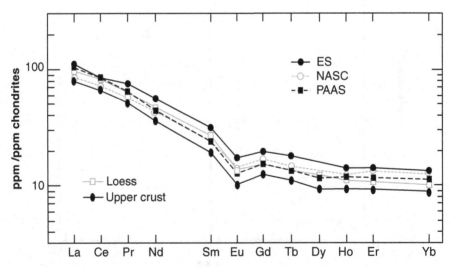

Fig. 11.3 The REE patterns for composite shale samples from Europe (European shale average = ES), North America (North American shale composite = NASC) and Australia (Post-Archean Australian shale = PAAS) and for average loess are effectively identical. Data from Table 11.2 and McLennan (1989) [15].

Table 11.2 *Rare earth element values for chondrites, the Post-Archean Average Australian shale (PAAS), North American shale composite (NASC) and the European shale composite (ES)*

Element	Chondrite	PAAS	NASC	ES
La	0.367	38.2	32	41.1
Ce	0.957	79.6	73	81.3
Pr	0.137	8.83	7.9	10.4
Nd	0.711	33.9	33	40.1
Sm	0.231	5.55	5.7	7.3
Eu	0.087	1.08	1.24	1.52
Gd	0.306	4.66	5.2	6.03
Tb	0.058	0.774	0.85	1.05
Dy	0.381	4.68	5.8	–
Ho	0.0851	0.991	1.04	1.20
Er	0.249	2.85	3.4	3.55
Tm	0.0356	0.405	0.50	0.56
Yb	0.248	2.82	3.1	3.29
Lu	0.0381	0.433	0.48	0.58
$\sum$REE	3.89	184.8	173	204
La_N/Yb_N	1.00	9.15	6.98	8.44
La_N/Sm_N	1.00	4.33	3.53	3.54
Eu/Eu*	1.00	0.65	0.70	0.70
Sc	8.64	16	14.9	–
Y	2.25	27	27	–

Data from McLennan, S. M. (1989) *Rare Earth Elements in Sedimentary Rocks*, Reviews in Mineralogy, vol. 36, p. 172, Table 2. Note the subscript N indicates chondrite-normalized values.

comes from a study of loess, that is derived from bedrock by glacial erosion during the Pleistocene. Neodymium isotopic model ages for loess give the age of extraction from the mantle. Samples from China, Europe, New Zealand and North America have depleted mantle model ages (T_{DM}) that range from 1060 to 1700 Myr. Thus the source material of the loess has been extracted from the mantle well back into the Proterozoic. However, the REE patterns from these samples of loess of widely varying ages (and locations) are similar, indicating uniformity of process [18]. Likewise the samples used to establish the PAAS REE patterns cover 1500 Myr in age but their REE patterns are identical (Fig. 11.3).

Other data for sedimentary rocks tell the same story. Thus sedimentary rocks that have neodymium model ages exceeding 2 Gyr all display the familiar PAAS upper-crustal REE patterns indicating the long continuity of the present crust-forming processes [19]. Thus the composition of the upper continental crust has been

Table 11.3 *The major element chemical composition (wt%) of the continental crust. Revisions from Taylor and McLennan (1985) are shown in underlined bold italic [30]*

Oxide	Upper continental crust	Bulk continental crust	Lower continental crust	Archean upper crust	Archean bulk crust
SiO_2	65.9	57.4	54.4	60.1	56.9
TiO_2	*0.65*	0.90	*0.99*	0.83	1.00
Al_2O_3	*15.2*	15.9	*16.1*	15.3	15.2
FeO	4.52	9.12	10.63	8.00	9.60
MgO	2.21	5.31	6.31	4.69	5.90
CaO	4.20	7.41	8.50	6.20	7.30
Na_2O	3.90	3.11	2.81	3.05	3.00
K_2O	3.36	*1.32*	*0.64*	1.81	*1.20*
Σ	**99.94**	**100.4**	**100.4**	**99.98**	**100.1**

Table 11.4 *The chemical composition of the continental crust. Revisions from Taylor and McLennan (1985) are shown in underlined bold italic [30]*

Element	Upper continental crust	Bulk continental crust	Lower continental crust	Archean upper crust	Archean bulk crust
Li (ppm)	20	13	11	–	–
Be (ppm)	3.0	1.5	1.0	–	–
B (ppm)	15	10	8.3	–	–
Na (wt%)	2.89	2.30	2.08	2.45	2.23
Mg (wt%)	1.33	3.20	3.80	2.83	3.56
Al (wt%)	8.04	8.41	8.52	8.10	8.04
Si (wt%)	30.8	26.8	25.4	28.1	26.6
P (ppm)	700	–	–	–	–
K (wt%)	2.80	*1.1*	*0.53*	1.5	*1.0*
Ca (wt%)	3.00	*5.29*	*6.07*	4.43	*5.22*
Sc (ppm)	*13.6*	30	*35*	14	30
Ti (ppm)	*0.41*	0.54	*0.58*	0.50	0.60
V (ppm)	*107*	230	*271*	195	245
Cr (ppm)	*83*	185	*219*	180	230
Mn (ppm)	*600*	1400	*1700*	1400	1500
Fe (wt%)	3.50	7.07	8.24	*6.22*	*7.46*
Co (ppm)	*17*	29	*33*	25	30
Ni (ppm)	*44*	128	*156*	105	130
Cu (ppm)	*25*	75	*90*	–	80
Zn (ppm)	71	80	83	–	–
Ga (ppm)	17	18	18	–	–
Ge (ppm)	1.6	1.6	1.6	–	–

Table 11.4 (*cont.*)

Element	Upper continental crust	Bulk continental crust	Lower continental crust	Archean upper crust	Archean bulk crust
As (ppb)	1.5	1.0	0.8	–	–
Se (ppm)	50	50	50	–	–
Rb (ppm)	112	*37*	*12*	50	28
Sr (ppm)	350	260	230	240	215
Y (ppm)	22	20	19	18	19
Zr (ppm)	190	100	70	125	100
Nb (ppm)	*12*	8.0	*6.7*	–	–
Mo (ppm)	1.5	1.0	0.8	–	–
Pd (ppb)	0.5	1	1	–	–
Ag (ppb)	50	80	90	–	–
Cd (ppb)	98	98	98	–	–
In (ppb)	50	50	50	–	–
Sn (ppm)	5.5	2.5	1.5	–	–
Sb (ppb)	0.2	0.2	0.2	–	–
Cs (ppm)	*4.6*	*1.5*	*0.47*	–	–
Ba (ppm)	550	250	150	390	300
La (ppm)	30	16	11	20	15
Ce (ppm)	64	33	23	42	31
Pr (ppm)	7.1	3.9	2.8	4.9	3.7
Nd (ppm)	26	16	12.7	20	16
Sm (ppm)	4.5	3.5	3.17	4.0	3.4
Eu (ppm)	0.88	1.1	1.17	1.2	1.1
Gd (ppm)	3.8	3.3	3.13	3.4	3.2
Tb (ppm)	0.64	0.60	0.59	0.57	0.59
Dy (ppm)	3.5	3.7	3.6	3.4	3.6
Ho (ppm)	0.80	0.78	0.77	0.74	0.77
Er (ppm)	2.3	2.2	2.2	2.1	2.2
Tm (ppm)	0.33	0.32	0.32	0.30	0.32
Yb (ppm)	2.2	2.2	2.2	2.0	2.2
Lu (ppm)	0.32	0.30	0.29	0.31	0.33
Hf (ppm)	5.8	3.0	2.1	3	3
Ta (ppm)	*1.0*	*0.8*	*0.73*	–	–
W (ppm)	2.0	1.0	0.7	–	–
Re (ppb)	0.4	0.4	0.4	–	–
Os (ppb)	0.05	0.05	0.05	–	–
Ir (ppb)	0.02	0.1	0.13	–	–
Au (ppb)	1.8	3.0	3.4	–	–
Tl (ppb)	750	360	230	–	–
Pb (ppm)	*17*	8.0	*5.0*	–	–
Bi (ppb)	127	60	38	–	–
Th (ppm)	10.7	*4.2*	*2.0*	5.7	*3.8*
U (ppm)	2.8	*1.1*	*0.53*	1.5	*1.0*

uniform and produced by similar geological processes for at least two billion years, since the Early Proterozoic.

11.3.1 Paradoxes: niobium and lead

Niobium forms a large cation and is a quintessential example of an incompatible element. So it might be expected to be enriched in the continental crust just as is uranium. Curiously the Nb/U ratio in the crust is only 7, compared to around 30 in the bulk Earth and to 40 in the depleted-mantle source of MORB. A frequently given answer to this "paradox" is that niobium (and tantalum) enter titanium minerals (or amphibole) and so are retained in these phases in the source regions of the arc lavas that erupt at subduction zones [20]. An alternative explanation is that it is the enrichment of other elements in arc lavas from fluids in the subduction zone that locally overwhelms the mantle signature in the overlying wedge and so causes an apparent depletion [21].

Lead presents us with the opposite problem. It is enriched in the continental crust with a complementary depletion in MORB. This occurs although the element is not particularly incompatible and has a strong tendency to form sulfide phases. Indeed uranium is notably more incompatible than lead but the $^{238}U/^{204}Pb$ ratio (or μ) is higher in MORB than in the crust or in the bulk Earth. This has been a long-term effect, as shown on a plot of $^{207}Pb/^{204}Pb$ versus $^{206}Pb/^{204}Pb$ (the well-known "lead evolution" diagram). Here the data for MORB are enriched in ^{206}Pb (derived from ^{238}U) and plot to the right of the isochron, although the opposite effect might have been predicted. This was long identified as the "lead paradox", although paradoxes arise from our imperfect understanding rather than from the operations of nature [22].

A more refined understanding of the geochemistry of lead has led to a consistent explanation of the paradox. The upper half of oceanic crust at the mid-ocean ridges is subject to massive alteration with seawater. Lead is leached and redeposited as sulfides and oxides near the seawater–crust interface. As the oceanic crust is subducted, heating and dehydration causes the sulfides to break down. Lead is more readily transported by hydrothermal fluids, compared to uranium, into the mantle wedge overlying the subduction zone. Here it is incorporated into the andesitic magmas and so into the crust. Thus the over-abundance of lead in the crust is a consequence of secondary processes caused by the geochemical peculiarities of its chalcophile character [23].

In summary, both these apparent paradoxes of niobium and lead abundances support the derivation of the continental crust from arc lavas produced in subduction zones. "The critical point is that fractionation of Nb/U and Ce/Pb appears to require the involvement of fluid ... and this is most likely to occur in the subduction zone" [24].

11.4 The lower crust (Post-Archean)

The Post-Archean upper crust is well understood. The problems in reaching some consensus become an order of magnitude more difficult when we attempt to estimate the composition of the lower crust of the continents. The problem has some parallels with the Archean. But the Archean crust possesses several advantages. It is relatively accessible: several percent of the Earth's surface is available for study; geological mapping is extensive as there are huge economic incentives to seek its rich mineral deposits; finally, intra-crustal differentiation processes became important only in the Late Archean. Even so its complex geological history has led to widely varying interpretations. Thus there were over 100 years of detailed investigations before even the Proterozoic Sudbury structure and its associated ore deposit were recognized as due to a very large impact (Chapter 10).

However, the lower crust lacks such benefits; uncertainty hangs over both its exposures and their interpretation [25]. It is mostly inaccessible. It is certainly as heterogeneous as the upper crust but unlike the upper crust, it lacks the possibility of comprehensive sampling to establish an overall composition except with geophysical tools. Thus our ideas about the composition of the lower crust are either model dependent, or rely on debatable interpretations of geophysics, scattered outcrops or xenoliths. To add further complications, some upper-crustal material has been dragged down to great depths and re-exhumed. Because of these uncertainties and the lack of anything even approaching a consensus, in this text we deal only briefly with these problems.

One of the few geophysical parameters that provides some broad overall average is Poisson's ratio, that is the ratio of seismic compressional to shear waves (V_p/V_s) in the lower crust. This indicates that the lower crust is significantly more basic than the upper crust [26]. The second geophysical constraint comes from heat-flow data that are consistent with low abundances of potassium, uranium and thorium below the enriched upper crust.

Granulite terrains seem to be at best debatable samples of the lower crust, although they are widely equated as representative examples [27]. Many such terrains appear to be upper-crustal material that has been buried in Himalayan-type collisions and so are irrelevant to the problem. For example Klaus Mezger [28] suggested that most regional granulite terrains represent the transitional region between upper and lower crust so that they represent mid crust rather than lower crust.

Among unequivocal samples that are available are xenoliths from the lower crust that occur in volcanic pipes and flows. These are mostly more basic than samples from granulite facies regions but are randomly and perhaps preferentially selected [29]. An undisputed fact in this problem is that the granitic rocks that dominate the upper crust are derived from melting from within the crust. However, although complementary positive Eu anomalies are common in granulite terrains and xenoliths [30] often these do not particularly resemble residual material from intra-crustal partial melting. Perhaps their Eu enrichment is inherited as we suspect happened with the Proterozoic anorthosites (next section). Thus another paradox exists in crustal geochemistry as intra-crustal partial melting must be a fundamental process governing the composition and chemical structure of the lower continental crust [31].

Here however we are less concerned with the petrological uncertainties of samples from this mostly inaccessible region and in the spirit of William of Ockham, consider the following factors:

   i. The upper crust is derived from the lower crust by intra-crustal melting as shown both experimentally and by the depletion in europium.
  ii. The upper crust is enriched in K, U, Th and many incompatible elements.
 iii. Heat-flow data demand a lower concentration of heat-producing elements with depth in the crust.
  iv. Poisson's ratio indicates a basic composition for the lower crust.
   v. Heat-flow and geological data assessed in Chapter 12 suggest an andesitic bulk crustal composition.
  vi. Accordingly we derive a lower crustal composition by subtracting the well-established upper-crustal composition from that of the bulk crust. The assumption here that the upper crust is derived by intra-crustal melting from the lower crust is supported by the experimental and geochemical data discussed above. We make no apologies for adopting this simple model approach. The alternatives involve questionable extrapolation from an uncertain database. Other solutions that have been proposed for this problem, such as delamination, are discussed in the next chapter.

11.4.1 Anorthosites

Anorthosite massifs form a significant fraction of the Mid-Proterozoic crust, covering, for example, twenty percent of the surface area of the Grenville Province in Canada. Accordingly they deserve comment here, in order to contrast them with their relatives that form the lunar highland crust. The terrestrial examples occur as slab- or sheet-like bodies 2 to 6 km thick and are dominated by an intermediate variety of plagioclase complete with an enrichment in europium [32]. A long-standing debate has occurred over whether these massive bodies of anorthosite were derived as plagioclase cumulates, from the crystallization of basaltic magmas derived from the mantle, or whether they are formed within the lower crust [33].

Musacchio and Mooney [34] consider that the massive anorthosites characteristic of Mid-Proterozoic time were derived from the mantle, a notion based on seismic evidence. This seems a peculiarly blunt tool to use to resolve a complex petrological and geochemical problem and the seismic data seem less decisive to us than the isotopic data.

A crucial observation in our opinion is that the Rb/Sr ratios are very low in the anorthosites, although their $^{87}Sr/^{86}Sr$ ratios are high. This indicates that the parent material of the anorthosites has previously resided in a high Rb/Sr environment (see below). However, the mantle has a low Rb/Sr ratio (and high-Ca feldspar) and so seems to be an unsuitable source. Neither is plagioclase of intermediate composition (around An_{50}) a product expected in mantle-derived magmas.

Multistage processes are indicated. In our model of crustal growth and evolution, the basic lower crust evolved by intra-crustal melting beginning in the Late Archean. Prior to that, intra-crustal differentiation was minimal so that the Rb/Sr ratios were high resulting in enhanced $^{87}Sr/^{86}Sr$ ratios. Intra-crustal melting then removed rubidium in granitic melts into the upper crust. The residue in the lower crust was thus plagioclase-rich material with low Rb/Sr ratios, but with inherited europium enrichment and high $^{87}Sr/^{86}Sr$ ratios. Subsequent melting of this material, perhaps induced by rising basaltic plumes in a dry environment, resulted in the generation of melts of plagioclase (An_{50}) with inherited high $^{87}Sr/^{86}Sr$ ratios that intruded the upper crust throughout the Mid Proterozoic. This model accounts for the restricted occurrence in time, the peculiar mineralogy, the strontium isotopic systematics and the chemistry of the Proterozoic massive anorthosites [35].

Thus there are similarities and differences between terrestrial and lunar anorthosites (Chapter 2). Both show enrichments in europium, albeit for different reasons. In the terrestrial case, this is inherited in part from a europium-rich lower crust, while the enrichment of europium (and strontium) in the lunar crust is a consequence of primary crystallization from the magma ocean. Their composition is also distinct (An_{50} versus An_{95}). Both are restricted in time, again for different reasons and reinforce the geological dictum that similar-looking rocks may form from independent causes.

Synopsis

Profound changes in crustal evolution occurred during the Late Archean and the Early Proterozoic, leading to a significant increase in the volume of the continental crust, while intra-crustal melting that produced a K-rich upper crust, became important for the first time. Large stable cratons appeared. These changes are recorded in the sedimentary rocks, whose K/Na and Th/Sc ratios and REE patterns with a significant depletion in europium, continued to characterize the upper crust throughout the Proterozoic and Phanerozoic. Enrichment of the upper crust in incompatible elements such as rubidium, that resulted in enhanced $^{87}Sr/^{86}Sr$ ratios in sediments, testifies to the magnitude of the change.

The composition of the upper crust is well established by widespread sampling and more efficiently by the analysis of clastic sediments. The processes responsible for producing the upper crust have been operating since the Early Proterozoic, as shown by the uniformity of the REE patterns and Sm–Nd isotopic systematics. The depletion of the incompatible element niobium in the crust is more apparent than real, while the enrichment of lead is due to hydrothermal processes associated with subduction-zone volcanism.

The lower crustal composition remains controversial. Difficult of access, the interpretation of all samples remains debatable and the composition is model dependent. We rely on heat-flow and bulk seismic data that indicate a much more basic composition than the upper crust. As the upper crust is derived by partial melting from within the crust, we interpret the lower crust as residual from the bulk crustal andesitic composition. Delamination may occur on local scales but seems an unlikely universal mechanism both to balance crustal growth and recycling and to account for the uniformity of crustal composition through Post-Archean time.

Anorthosites, that differ in composition from their lunar counterparts, form a significant component of the Mid-Proterozoic crust and are here interpreted as arising from lower crustal melting following the extraction of the upper crust. The appendices provide basic information about the area, volume, age of the present continental crust and the use of sedimentary rocks to establish crustal composition, an approach that we favor.

Appendices

Area, thickness and density of the present continental crust

The continental crust presently occupies 41.2% of the surface area of the Earth or 2.10×10^8 km^2; 71.3% or 1.50×10^8 km^2 lies above sea level. There are at least ten major continental blocks and four submerged microcontinents. The mean elevation of the continental crust above present sea level is only 125 m, but the elevation of the area above the shelf/slope break (the 200 m isobath) is 690 m. Crustal thickness varies between 10 and 80 km from an average of 41 km [36]. The thickness correlates with the size of the continental block and the age of the last tectonic event. The crustal volume is 7.35×10^9 km^3. This estimate includes the volumes of the submerged continental masses but depends on the value adopted for crustal thickness and has an error of ±5%.

Archean terrains tend to have thinner crust, with crustal thicknesses between 27 and 40 km while Proterozoic crust is about 40–55 km thick, with higher P-wave velocities (up to 7 km/s) at the base [37]. This does not seem to be due to greater erosion of the earlier terrains.

The major seismic discontinuity at the Mohorovicic Discontinuity or Moho, where compressional wave velocities (V_p) rise from about 7 to about 8 km/s is conventionally taken as the base of the crust. However, the Moho is sometimes not sharp and may be absent. The crust can be defined as material with $V_p < 7.8$ km/s or $V_s < 4.3$ km/s. These seismic velocities are consistent with densities less than 3.1 g/cm^3. Probable interlayering of mantle material, underplating by basalt and the presence of sub-lithospheric mantle keels are additional complications for the base of the crust.

The discontinuous Conrad Discontinuity sometimes separates upper and lower crust at a depth of 10–20 km, although super-deep drill holes rarely identify discontinuities based on interpretations of the geophysical data.

Estimates of the density of the crust range from 2.7 to 2.9 g/cm^3. The great geological complexity of the crust and the uncertain nature of the lower crust make this calculation difficult. The mass of the continental crust is 2.06×10^{25} g ($\pm 7\%$) for an average density of 2.8 g/cm^3. The continental crust forms 0.54% of the mass of the crust–mantle system and only 0.35% of the mass of the whole Earth from this calculation.

Age of the continental crust

One of the great contrasts between the oceanic and continental crust is the great antiquity of the latter. However, the mean age of the continents, although important for geochemical and geophysical models, is difficult to estimate. The most reliable estimates are based on the Sm–Nd isotope system as this seems to be among the least prone of the several radiogenic systems to resetting during metamorphism. Its most important characteristic is that it dates the separation of Sm from Nd at the time of formation of the crust from the mantle. The Nd model age of crustal igneous rocks is thus thought to reflect the time of their derivation from the mantle. Those of sedimentary rocks in general give the average time of formation of the various sources of the sediments rather than the age of deposition. This assumes that Sm and Nd are not fractionated during sedimentation or diagenesis.

This simple view is complicated by a number of processes that may later change (usually decrease) the Sm/Nd ratio or underestimate the importance of older crustal rocks. Various processes such as intra-crustal melting, metamorphic resetting and assimilation of older material tend to decrease the Sm/Nd ratio and so complicate this simple picture. Although intra-crustal melting events lead to further fractionation of Sm and Nd, this generally follows crust formation within 100–200 Myr and so has a comparatively minor effect on dating processes that operate over much of the history of the Earth.

Estimates of the average crustal age based on the Nd isotope system thus usually represent a minimum [38], so the average age of continental crust, based on the Sm/Nd system, is about 2.0 Gyr [39]. An age of about 2.4 Gyr for the continental crust is obtained if about 60% of the crust was in place by 2.7 Gyr, a model that we favor here.

Sedimentary rocks as crustal samples

The upper continental crust is heterogeneous in detail with changes in composition occurring sometimes on scales of meters while geological maps display complexity at all scales. How can this bewildering mixture of rock types be sampled to yield a bulk composition? Although this has led to a perception that solving this problem is very difficult or perhaps impossible, the processes of erosion and sedimentation have simplified this task. The history of the crust is to a major extent preserved in the record in the sedimentary rocks. Just as the fossils entombed in the rocks have enabled us to understand the evolution of life and to establish the sequence of geological periods, so the sediments themselves also record the chemical and isotopic evolution of their crustal sources.

The first suggestion that sedimentary processes might homogenize the diversity of compositions observed in the crust seems to be due to Goldschmidt [40]. His prediction has proven robust with respect to fine-grained clastic sedimentary rock. Elements such as the REE are very insoluble and thus have a short residence time in the oceans. They are thus transferred rapidly from the upper crust exposed to weathering into the sedimentary record. This is shown clearly by the uniform REE patterns in sedimentary rocks in the Post-Archean crust (Table 11.2).

Many other less-soluble elements such as Y, Sc and Th are little affected by the processes of erosion and sedimentation so that their abundances in shales and sandstones reflect those of their source. Extremely insoluble elements such as Zr, Hf and Sn are concentrated in minerals such as zircon and cassiterite in sandstones. Anomalous REE patterns with cerium anomalies or severe LREE depletion may form in unusual sedimentary environments but these effects are rare [41].

In sedimentary rocks, REE occur in the clay and silt fraction as well as in accessory minerals such as zircon, monazite and apatite. There is no correlation with individual clay minerals in which some fraction of the REE may be present as micro-inclusions of trace-element-rich accessory minerals such as apatite. Lower concentrations of REE occur in the sand fractions. Carbonates and coarse-grained sedimentary rocks typically have REE patterns essentially parallel (Eu/Eu^*, La_N/Yb_N) to those of shales, but with lower total abundances than shales.

Fractionation of heavy minerals such as zircon and monazite during sedimentary transport can affect REE patterns particularly in clastic sediments (e.g. quartzites) that have low concentrations of the REE. Thus the presence of monazite in Archean metaquartzites from the Western Gneiss Terrain, Australia causes an enrichment in light REE. Mostly such minerals are rarely concentrated in amounts to cause significant or even visible effects on the REE patterns. Thus the resistant mineral zircon, typically enriched in heavy REE, affects the bulk rock patterns only when zircon constitutes more than about 0.06% (or Zr abundances exceed about 300 ppm) Thus 100 ppm of Zr present as zircon, enriched in HREE, adds only 0.25 times chondritic levels of Yb to patterns that typically contain 10–15 times chondritic levels.

Diagenesis may also affect the REE patterns. However, newly formed diagenetic minerals mostly just redistribute the REE without significant transport. During erosion and sedimentation, Ce^{3+} may be oxidized to Ce^{4+}. It then separates as an insoluble phosphate in seawater. The depletion or enrichment of Eu from the other REE, due to its reduction to Eu^{2+}, does not occur in surficial environments, where Eu is present as the trivalent ion. Thus the enrichment or depletion of Eu records a previous signature of crystal fractionation and removal of Eu by plagioclase.

Finally it should be noted that the sedimentary record is intrinsically cannibalistic and, on average, 70% of most sedimentary rocks are derived from the erosion of pre-existing sedimentary rocks. This recycling has a strong buffering effect on the ability of the sedimentary record to record changes in upper-crustal composition and only major changes are likely to be observed [42].

Pleistocene loess covers about 10% of the land surface and provides yet another sampling of the exposed upper crust. Loess originates by wind transport from glacial outwash, mostly during cold dry climatic intervals. Silt-sized rock flour is produced by glacial erosion and so samples essentially unweathered upper-crustal material. Uniform REE patterns, similar to PAAS, ES and NASC, although sometimes diluted by carbonate or quartz, are seen in samples of loess from China, Europe, New Zealand and North America [43]. Thus these widely scattered loess deposits provide another sample of the composition of the upper crust. Their derivation from unweathered crust demonstrates that any effects of sedimentary processes during sedimentation for the insoluble elements are overshadowed by upper-crustal provenance.

Notes and references

1. Sorby, H. C. (1908) On the application of quantitative methods to the structure and history of rocks. *Quart. J. Geol. Soc.* **64**, 190.
2. See Frarey, M. J. and Roscoe, S. M. (1970) *Geol. Surv. Can. Paper* **70–40**, 143.
3. It is interesting that the book on this topic by Rogers, J. J. W. and Santosh, M. (2004) *Continents and Supercontinents*, Oxford University Press, deals almost exclusively with Post-Archean time in line with our perception of continental evolution. The complexities involved in continental reconstructions are well summarized by Windley, B. F. *et al.* (2007) Tectonic models for accretion of the Central Asian orogenic belt. *J. Geol. Soc. London* **164**, 31–47. See also Cloud, P. (1988) *Oasis in Space: Earth History from the Beginning*, W.W. Norton, for an excellent stratigraphic account; Pollack, H. N. (1986) Cratonization and thermal evolution of the mantle. *EPSL* **80**, 175–82; Wyllie, P. J. (1977) Crustal anatexis: An experimental view. *Tectonophysics* **43**, 41–71; Clemens, J. D.

(2005) Melting of the continental crust: Fluid regimes, melting reactions and source rock fertility, in *Evolution and Differentiation of the Continental Crust* (eds. M. Brown and T. Rushmer), Cambridge University Press, pp. 296–330.

4. See review in Veizer, J. and Mackenzie, F. T. (2003) Evolution of sedimentary rocks, in *Treatise on Geochemistry* (eds. H. D. Holland and K. K. Turekian), Elsevier, vol. 7, pp. 367–407. The study of secular changes in sediment composition has a long history, dating back to Daly, R. A. (1909) First calcareous fossils and evolution of limestones. *GSA Bull.* **86**, 1085–8. Interpretations fell into two basic categories. Ronov and co-workers in the USSR, in line with the classic studies of Goldschmidt, suggested that sedimentary rocks provided a sampling of exposed crust and largely tracked crustal evolution. Garrels and Mackenzie noted that secular changes in sediment composition mimicked changes observed during burial diagenesis and proposed that cannibalistic sedimentary recycling, with successive cycles of burial and diagenesis, imposed long-term chemical changes. Garrels, R. M. and Mackenzie, F. T. (1971) *Evolution of Sedimentary Rocks*, Norton. Engel, A. *et al.* (1974) Crustal evolution and global tectonics: A petrogenic view. *GSA Bull.* **85**, 843–58 were the first clearly to document that rather than gradual changes, there was an abrupt increase in K_2O/Na_2O ratios of sedimentary rocks and continental basement at the Archean–Proterozoic transition. At about the same time, abrupt changes in a variety of sedimentary trace-element (notably REE) and isotopic compositions were also correlated with the Archean–Proterozoic transition. See review in Taylor, S. R. and McLennan, S. M. (1985) *The Continental Crust: Its Composition and Evolution*, Blackwell. Although suggestions that such changes either do not exist or are related to sampling biases continue to appear, it is now well established that these changes at the Archean–Proterozoic transition result from the chemical evolution of the Late Archean crust.

5. The depletion in europium relative to neighboring samarium and gadolinium is commonly referred to as a "negative Eu anomaly" while a reciprocal enrichment in Eu is referred to as a "positive Eu anomaly". The REE patterns in clastic sedimentary rocks show rapid changes from Archean to Post-Archean patterns during the interval from 3.2 to 2.5 Gyr in South Africa. See McLennan, S. M. *et al.* (1983) Geochemical evolution of Archean shales from South Africa. *Precambrian Res.* **22**, 93–124. Similar changes from Archean to Post-Archean REE patterns are recorded from the Early Proterozoic successions of the Pine Creek Geosyncline and Hamersley Basin, Australia. Taylor, S. R. *et al.* (1983) Geochemistry of Early Proterozoic sedimentary rocks and the Archean/Proterozoic boundary. *GSA Memoir* **161**, 119–31. In a particularly striking example, the REE patterns in the early Proterozoic (2.5–2.2 Gyr) Huronian sedimentary succession of Canada change from Archean-like REE patterns at the base to Post-Archean patterns at the top of the sequence, as shown by McLennan, S. M. *et al.* (1979) Rare earth elements in Huronian (Lower Proterozoic) sedimentary rocks. *GCA* **43**, 375–88. This trend is also shown by the Sm–Nd isotopic system: McLennan, S. M. *et al.* (2000) Nd and Pb isotopic evidence for provenance and post-depositional alteration of the Paleoproterozoic Huronian Supergroup, Canada. *Precambrian Res.* **102**, 263–78.

6. e.g. Veizer, J. (1983) Trace elements and isotopes in sedimentary carbonates. *Rev. Mineral.* **11**, 265–99. The element rubidium has a nearly identical geochemical behavior to potassium on account of its similarity in ionic radius and valency. Thus no independent rubidium minerals are known, as rubidium is so mildly fractionated from potassium that its concentration never reaches the point at which a separate phase could be precipitated. This behavior, the "camouflage" of rubidium by potassium, contrasts with the much less abundant, but larger alkali element ion, cesium, that commonly is concentrated in late-

stage differentiates such as pegmatites to a sufficient degree that independent cesium minerals form.

7. Scandium, although trivalent and a member of the same group (3) of the periodic table as the REE, is a much smaller ion and enters early pyroxenes during magmatic crystallization. In contrast, tetravalent thorium is a large incompatible ion that is concentrated in residual melts in igneous processes. For these reasons, thorium is concentrated in granitic rocks, and scandium is concentrated in basic rocks. As both elements are transported into sediments with little loss to seawater, the Th/Sc ratio observed in fine-grained sedimentary rocks provides a ratio of the relative proportions of the granitic and basic parents in their source rocks. McLennan, S. M. and Hemming, S. (1992) Samarium/neodymium elemental and isotopic systematics in sedimentary rocks. *GCA* **56**, 887–98. For a number of trends (e.g. Sm/Nd, Th/Sc, K_2O/Na_2O) there is an apparent "reversal" during the Mesozoic and Tertiary. Such "reversals" appear to be statistically significant but rather than reflecting long-term chemical evolution of the upper continental crust instead may be a transient feature due to first-cycle sediments sampling undifferentiated crustal additions. Veizer, J. and Mackenzie, F. T. (2003) Evolution of sedimentary rocks, in *Treatise on Geochemistry* (eds. H. D. Holland and K. K. Turekian), Elsevier, vol. 7, pp. 367–407.

8. Hemming, S. and McLennan, S. M. (2001) Pb isotopic composition of modern deep-sea turbidites. *EPSL* **184**, 489–503. There is an increase in Th/U ratios in fine-grained sedimentary rocks over geological time, related to a decrease in uranium, notably for samples of Post-Archean age. This in turn is due to progressive loss of uranium from the fine-grained sedimentary record associated with repeated recycling of sedimentary material resulting in the ultimate loss of uranium to the mantle during hydrothermal alteration of ocean crust, as well as to continental ore deposits. Collerson, K. D. and Kamber, B. S. (1999) Evolution of the continents and the atmosphere inferred from Th–U–Nb systematics of the depleted mantle. *Science* **283**, 1519–22.

9. This is consistent with the massive increase in continental area beginning in the Late Archean that exposes broad expanses of cratonic sediments to weathering by the terrestrial atmosphere. Valley, J. W. *et al.* (2005) 4.4 billion years of crustal maturation: Oxygen isotope ratios of magmatic zircon. *Contrib. Min. Petrol.* **150**, 561–80. We note that the $\partial^{18}O$ variation through time parallels our crustal growth curve (the $\partial^{18}O$ value for the mantle is 5.5‰). The Archean–Proterozoic transition has also been linked to the rise in atmospheric oxygen at that time. Kump, L. H. and Barley, M. E. (2007) Increased subaerial volcanism and the rise of atmospheric oxygen 2.5 billion years ago. *Nature* **448**, 1033–6.

10. e.g. Unrug, R. (1992) The supercontinent cycle and Gondwanaland assembly. *J. Geodynam.* **16**, 215–40; Rogers, J. J. W. and Santosh, M. (2004) *Continents and Supercontinents*, Oxford University Press.

11. See also Cloud, P. (1988) *Oasis in Space: Earth History from the Beginning*, W.W. Norton, for much stratigraphic information. Although the interpretations of the changes at the Archean–Proterozoic transition appear to be generally agreed upon, there has been debate over changes in sedimentary REE patterns across this boundary. The more recent papers include Gao, S. and Wedepohl, K. H. (1995) The negative Eu anomaly in Archean sedimentary rock – Implications for decomposition, age and importance of their granitic sources. *EPSL* **133**, 81–94 and Cox, R. and Lowe, D. R. (1995) A conceptual review of regional scale controls on the composition of clastic sediment and the coevolution of continental blocks and their sedimentary cover. *J. Sed. Res.* **65**, 1–12. These concerns fail on several accounts. Early studies of Archean sedimentary rocks indeed relied almost exclusively on samples from relatively

low-grade greenstone belts but it is now understood that Archean greenstone belts represent a variety of tectonic settings. Although modern sediments found in purely oceanic island arcs may lack Eu anomalies and have other features superficially similar to Archean greenstone sediments, most tectonically active settings contain Post-Archean-type (PAAS) REE patterns. Where stabilized Archean crust existed, the associated sediments may have Post-Archean-type REE patterns with depletions in europium. However, these terrains appear to make up a relatively small proportion of the preserved Archean crust. Cratonic terrains are preserved and volcanically active tectonic terrains are preferentially lost over geological time due to intra-crustal and mantle–crust recycling processes. This is the opposite of that required to explain the difference by preservation bias. Accordingly the distinctions in Eu anomalies and REE patterns reflect the differences in average upper crustal REE patterns for the Archean and Post Archean, e.g. Taylor, S. R. *et al.* (1986) Rare earth elements in Archean high-grade metasediments and their tectonic significance. *GCA* **50**, 2267–79; Taylor, S. R. and McLennan, S. M. (1995) The geochemical evolution of the continental crust. *Rev. Geophys.* **33**, 241–65; Taylor, S. R. and McLennan, S. M. (1985) *The Continental Crust: Its Composition and Evolution*, Blackwell; McLennan, S. M. *et al.* (2003) The role of provenance and sedimentary processes in the geochemistry of sedimentary rocks, in *Geochemistry of Sediments and Sedimentary Rocks* (ed. D. R. Lentz), Geological Association of Canada, Geotext 5, pp. 1–31; Veizer, J. and Mackenzie, F. T. (2003) Evolution of sedimentary rocks, in *Treatise on Geochemistry* (eds. H. D. Holland and K. K. Turekian), Elsevier, vol. 7, pp. 367–407.

12. Shaw, D. M. *et al.* (1986) Composition of the Canadian Precambrian shield and the continental crust of the Earth. *Geol. Soc. Canada Spec. Pub.* **24**, 275–82; Eade, K. E. and Fahrig, W. F. (1973) Regional, lithological and temporal variation in the abundance of some trace elements in the Canadian shield. *Geol. Surv. Canada Paper* **72–46**. A USSR program involved 40 780 samples: Ronov, A. B. *et al.* (1992) General trends in the evolution of the chemical composition of sedimentary and magmatic rocks of the continental Earth crust. *Sov. Sci. Rev. Sect. G. Geology* **1**, 1–3. Other examples include Wedepohl, K. H. (1991) Chemical composition and fractionation of the continental crust. *Geol. Rundsch.* **80**, 207–23 and Condie, K. C. (1992) Chemical composition and evolution of the upper continental crust: Contrasting results from surface samples and shales. *Chem. Geol.* **104**, 1–37.

13. e.g. Gao, S. *et al.* (1998) Chemical composition of the continental crust as revealed by studies in East China. *GCA* **62**, 1959–75.

14. See Rudnick, R. L. and Gao, S. (2003) Composition of the continental crust, in *Treatise on Geochemistry* (eds. H. D. Holland and K. K. Turekian), Elsevier, vol. 3, pp. 1–64 for earlier references.

15. This is borne out by the excellent record of crustal composition that resides in the sedimentary record. McLennan, S. M. (2001) Relationship between the trace element composition of sedimentary rocks and upper continental crust. *Geochem. Geophys. Geosystems* **2**, doi: 10.1029/2000GC000109; McLennan, S. M. (1989) Rare earth elements in sedimentary rocks: Influence of provenance and sedimentary processes, in *Geochemistry and Mineralogy of Rare Earth Elements* (eds. B. R. Lipin and G. A. McKay), Reviews in Mineralogy 21, pp. 169–200; Taylor, S. R. and McLennan, S. M. (1995) The geochemical evolution of the continental crust. *Rev. Geophys.* **33**, 241–65; Taylor, S. R. and McLennan, S. M. (1985) *The Continental Crust: Its Composition and Evolution*, Blackwell.

16. The REE patterns in the Post-Archean crust are especially instructive. Relative to chondritic patterns, these patterns have flat heavy REE abundances at about 10 times

chondritic, a smooth light REE enrichment (La = 100 × chondritic) and a uniform depletion in Eu (Eu/Eu* = 0.65 ± 0.05) (Fig. 11.3, Table 11.2). The REE patterns of chondritic meteorites are considered to be representative of the primordial solar nebula and, as refractory elements, parallel to that of the whole Earth (see Chapter 1). Except for first-cycle sediments, REE patterns of modern sediments are close to those of Post-Archean shales. The particulate matter in major rivers is likewise similar to PAAS except that due to its finer grain size, it is slightly enriched in total REE. Goldstein, S. J. and Jacobsen, S. B. (1988) Rare earth elements in river waters. *EPSL* **89**, 35–47. Even in modern volcanic provenances, such as in sediments derived from island-arc volcanoes or in the Pacific sea-floor sediments, REE patterns display the europium depletion characteristic of the upper crust. e.g. Lin, P. -N. (1992) Trace element and isotopic characteristics of western Pacific pelagic sediments. *GCA* **56**, 1641–54.

17. See McLennan, S. M. (2001) [15]. Each average is based on a relatively large number of samples for a variety of clastic lithologies (shales, loess, sandstones, tillites) from a variety of settings (active margins, passive margins, cratonic, deep marine).
18. Taylor, S. R. *et al.* (1983) Geochemistry of loess, continental crustal composition and crustal model ages. *GCA* **47**, 1897–905.
19. McLennan, S. M. and Hemming, S. R. (1992) Samarium/neodymium elemental and isotopic systematics in sedimentary rocks. *GCA* **56**, 887–98.
20. Hofmann, A. W. *et al.* (1986) Nb and Pb in oceanic basalts: New constraints on mantle evolution. *EPSL* **79**, 33–45; Ionov, D. A. and Hofmann, A. W. (1995) Nb–Ta-rich mantle amphiboles and micas: Implications for subduction-related metasomatic trace element fractionations. *EPSL* **131**, 341–56. Kamber, B. S. *et al.* (2003) A refined solution to Earth's hidden niobium. *Precambrian Res.* **126**, 289–308. Hofmann, A. W. (2004) Sampling mantle heterogeneity through oceanic basalts: Isotopes and trace elements, in *Treatise on Geochemistry* (eds. H. D. Holland and K. K. Turekian), Elsevier, vol. 2, pp. 61–101.
21. Pearce, J. A. *et al.* (2006) Geochemical mapping of the Mariana arc-basin system: Implications for the nature and distribution of subduction components. *Geochem. Geophys. Geosystems* **6**, doi: 10.1029/2004GC000895; Arculus, R. J. (1994) Aspects of magma genesis in arcs. *Lithos* **33**, 189–208.
22. Allègre, C. L. (1969) Comportement des systèmes U-Th-Pb dans le manteau supérieur et modèle d'évolution de ce dernier au cours des temps géologiques. *EPSL* **5**, 261–9. Numerous explanations have appeared, one of the most popular being that lead was secreted in the core that was supposed to grow slowly through geological time. Vidal, P. and Dosso, L. (1978) Core formation: Catastrophic or continuous? – Sr and Pb isotopic geochemistry constraints. *GRL* **5**, 169–72; Allègre, C. J. *et al.* (1982) Chemical aspects of the formation of the core. *Phil. Trans. Royal Soc. London* **306**, 49–59. These notions have not survived the current recognition that core formation is effectively coincident with the accretion of the planet (Chapter 1).
23. Newsom, H. E. *et al.* (1986) Siderophile and chalcophile element abundances in oceanic basalts, Pb isotope evolution and the growth of the Earth's core. *EPSL* **80**, 299–313; Chauvel, C. *et al.* (1995) Hydration and dehydration of oceanic crust controls Pb evolution of the mantle. *Chem. Geol.* **126**, 65–75.
24. Davidson, J. P. and Arculus, R. J. (2005) The significance of Phanerozoic arc magmatism in generating continental crust, in *Evolution and Differentiation of the Continental Crust* (eds. M. Brown and T. Rushmer), Cambridge University Press, p. 139.
25. The most thorough study of the lower crust was by Rudnick, R. L. and Fountain, D. M. (1995) Nature and composition of the continental crust: A lower crustal perspective.

Rev. Geophys. **33**, 267–309. Also see an extended earlier discussion in Taylor, S. R. and McLennan, S. M. (1985) *The Continental Crust: Its Composition and Evolution*, Blackwell, Oxford, ch. 4.

26. Zandt, G. and Ammon, C. J. (1995) Continental crust composition constrained by measurements of crustal Poisson's ratio. *Nature* **374**, 152–4; Musacchio, G. and Mooney, W. D. (2002) Seismic evidence for a mantle source for mid-Proterozoic anorthosites and implications for models of crustal growth, in *The Early Earth: Physical, Chemical and Biological Development* (eds. C. M. R. Fowler *et al.*), Geological Society of London Special Publication 199, pp. 125–34.

27. e.g. Ellis, D. J. (1987) Origin and evolution of granulites in normal and thickened crust. *Geology* **15**, 167–70; Bohlen, S. R. and Mezger, K. (1989) Origin of granulite terrains and the formation of the lowermost continental crust. *Science* **244**, 326–9.

28. Mezger, K. (1992) Temporal evolution of regional granulitic terrain: implications for the formation of lowermost continental crust, in *Continental Lower Crust* (eds. D. M. Fountain *et al.*), Elsevier, pp. 447–74. An excellent example of the complexity involved in studying the lower crust appears in this example from two workers of wide experience in Archean terrains. Rollinson, H. R. and Tarney, J. (2005) Adakites – the key to understanding LILE depletion in granulites. *Lithos* **79**, 61–81. They propose that the depletion of incompatible elements in granulites, commonly attributed to some signature of partial melting within the crust, is instead inherited from the crustal forming material. A basic assumption that runs through their paper is that granulites represent lower crustal samples but they conclude that there are "granulites and granulites" and that "there is a great variation in the composition of granulite protoliths (so) there are many stages of melt extraction preserved". They propose a complex model but admit that "why LILE-depleted melts should be restricted to the lower crust and preserved as granulites is something not yet fully resolved". Here we attribute any depletion in incompatible elements in the lower crust to intra-crustal melting processes.

29. Rudnick, R. L. and Taylor, S. R. (1987) The composition and petrogenesis of the lower crust: a xenolith study. *JGR* **92**, 13,981–4,005; Rudnick, R. L. (1992) Restites, Eu anomalies and the lower continental crust. *GCA* **56**, 963–70.

30. See e.g. Taylor, S. R. and McLennan, S. M. (1985) *The Continental Crust: Its Composition and Evolution*, Blackwell, ch. 4.

31. Wyllie, P. J. (1977) Crustal anatexis: An experimental view. *Tectonophysics* **43**, 41–71; Clemens, J. D. (2005) Melting of the continental crust: Fluid regimes, melting reactions and source rock fertility, in *Evolution and Differentiation of the Continental Crust* (eds. M. Brown and T. Rushmer), Cambridge University Press, pp. 296–330 in which he concludes that intra-crustal melting of rocks of andesitic composition can yield "10 to 50 volume percent of H_2O-undersaturated melt at attainable crustal temperatures" (p. 322) although basaltic underplating is mostly required as a heat source. See also Rudnick, R. L. (1992) Restites, Eu anomalies and the lower continental crust. *GCA* **56**, 963–70. Many residual crystal cumulates have positive Eu-anomalies. Other alternatives include the possibility that intra-crustal melting takes place under wet conditions at low temperatures so that the residual materials (restites) are present in amphibolite rather than granulite terranes.

32. Typically An_{50} plagioclase or andesine–labradorite.

33. Ashwal, L. (1993) *Anorthosites*, Springer-Verlag.

34. Musacchio, G. and Mooney, W. D. (2002) Seismic evidence for a mantle source for mid-Proterozoic anorthosites and implications for models of crustal growth, in *The Early Earth: Physical, Chemical and Biological Development* (eds. C. M. R. Fowler *et al.*), Geological Society of London Special Publication 199, pp. 125–34.

35. Taylor, S. R. *et al.* (1984) A lower crustal origin for massif-type anorthosites. *Nature* **311**, 372–4; Duchesne, J. C. *et al.* (1999) The crustal tongue melting model and the origin of massive anorthosites. *Terra Nova* **11**, 100–5.

36. Christensen, N. I. and Mooney, W. D. (1995) Seismic velocity structure and composition of the continental crust: A global view. *JGR* **100**, 9761–88.

37. Durrheim, R. J. and Mooney, W. D. (1994) Evolution of the Precambrian lithosphere: Seismological and geochemical constraints. *JGR* **99**, 15,359–74.

38. McLennan, S. M. (1988) Recycling of the continental crust. *Pure Appl. Geophys.* **128**, 683–724.

39. e.g. DePaolo, D. J. *et al.* (1991) The continental crustal age distribution. *JGR* **96**, 2071–88.

40. Goldschmidt, V. M. (1938) Geochemische verteilungsgesetze der elemente IX. *Skr. Nor. Vidensk. Akad. Kl. 1 Mat. Naturvidensk Kl.* **4**, 1–148. See the following for details of this approach to crustal sampling: McLennan, S. M. (2001) Relationship between the trace element composition of sedimentary rocks and upper continental crust. *Geochem. Geophys. Geosystems* **2**, doi: 10.1029/2000GC000109; McLennan, S. M. (1989) Rare earth elements in sedimentary rocks: Influence of provenance and sedimentary processes, in *Geochemistry and Mineralogy of Rare Earth Elements* (eds. B. R. Lipin and G. A. McKay), Reviews in Mineralogy 21, pp. 169–200; Taylor, S. R. and McLennan, S. M. (1995) The geochemical evolution of the continental crust. *Rev. Geophys.* **33**, 241–65; Taylor, S. R. and McLennan, S. M. (1985) *The Continental Crust: Its Composition and Evolution*, Blackwell.

41. Reviews of the influence of various sedimentary processes, including transport, weathering, diagenesis and recycling, on the chemical and isotopic composition of sedimentary rocks can be found in Veizer, J. and Mackenzie, F. T. (2003) Evolution of sedimentary rocks, in *Treatise on Geochemistry* (eds. H. D. Holland and K. K. Turekian), Elsevier, vol. 7, pp. 367–407; McLennan, S. M. *et al.* (2003) The roles of provenance and sedimentary processes in the geochemistry of sedimentary rocks, in *Geochemistry of Sediments and Sedimentary Rocks* (ed. D. R. Lentz), Geological Association of Canada, Geotext 5, pp. 7–38.

42. McLennan, S. M. (2001) Relationship between the trace element composition of sedimentary rocks and upper continental crust. *Geochem. Geophys. Geosystems* **2**, doi: 10.1029/2000GC000109; McLennan, S. M. (1989) Rare earth elements in sedimentary rocks: Influence of provenance and sedimentary processes, in *Geochemistry and Mineralogy of Rare Earth Elements* (eds. B. R. Lipin and G. A. McKay), Reviews in Mineralogy 21, pp. 169–200. Veizer, J. and Mackenzie, F. T. (2003) Evolution of sedimentary rocks, in *Treatise on Geochemistry* (eds. H. D. Holland and K. K. Turekian), Elsevier, vol. 7, pp. 367–407.

43. Taylor, S. R. *et al.* (1983) Geochemistry of loess, continental crustal composition and crustal model ages. *GCA* **47**, 1897–905; Liu, T. S. *et al.* (1993) A geochemical study of loess and desert sand in Northern China: Implications for continental crust weathering and composition. *Chem. Geol.* **196**, 359–74; Gallet, S. *et al.* (1998) Loess geochemistry and its implications for particle origin and composition of the upper continental crust. *EPSL* **156**, 157–72.

44. Bowes, D. R. (1972) Geochemistry of Precambrian crystalline basement rocks, north-west highlands of Scotland. Proceedings of the 24th International Geology Congress, Montreal, Canada, August 21–30, 1972, sect. 1, pp. 97–103. Condie, K. C. (1993) Chemical composition and evolution of the upper continental crust: Contrasting results from surface samples and shales. *Chem. Geol.* **104**, 1–37. Condie, K. C. and Brookins, D. G. (1980) Composition and heat generation of the Precambrian crust in

New Mexico. *Geochem. J.* **14**, 95–9. Condie, K. C. and Selverstone, J. (1999) The crust of the Colorado Plateau: New views of an old arc. *J. Geol.* **107**, 387–97. Eade, K. E. and Fahrig, W. F. (1971) Geochemical trends of continental plates: A preliminary study of the Canadian shield. *Geol. Surv. Canada Bull.* **71**. Eade, K. E. and Fahrig, W. F. (1973) Regional, lithological, and temporal variation in the abundances of some trace elements in the Canadian shield. *Geol. Surv. Canada Paper* **72–46**. Fahrig, W. F. and Eade, K. E. (1968) The chemical evolution of the Canadian shield. *Canad. J. Earth. Sci.* **7**, 1247–51. Gao, S. *et al.* (1998) Chemical composition of the continental crust as revealed by studies in East China. *GCA* **62**, 1959–75. McLennan, S. M. (2001) Relationships between the trace element composition of sedimentary rocks and upper continental crust. *Geochem. Geophys. Geosystems* **2**, doi: 10.1029/2000GC000109. Plank, T. and Langmuir, C. H. (1998) The chemical composition of subducting sediment and its consequences for the crust and mantle. *Chem. Geol.* **145**, 325–94. Rudnick, R. L. and Gao, S. (2003) Composition of the continental crust, in *Treatise on Geochemistry* (eds. H. D. Holland and K. K. Turekian), Elsevier, vol. 3, pp. 1–64. Shaw, D. M. *et al.* (1986) Composition of the Canadian Precambrian shield and the continental crust of the earth, in *The Nature of the Lower Continental Crust* (eds. J. B. Dawson *et al.*), Geological Society of London Special Publication 24, pp. 275–82. Taylor, S. R., and McLennan, S. M. (1985) *The Continental Crust: Its composition and Evolution*, Blackwell. Taylor, S. R., and McLennan, S. M. (1995) The geochemical evolution of the continental crust. *Rev. Geophys.* **33**, 241–65. Wedepohl, K. H. (1995) The composition of the continental crust. *GCA* **59**, 1217–32.

12

Composition and evolution of the continental crust

It is difficult to calculate what the composition of the crust of the Earth is in any reliable way

(Harold Urey) [1]

The composition of the upper part of the continental crust is well established, but it is so enriched in incompatible elements and the heat-producing elements K, U and Th in particular, that it cannot be representative of the entire crust. Unfortunately the inaccessible and largely unknown nature of the lower continental crust makes it more difficult to determine the overall crustal composition so that elements of model-dependency enter the discussion. Because the crust is a significant reservoir for many elements, understanding its overall chemical composition is of fundamental importance to geochemistry as these data place constraints on the basic processes of crustal growth, differentiation and evolution of the mantle.

Because of these restrictions, indirect evidence from the geophysical disciplines (e.g. heat flow, seismology) has to be employed mostly to obtain the bulk composition of the continental crust. So in contrast to upper crustal abundances where there is a consensus, the chemical composition of the bulk crust is much more controversial, with recent models covering a broad range from basalt through to dacite [2] (Fig. 12.1).

However, compositions at both extremes encounter a variety of problems that are difficult to reconcile with known crustal characteristics. In our opinion, the combination of constraints imposed by the upper crustal composition, heat flow and geochemistry yields reliable compositions for the bulk crust. We regard these as inherently more reliable than attempts to calculate bulk crustal compositions from geological assumptions. The latter are based inevitably on local areas, from debatable exposures of lower crustal material or from extrapolations from the seismological structure of the crust to its overall chemical composition.

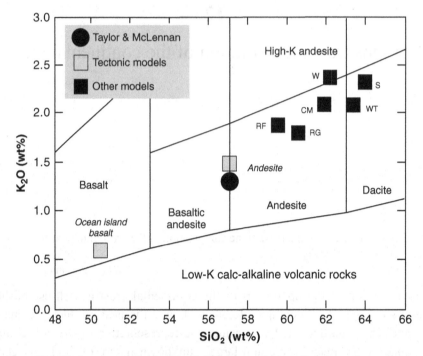

Fig. 12.1 Various models for the bulk composition of the continental crust superimposed on the K_2O–SiO_2 classification of subduction-zone volcanic rocks, compared with our estimate – see Taylor and McLennan (1985, 1995) and Table 11.3. CM = Christensen and Mooney (1995); RF = Rudnick and Fountain (1995); RG = Rudnick and Gao (2003); S = Shaw *et al.* (1986); W = Wedepohl (1995); WT = Weaver and Tarney (1984). See Note 49 for full references.

12.1 Heat flow constraints

Continental heat flow data provide important information on bulk crustal models. The average heat flow from the continents remains as one of the few independent constraints on all models of crustal composition because of the limits it places on the abundances of the heat-producing elements (K, U and Th). These are so strongly enriched in the upper crust that if these concentrations were typical of the whole crust, they would demand that the entire complement of these elements in the Earth be present in the continental crust [3].

Continental heat flow may be divided into three major components

i. Heat derived from radioactive decay within the crust (so-called crustal heat) and thus providing a measure of the average levels of heat-producing elements (K, Th, U) by assuming $K/U = 10^4$ and $Th/U = 3.8$. The average heat flow from stabilized continental crust is well established at 48 ± 1 mW/m^2 [4].

ii. Heat derived from the mantle beneath the continents (the so-called reduced heat flow). This component of heat may also vary as a function of crustal age and be influenced by mantle convection regimes. The best estimates of mantle contribution of overall heat flow are in the range of 10–18 mW/m² with an average of 14 ± 3 mW/m².

iii. Heat derived from various processes associated with orogenic activity (the so-called tectonothermal heat). This component decays on timescales of several hundred million years following the last tectonothermal event [5].

The crucial point is that crustal composition models that predict crustal radiogenic contributions to heat flow greater than 38 mW/m² (48 ± 1 minus 14 ± 3 mW/m²) are inconsistent with the global heat flow data [6] (Fig. 12.2).

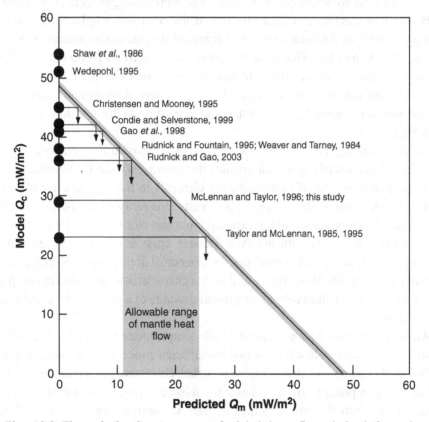

Fig. 12.2 The calculated component of global heat flow derived from the continental crust (Q_c) versus the predicted mantle component (Q_m) of heat flow for various models of crustal composition. The diagonal line represents the constraint on total crustal heat flow of 48 ± 1 mW/m². The shaded area represents the most likely range of average heat flow from the mantle. See Note 49 for references.

12.2 Composition of the bulk crust

The bulk composition of the continental crust has been examined in exhaustive detail by us in our 1985 book and in several subsequent review papers [7]. Rather than repeat these discussions, here we summarize our conclusions that suggest the overall composition of the continental crust is similar to that of the arc-derived igneous rock, andesite. We base this concept on the following considerations.

Firstly the well-documented upper crust is derived from the lower crust dominantly by intra-crustal melting. It is so enriched in potassium, uranium and thorium and many incompatible elements so that it can only represent 25–30% of the total crust. Poisson's ratio and such reliable samples as are available, from xenoliths and exposed granulite terrains, indicate a basic composition for the lower crust. Heat flow data also require substantially lower concentrations of heat-producing elements with depth.

By the Late Archean, at least 60–70% of the crust was emplaced. The overall composition of the Archean upper crust is recorded in the compositions of Archean sediments (Chapter 10). That crust was composed effectively of similar amounts of basalt and Na-rich granites (tonalite–trondhjemite–granodiorite or TTG) and their extrusive equivalents. The average of these diverse compositions mimics that of andesite as is shown by REE patterns preserved in Archean sedimentary rocks (Fig 10.2).

However, during the Archean, only minor intra-crustal differentiation occurred, with few intrusions of the K-rich granites that now typify the Post-Archean crust. Thus the concentration of heat-producing elements in the Archean upper crust is much less than that of the present upper crust. This rarity of intra-crustal melting means that the composition of the exposed crust was close to that of the bulk crust. This situation contrasts with the Post-Archean crust in which such intra-crustal melting was ubiquitous. This produced an upper crust dominated by K-rich granites, also rich in incompatible elements and with a characteristic depletion in europium. This process has led ultimately to the present structure of a granitic upper and a basic lower crust.

An interesting conclusion is that the bulk compositions of both the Archean and Post-Archean crusts, although formed by different processes, are broadly similar (although the Archean crust was higher in nickel, cobalt and chromium derived from the basaltic component). Both the basic and felsic components of the Archean crust were derived from the mantle in the Archean. In contrast, andesites derived more indirectly from the mantle via subduction-zone volcanism have been the main source of new continental material during the Proterozoic and Phanerozoic. The difference in the upper crusts between the Archean and later times lies mainly in the amount of differentiation within the crust. This was minimal in the Archean, became more common in the Late Archean and was dominant in the Post-Archean crust.

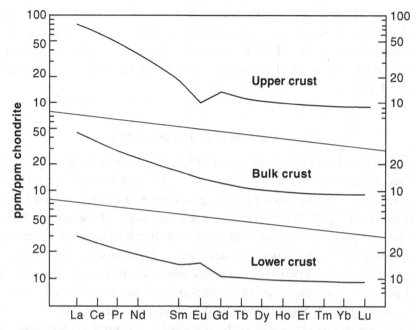

Fig. 12.3 Rare earth element patterns for the upper crust (Eu/Eu* = 0.65), bulk crust (Eu/Eu* = 1.0) and the lower crust (Eu/Eu* = 1.14). Data from Table 11.4.

The present crust is estimated to be a 60/40 mixture of the Archean bimodal and the Post-Archean andesitic compositions. The contribution of higher Ni, Co, Cr and Mg# from the Archean component in the crust, leading to the values for these elements exceeding those currently supplied by andesites, is sometimes raised as a difficulty with the andesite model. The concentrations of 1.1% K, 4.2 ppm Th and 1.1 ppm U produce a crustal contribution to heat flow of 29 mW/m^2; this allows for an average mantle contribution of about 18–20 mW/m^2. In Tables 11.3. and 11.4 our crustal abundances are given for the total continental crust and for the Archean continental crust with revisions from our previous estimates highlighted (Fig. 12.3).

12.3 The andesite model

The bulk composition of the continental crust from this assessment is close to that of andesite, a major product from subduction-zone volcanism. This similarity has led to the concept of continental growth by this process, the "Andesite Model", as a viable candidate for the generation of continental crust, although only following the Archean–Proterozoic transition [8].

However, the upper 10 km or so of the present continental crust has the average composition of granodiorite that originates by melting within the crust. Two decisive facts inform us of such an origin for the upper crust: the experimental studies on the origin of granites; and the ubiquitous depletion in europium that is retained in residual phases (such as plagioclase) in the lower crust [9]. The upper crust is thus characterized by an average REE pattern with a distinctive depletion in europium (Eu/Eu* = 0.65). In our model, europium is retained in the lower crust that has Eu/Eu* = 1.14.

12.3.1 Delamination and its problems

An apparent paradox disturbs this model of an andesitic bulk crustal composition. It has become an article of faith among petrologists that the bulk composition of subduction-zone volcanics, as for any magmas derived ultimately from the mantle, should be basaltic [10]. But if island arcs are the source of continental material, how does one produce siliceous continents from basalt? The problem has led to the popularity of "delamination" models. In these, basalts are the fundamental building blocks of the continents, an upper siliceous crust is produced by intra-crustal differentiation and much of a residual dense ultrabasic lower continental crust separates and subsides (delaminated) back into the mantle [11].

There are several problems with such a non-observable process. There needs to be a near-complete balance between crustal additions and subcrustal subtractions, for we see no evidence of changing upper (or indeed bulk) crustal compositions in Post-Archean time. Although a few localized occurrences of delamination have been reported, no process that occurs on the required massive scale has been identified [12]. Thus delamination of the lower crust must be a major continuing, rather than sporadic, process. It needs to be as efficient in removing, as subduction-zone volcanism is in adding material to the continents. These two processes are unrelated both geographically and tectonically, but need to be coordinated in some fashion in the delamination model. If under-crustal delamination follows the same volume and timescales as crustal growth, unlikely coincidences are involved.

Oceanic island basalts (OIB) display chemical and isotopic signatures of ancient subducted oceanic (HIMU) and of upper continental crust (as in the EM2 species). Significantly, no evidence of the lower crust, such as positive Eu anomalies, appears in mantle-derived rocks. This is surprising because if delamination has occurred on the scale required, a mass the order of two thirds of the continental crust would have to have subsided into the mantle and so some mantle-derived lavas might confidently be expected to show lower crustal signatures. Fermi's famous paradox might be recast as "Where's the europium?"

Among minor problems, buoyant lithospheric keels (Mg-rich and Fe-poor) appear to be attached underneath cratons and so might get in the way. Finally, there may not be a sufficient density contrast between the lower crust and the mantle to make delamination a likely process.

We regard the evocative term delamination as another unfortunate example of jargon that casts a cloak of understanding over an unobservable process. Here we restrict the term to its entrenched usage (the removal of the basal sections of cratons). We prefer to use the simpler words "foundering" or "sinking" to describe the apparent loss of gabbroic cumulates in island arcs, a topic to which we now turn.

12.3.2 The formation of andesites

Andesites are a common product of subduction-zone volcanism. Although ultimately derived from the mantle, they differ from those other common mantle melts, MORB and OIB. Notably their derivation through differentiation from island-arc basalts follows the famous (or infamous) "calc-alkaline" trend [13]. The significant difference is that andesites and their relatives lack the enrichment in iron that is characteristic of other differentiates of mantle-derived lavas. This low-Fe trend that is a characteristic signature of wet arc-derived lavas, is due to early precipitation of Ti-rich magnetite.

This process is a consequence of the unique terrestrial environment associated with subduction zones. As the slab carrying the oceanic crust descends into the mantle, rising temperatures cause it to dehydrate. Large wet oxidized fluxes carrying many incompatible elements infiltrate into the mantle wedge overlying the slab. Melting in this wet oxidizing environment in the wedge occurs at low temperatures. This leads to the early separation of magnetite and depletion of iron so that the subsequent crystallization history of such magmas differs from that of MORB and OIB [14]. As the magmas differentiate, sinking of gabbroic cumulates results in the production of andesites that are erupted, mostly in spectacular fashion.

However, the genesis of andesites at subduction zones is complicated in most regions by the presence of continental crust. So it is difficult to disentangle the contributions of recycled crustal contributions from those coming from the mantle and wedge. The many andesitic volcanoes that crown the Andes form the archetypical example [15].

In the Izu-Bonin arc in the Western Pacific, oceanic crust is subducted without any significant contribution from the continents. It provides a "clean" example where continental crustal material is minimal. Detailed seismic studies of the Izu-Bonin island arc indicate that much of it has an overall andesitic composition, that would provide suitable material to make continental crust of our bulk composition. This andesitic composition is thought to have originated through differentiation from basalts through the sinking or foundering of dense olivine–pyroxene cumulates. Similar conclusions have been reached from a study of the Talkeetna arc section in Alaska [16] (Fig. 12.4).

Foundering of dense gabbroic cumulates in subduction zones remains a currently unobservable process but seems inherently more viable than delamination beneath

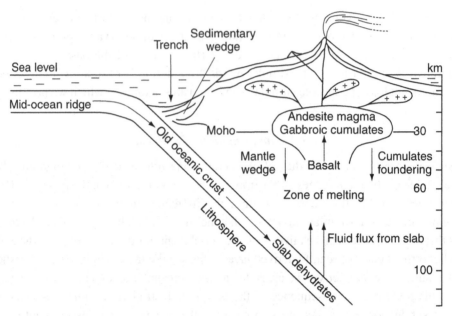

Fig. 12.4 Sketch of the formation of andesites in subduction zones during the current plate-tectonic regime. The flux of fluids from the dehydration of the down-going slab, containing many incompatible elements, induces melting in the overlying mantle. As the wet basaltic melts rise, they differentiate, forming dense gabbroic cumulates that founder. The volatile-rich andesitic magmas either intrude the crust or erupt as explosive volcanics, in either case adding new continental material. Based on Ref. 16.

the stable cratons and mostly avoids the problems discussed above. It receives additional support from the interpretation of seismic data at arcs [17]. A seismic discontinuity at a depth of 300 km attributed to sinking of eclogite appears under many island arcs but is rarely observed under cratons. It is clear that the discontinuity is concentrated beneath arcs, rather than under continents. The only two occurrences that have been detected beneath cratons may be residual from earlier subduction episodes. These data certainly provide no evidence for the ubiquitous delamination beneath cratons as required by other models of the continental crust.

Such foundering of basal sections of island arcs greatly alleviates the requirement for the necessity of delaminating the lower continental crust, more or less completely, long after its formation; and for significant amounts of ultramafic cumulate to be present in the sub-continental lithospheric mantle. If the foundering of cumulates occurs in the arc, the ultramafic cumulates sink as a consequence of the subduction process itself.

Accordingly the Andesite Model has been revived [18]. This model overcomes the basaltic nature of arcs by proposing that the removal of cumulate residues occurs in the arc, rather than at the base of the continental crust. The resulting arc material, now of andesitic composition, is added to the continents in line with the Lyellian doctrine of currently observable processes [19].

Can arc-related volcanism supply enough material, not only to grow the continents, but also to balance that removed by erosion and lost to the oceans or subduction zones? A linear growth rate for the continents requires an addition of over 1 km^3 per year which is about the same as a popular estimate of the net magma flux from arcs [20]. However this does not allow for losses through erosion that are also about this value. So if these estimates are correct, the continents might be shrinking with time [21]. However, the contribution from arc volcanism based on surface studies seems to have been seriously underestimated. Sea-floor mapping has revealed that arc volcanism is much more extensive than that which is so dramatically obvious above sea level. The amount of arc volcanism is now revealed to be between three and six times more voluminous than the earlier estimates [22]. The consequence is that the magmatic flux from arcs is adequate to sustain continental growth.

12.3.3 Granites and granites

The origin of granites continues to arouse controversy. The arguments over whether they originated as magmas or through the alteration of pre-existing rocks by assorted fluids (or "granitization") continued through the first half of the twentieth century. These debates were resolved in favor of the former view by experimental work [23] that demonstrated an origin through intra-crustal melting for the K-rich granites that now typify the Post-Archean upper continental crust. These granites all show REE patterns with a rather uniform depletion in europium, indicating equilibration with plagioclase, stable only at crustal depths on Earth. A clear distinction was established from the older Na-rich granites (the TTG suite) that originated at mantle depths and that dominate the Archean crust. Their steep REE patterns, showing no depletion in europium, indicate equilibration of the melt with garnet, generally stable only at depths greater than 30 km.

However, the K-rich granites, typical of the present upper crust, display a subtle variety of types, that initially were labeled as I and S types of granite [24]. These showed differences reflecting derivation from or interaction with "igneous" (I) or "sedimentary" (S) source rocks. The popularity of this classification led to further subdivisions so that an alphabet soup of granite types soon arose.

A serious debate has turned on the origin of the I and S types [24, 25]. Both are similar in major and trace element compositions. The S type have elevated $^{87}Sr/^{86}Sr$, K/Na ratios and $\partial^{18}O$ values, characteristics of interaction with sedimentary rocks so that their origin by intra-crustal melting is not seriously disputed. Although the I type show a wider spread in composition, their average compositions and REE patterns can scarcely be distinguished from those of the S-type suite and their

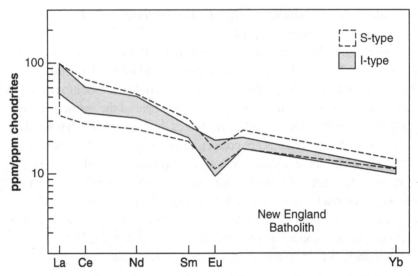

Fig. 12.5 The similar REE patterns for I- and S-type granites from the Palaeozoic New England batholith of eastern Australia. Data from Shaw and Flood (1981) [25].

overall composition is also typical of the granites that dominate the upper crust (Fig. 12.5). However I-type granites tend to have lower $^{87}Sr/^{86}Sr$, K/Na ratios and $\partial^{18}O$ values compared to S-type, indicative of an "igneous" signature.

These rocks are sometimes considered to be residues from melting (so-called "restites") or alternatively have been derived from basalts. While the S type reflect interaction with material that has been exposed to sedimentary processes, I-type granites reflect the overall andesitic composition of the unfractionated crust and accordingly signatures of this source, such as low $^{87}Sr/^{86}Sr$, K/Na ratios and $\partial^{18}O$ values are to be expected.

However, behind this debate lurks a larger question: the very nature of the growth of the continents themselves. Is the crust and the granites it contains ultimately derived from basalt, that itself is directly derived from the mantle? So is basalt the parent of the I-type granites or mainly the source of heating of lower crustal material? Much of the debate turns on a question of scale and perspective and is typical of many geological dilemmas. It also suffers from the classic problem of looking at the end product and trying to deduce the pathway. Identification of minor components with mantle signatures within the I-type granites does not constitute evidence for deriving the crust solely from mantle basalt via Bowen-type differentiation. Such a process would require huge amounts of intermediate lithologies in the differentiation process leading from basalt to granite.

In our opinion, granites result from lower crustal melting, perhaps induced by basaltic underplating which adds little directly to their overall volume or bulk composition. If the heat source is from the underplating of basaltic magma, incorporation of minor amounts of this material can explain some of the isotopic signatures in the granites noted above, as well as hafnium and oxygen isotope mantle signatures in zircons [25].

Thus the I-type composition is similar to S-type (identical REE patterns with europium depletion) but with variable admixtures of basaltic components and their derivatives; therefore we can rule out derivation of the continental crust from the mantle in a one-step process. If granites were derived by such a direct process, this might result in the development of granitic crusts wherever basalt eruptions are common, so that granites might abound on Venus, the Moon and Mars or even at mid-ocean ridges. Instead we favor the model in which the continental crust is derived in a multistage process, unique to the Earth as described above. This process of remelting within the "andesitic" crust produces the upper granitic crust and the more basic lower crust.

12.4 Alternatives: basaltic compositions

One class of crustal models proposes that the continental crust grows predominantly by basaltic volcanism. This is driven in part by models that suggest that the principal product of island-arc volcanism is basaltic in composition. Others suppose that mantle plumes or accreted oceanic plateaus are the main source [26]. However continents of basaltic composition, even if relatively enriched in incompatible elements such as ocean island basalts, encounter at least two difficulties. The fundamental problem is that it is necessary to produce the observed granodioritic upper crust that comprises between about one-quarter and one-third of the entire crust. So basaltic crustal models require a complementary ultramafic composition for the lower crust to balance the granodioritic composition of the upper crust. But no evidence appears for a lower crust of overall ultramafic composition from seismology, xenoliths or from the debatable exposures [27].

A second problem is that the level of continental heat flow is too low to match the observed values in a crust of overall basaltic composition. If continents were built from OIB, the most favorable case, they would have maximum K_2O contents of 0.5–0.7%. A more likely MORB-dominated composition would contain only about 0.3% K_2O. Such compositions do not provide enough potassium to account for that observed in the upper crust and predict a present-day crustal radioactive contribution to heat flow of less than about 15 mW/m^2, far below the observed value.

These difficulties are generally recognized by most proponents of such models and are solved by proposing additional (but unspecified) sources of potassium or loss of ultramafic residues back into the mantle by delamination or subduction erosion [28].

Basaltic underplating at convergent margins and magmatism and underplating associated with mantle plumes may also add material to continents, but the scale of such processes is uncertain [29]. Lower crustal xenoliths provide some evidence of underplating; but such processes reflect inherently random samples, as are all the interpretations based on xenoliths. Therefore these are not relevant to our problem of arriving at a bulk crustal composition [30].

In summary, the amount of material added to continents through intra-plate mantle plumes and oceanic plateaus probably amounts only to a few percent [31].

12.5 Alternatives: felsic compositions

Many bulk crustal composition models use a variety of constraints from seismic velocity profiles and geochemistry of deep crustal cross-sections and lower crustal xenoliths. These are dependent on the vagaries of surface exposures and the anecdotal evidence from xenoliths and so suffer from sampling uncertainties [32].

Although there is considerable variability among these models in detail as might be expected, they are all similar in that they predict bulk crustal compositions that are significantly more felsic than average andesite. Estimates such as those of Shaw, Wedepohl, Rudnick, Gao, Condie and others (Fig. 12. 2) are usually biased toward upper crustal compositions due to reliance on local areas as shown in Table 11.1. Only the data provided from the widespread sampling by sedimentary rocks or the huge continent-wide sampling by the USSR are likely to provide a useful average. At the extreme, some of these models predict a crustal radiogenic component of heat flow that is in excess of overall average continental heat flow. Their compositions also contain concentrations for many elements in excess of their bulk Earth abundances. We have used the heat flow data to argue that crustal composition models with K_2O greater than about 1.6% (and appropriate corresponding thorium and uranium abundances) cannot be reconciled with the current state of knowledge of average continental heat flow. This is because they predict continental heat flow in excess of 48 ± 1 mW/m^2 [33].

Accordingly, all of the relatively felsic crustal models encounter difficulties and at best can only be considered to be representative of local regions of the crust. They mostly appear to be overweighted towards the more readily accessible upper crust with its high concentrations of incompatible elements, a well-known trap in estimating the composition of the bulk crust [34] (Fig. 12.6).

Most workers appear to agree that primary melts derived from the mantle are dominated by basaltic compositions. Accordingly, to overcome this problem, felsic models for the crust typically predict that a lower crust of ultramafic–mafic composition must be returned to the mantle subsequent to crust formation by processes such as delamination of the lower crust or subduction erosion [35, 36]. However,

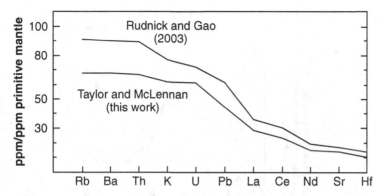

Fig. 12.6 The enrichment of incompatible elements in the continental crust according to Rudnick and Gao [32] and this work (Table 11.4) relative to values for the primitive mantle. (Primitive mantle data from Table 8.3, column B.)

such lower crustal material has not been identified in samples from the mantle, despite the chemical and isotopic evidence both of recycled oceanic and upper continental crust that appears in OIB, as discussed above.

12.6 Crustal growth and its episodic nature

The standard model, already adopted in our 1985 book, is that the crust grew throughout geological time mostly through major episodic pulses [37]. No subsequent evidence has appeared to change this model except in minor detail (Fig. 12.7).

The alternative steady-state model of crustal growth proposes that the crust formed "at the beginning" and has been recycled through the mantle so that new additions are balanced by losses. This view of an early "granitic" crust has been discussed extensively in Chapters 9 (Hadean) and 10 (Archean) [38] and we find no evidence to support it.

The REE evidence records a major change in the sedimentary record around the Archean–Proterozoic boundary (*c.* 2500 Myr) as does the oxygen isotope record in zircons [39]. This episodic growth of the crust must be due to as yet little-understood intermittent periods of major production of magma in the mantle. It raises the question whether the resurfacing of Venus 750 Myr ago was another such example.

12.6.1 *The freeboard constraint*

The widespread occurrence of flat-lying Proterozoic sediments on cratons shows that sea level has not varied relative to the continents by more than about 1 km for the past two billion years. This fact, referred to as the "freeboard constraint",

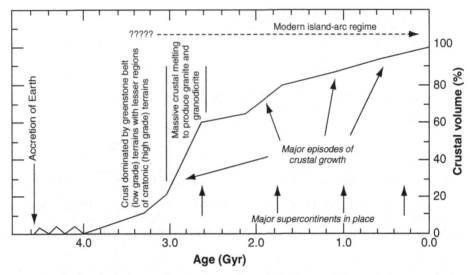

Fig. 12.7 The growth rate of the continental crust.

indicates that the relative volumes of oceans and continents have been broadly similar over that period [40]. The observation is relevant as far back as the Early Proterozoic, halfway through the geological record but does not extend into the Archean as the sedimentary record in the Archean is too incomplete to make freeboard a restriction on crustal volume [40].

The freeboard constraint indicates that most crustal growth was completed by the Late Archean and that additions to the continents since the Early Proterozoic have been minor. Those models that require the amount of crust present at 2.5 Gyr to be under 50% and that call for significant crustal growth at 2.1–1.8 Gyr do not meet the freeboard constraint. The confusion is in part due to interpreting easily accessible granitic intrusions in the upper crust as additions to the crust rather than being the products of intra-crustal differentiation.

12.6.2 Recycling

While the geological, geochronological and geochemical evidence supports a model for the episodic growth of the continental crust through geological time, with a major increase in the Late Archean, how much recycling of the continental crust does occur? In the modern subduction zones, a few percent of upper crustal sediments have been recycled into the mantle as shown by evidence from ^{10}Be and Pb isotopes. The amount of sediment available for subduction is less than 1.6×10^{15} g/yr (about $0.5 \, km^3$/yr). How much of the sediment cover reaches the mantle or is scraped off in the forearc remains unclear. Signatures of recycled sediment in OIB such as the

EM2 example at Samoa, are rare. These amounts do not seem sufficient to support models of a steady-state crustal mass, continuously recycled through the mantle such as that proposed most famously by Armstrong [41]. The buffering effects of cannibalistic sedimentary recycling and the efficient mixing during sedimentation doubtless conceal variations in the rate of continental growth and perhaps in upper crustal composition. Probably for these reasons, a period of growth around 1800 Myr ago is not distinguishable in the geochemical record in sedimentary rocks [42].

So in summary, the growth of the continental crust appears to have proceeded in an episodic fashion throughout geological time. The period of crustal growth during the Late Archean was by far the most important single episode in Earth history, followed by that at about 2.0–1.7 Gyr.

12.6.3 Continental break-up and assembly

Although somewhat peripheral to our main thesis, this topic warrants mention. At the present time, the continents are fragmented but periodically they unite to form supercontinents [43]. Such events occurred during the Late Paleozoic (Pangea; 0.3 Gyr), Late Proterozoic (0.9 Gyr), possibly at about 1.8 Gyr (for North America/ Baltica at least) and perhaps during the Late Archean (*c.* 2.7 Gyr) [44]. The apparent periodicity of crust-forming events during the Post Archean may be related to these supercontinental cycles. These episodes that occurred around 2.9–2.7, 2.0–1.7, 1.3–1.1 and 0.5–0.3 Gyr correlate with supercontinental assembly phases. At these times, subduction should increase with enhanced arc volcanism. The collisions between continents in these episodes would promote crustal thickening, granulite facies metamorphism and continental underplating.

The causes of continental break-up are much debated [45]. Large-scale igneous provinces are associated, for example, with the opening of the Atlantic Ocean. These include, from south to north, the Parana and Etendeka flood basalts, about 120 Myr old, the large igneous province, 200 Myr old, on the US East Coast and the Greenland–British tertiary province that formed 56 Myr ago [46]. Are such vast outpourings of basaltic lavas that accompany such break-ups, the cause of the split or only a consequence of decompression melting of the mantle following the rifting of the crust [47]?

What role, if any, do deep mantle plumes play? Perhaps by doming the crust, they might initiate splitting. But there are perhaps differences between cold and hot rifts. In cold rifting, the continental rift is "stretched like gum" eventually forming a new oceanic spreading center. In contrast, massive volcanism from mantle plumes causes hot rifts in which very large volumes of lavas are extruded [48]. So mantle plumes may provide a trigger in the shape of additional heat that boosts the splitting and break-up of the continents.

Synopsis

The bulk composition of the crust is an important geochemical parameter because it contains a sizeable fraction of the whole Earth budget for many incompatible elements (Fig. 1.6). However, in contrast to the upper crust, the composition of the bulk crust remains model-dependent and controversial. We employ the constraints from heat flow and seismic data, both of which indicate a basic lower crust as well as the established incompatible-element-rich upper crustal composition, that is derived from intra-crustal melting, to arrive at a bulk crustal composition. This is close to that of andesite. Critical components are 1.1% K, 4.2 ppm Th and 1.1 ppm U.

We conclude that the Archean (Chapter 10) and Post-Archean bulk crusts (Chapter 11), although formed by different processes, are broadly similar in composition. The most important differences lie in the amount of intra-crustal differentiation that leads to distinct upper crustal compositions. This process was rare during the Archean but became dominant in Post-Archean time.

We examine the "Andesite Model" for Post-Archean crustal growth and the formation of andesites in subduction zones, particularly in intra-oceanic arcs where continental crustal contamination is minimal. In the subduction-zone environment, the crucial component is water. This allows for magnetite crystallization and for the iron-depleted calc-alkaline trend to develop, that contrasts with the iron-enriched MORB and OIB compositions. Sinking or foundering of basic material in subduction zones is probably required to produce the andesitic compositions in the upper regions of arcs. We conclude that andesites form the major contribution to crustal growth during Post-Archean time.

The familiar granites originate within the crust. The origin of the upper crustal granites, including the I and S types, is assessed, and we conclude that both arise through intra-crustal melting, sometimes assisted by basaltic underplating that may impart some mantle signatures.

The many alternative bulk crustal compositions that have been proposed, ranging from basaltic through to granitic, fail either to meet heat flow constraints or result in concentrations of incompatible elements that exceed the limitations imposed by the bulk Earth composition or both. Delamination of the lower crust, a process required for the more felsic models, encounters several problems. The wholly unrelated processes of addition of andesites at subduction zones and the removal of the lower crust need to balance, in order to maintain a uniform crustal composition over Post-Archean time, while low-density keels lie under the continents.

The crust grew in an episodic manner from about 3800 Myr, initially through basaltic and TTG-suite production, with a major increase in the Late Archean. This resulted in 60–70% of the crust being in place by the end of the Archean consistent with the "freeboard" constraint. Recycling of the crust back into the mantle is too limited in volume to support constant volume models for the crust. There is much break-up and reassembly of the cratons, episodically forming supercontinents, that may correlate with episodes of crustal growth.

Notes and references

1. Urey, Harold C. (1966) Personal communication.
2. McLennan, S. M. *et al.* (2006) Composition, differentiation and evolution of continental crust: Constraints from sedimentary rocks and heat flow, in *Evolution and Differentiation of the Continental Crust* (eds. M. Brown and T. Rushmer), Cambridge University Press, pp. 92–134; Rollinson, H. (2006) Crustal generation in the Archean, in *Evolution and Differentiation of the Continental Crust* (eds. M. Brown and T. Rushmer), Cambridge University Press, pp. 173–230; Davidson, J. P. and Arculus, R. J. (2006) The significance of Phanerozoic arc magmatism in generating continental crust, in *Evolution and Differentiation of the Continental Crust* (eds. M. Brown and T. Rushmer), Cambridge University Press, pp. 135–72.
3. The radioactive decay of ^{40}K, ^{238}U, ^{235}U and ^{232}Th contributes up to 80% of the terrestrial heat flow, with the remainder from the secular cooling of the Earth. The relative proportions are a matter of debate. The average heat flow from continental areas at present is about 50 mW/m^2 compared to 100 mW/m^2 for the oceans and 84 mW/m^2 for the whole Earth. At the base of the crust, temperatures are typically 300–400 °C in shield areas, 500–600 °C in young orogenic areas and 650–750 °C in rift regions. Nyblade, A. A. and Pollack, H. N. (1993) A global analysis of heat flow from Precambrian terrains. *JGR* **98**, 12,207–18; Jaupart, C. and Mareschal, J. -C. (2003) Constraints from crustal heat production from heat flow data, in *Treatise on Geochemistry* (eds. H. D. Holland and K. K. Turekian), Elsevier, sect. 3.2, pp. 65–84.
4. Nyblade, A. A. and Pollack, H. N. (1993) A global analysis of heat flow from Precambrian terrains. *JGR* **98**, 12,207–18.
5. Bickle, M. J. (1978) Heat loss from the Earth. *EPSL* **40**, 301–15. See also Jaupart and Mareschal (2003) in Note 3.
6. Rudnick *et al.* (1998) argued that crustal contributions to heat flow for individual crustal sections are variable, ranging from about 20 to 45 mW/m^2 and concluded from this observation that there was too much uncertainty to use heat flow to constrain crustal composition models. Rudnick, R. L. *et al.* (1998) Thermal structure, thickness and composition of continental lithosphere. *Chem. Geol.* **145**, 395–411. But this reflects the inherent difficulties in dealing with geological problems. Nature is rarely uniform or accommodating and heat flow remains, like Poisson's ratio, among the few independent constraints. We consider the opinions of Rudnick and co-workers to be overly pessimistic because they essentially neglect the well-constrained value of average heat flow from tectonically stabilized Precambrian continents of 48 ± 1 mW/m^2 based on more than 1600 locations. Nyblade, A. A. and Pollack, H. N. (1993) A global analysis of heat flow from Precambrian terrains. *JGR* **98**, 12,207–18; Jaupart, C. and Mareschal, J. -C. (2003) Constraints from crustal heat production from heat flow data, in *Treatise on Geochemistry* (eds. H. D. Holland and K. K. Turekian), Elsevier, sect. 3.2, pp. 65–84. Regional variations in heat flow are well documented (e.g. Jaupart and Mareschal, 2003). They likely reflect variations in the mantle heat flux as well as variations in crustal composition and do not seem to us to be particularly surprising. Jaupart and Mareschal (2003) adopted a value of 51 mW/m^2 for continental heat flow from stabilized regions. This value is higher than that of Nyblade and Pollack (1993) because it also includes

Phanerozoic regions older than about 200 Myr, and which have relatively high heat flow that Jaupart and Mareschal (2003) interpret to be due to high crustal heat production (i.e., high abundances of K, Th, U). We are unconvinced by this interpretation because it implies a significant change in crustal composition (e.g. in arc magmatism) across the Precambrian–Phanerozoic transition. See also Hofmeister, A. M. and Criss, R. E. (2005) Earth's heat flux revised and linked to chemistry. *Tectonophysics* **395**, 159–77; Von Herzen, R. *et al.* (2005) Comments on "Earth's heat flux revised and linked to chemistry". *Tectonophysics* **409**, 193–8; Hofmeister, A. M. and Criss, R. E. (2005) Reply. *Tectonophysics* **409**, 199–203. For further discussion, see McLennan, S. M. *et al.* (2006) Composition, differentiation and evolution of continental crust: Constraints from sedimentary rocks and heat flow, in *Evolution and Differentiation of the Continental Crust* (eds. M. Brown and T. Rushmer), Cambridge University Press, pp. 92–134. Archean terrains are characterized by significantly lower heat flow that seems to be related to a greater thickness of subcontinental lithosphere beneath these terrains with perhaps some component of the difference being due to differing crustal composition. Nyblade, A. A. and Pollack, H. N. (1993) A global analysis of heat flow from Precambrian terrains. *JGR* **98**, 12,207–18.

7. For the latest most comprehensive assessment, see McLennan, S. M. *et al.* (2006) Composition, differentiation and evolution of continental crust: Constraints from sedimentary rocks and heat flow, in *Evolution and Differentiation of the Continental Crust* (eds. M. Brown and T. Rushmer), Cambridge University Press, pp. 92–134. Earlier major reviews include Taylor, S. R. and McLennan, S. M. (1985) *The Continental Crust: Its Composition and Evolution*, Blackwell; Taylor, S. R. and McLennan, S. M. (1995) The geochemical evolution of the continental crust. *Rev. Geophys.* **33**, 241–65; McLennan, S. M. and Taylor, S. R. (1996) Heat flow and the chemical composition of the continental crust. *J. Geol.* **104**, 377–96; McLennan, S. M. (2001) Relationship between the trace element composition of sedimentary rocks and upper continental crust. *Geochem. Geophys. Geosystems* **2**, doi: 10.1029/2000GC000109; Taylor, S. R. and McLennan, S. M. (2002) Chemical composition and element distribution in the Earth's crust. *Encyclopedia of Physical Science and Technology* (ed. R. Meyers), Academic Press, vol. 2, pp. 697–719. Our revisions to the bulk crustal abundances have a significant effect only on the most incompatible elements Cs, Rb, K, Th and U. To maintain a K/Rb of 300, Rb has been revised upwards to 37 ppm (from 32 ppm) and Cs is estimated from the average island-arc volcanic Rb/Cs ratio of 25 (there are insufficient high-quality Archean data for Cs) at 1.5 ppm.

8. Taylor, S. R. (1967) The origin and growth of continents. *Tectonophysics* **4**, 17–34. Taylor, S. R. (1977) Island arc models and the composition of the continental crust. *Amer. Geophys. Union Maurice Ewing Series* **I**, 325–35; Taylor, S. R. (1979) The composition and evolution of the continental crust: The rare earth element evidence, in *The Earth: Its Origin, Structure and Evolution* (ed. M. W. McElhinny), Academic Press, ch. 11, pp. 353–76. The crust clearly bears subduction-zone signatures. The most dramatic are the depletion in niobium and the enrichment in lead relative to MORB or OIB discussed in Chapter 11. As we have seen, while the Archean crust is a mixture of basaltic rocks and the TTG suite, both derived from the mantle, it has an average composition close to that of andesites (albeit with higher Ni, Cr and Mg#); TTG rocks also display niobium depletion (Chapter 10). For an excellent earlier discussion about island arcs before plate tectonics, see Umbgrove, J. H. F. (1947) *The Pulse of the Earth*, 2nd edn., Nijhoff, ch. 3, pp. 144–216. The high Mg#, Cr and Ni content of the bulk crust compared to that of andesites is sometimes cited as a problem with the andesite model. However, it should be recalled that the bulk crust contains a substantial fraction with high Ni, Cr and

Mg# that was contributed from basaltic rocks in the Archean, obviating the necessity to generate the crust from the rare high-Mg# andesites. Kelemen, P. B. (1995) Genesis of high Mg# andesites and the continental crust. *Contrib. Mineral. Petrol.* **120**, 1–19.

9. Wyllie, P. J. (1977) Crustal anatexis: An experimental view. *Tectonophysics* **43**, 41–71; Clemens, J. D. (2006) Melting of the continental crust: Fluid regimes, melting reactions and source rock fertility, in *Evolution and Differentiation of the Continental Crust* (eds. M. Brown and T. Rushmer), Cambridge University Press, pp. 296–330. Thus the europium-depleted K-rich granites and granodiorites, that are now dominant in the upper crust, are formed mainly through intra-crustal melting. During these intra-crustal melting events, divalent Eu (and Sr that has a closely similar divalent ion radius) was retained in the lower crust in residual plagioclase. Plagioclase is not stable at pressures greater than 10 kbar, a pressure that is reached at a depth of 40 km on the Earth. Thus the presence of the depletion in europium in igneous rocks tells us that the rocks originated within the crust. This observation is in contrast to the REE patterns observed in the TTG suites of igneous rocks that are typical of the Archean crust. Their lack of signatures of a depletion in europium and their steep patterns denote a deeper source where garnet is a residual phase. As pointed out in Chapter 11, the depletion in europium that is observed in Post-Archean sedimentary rocks was not caused by surficial processes of oxidation or reduction near the Earth's surface. Europium is present as the trivalent ion in sediments; but under reducing conditions it is divalent. Thus the depletion in europium that is observed in the present-day upper crust is a memory of a previous igneous event when europium was divalent under reducing conditions.

10. e.g. Arculus, R. J. (1981) Island arc magmatism in relation to the evolution of the crust and mantle. *Tectonophysics* **75**, 113–33; Gill, J. B. (1981) *Orogenic Andesites and Plate Tectonics*, Springer-Verlag; Pearcy, L. G. *et al.* (1990) Mass balance calculations for two sections of island arc crust and implications for the origin of continents. *EPSL* **96**, 427–42; Rudnick, R. L. (1995) Making continental crust. *Nature* **378**, 571–8.

11. Kay, R. W. and Kay, S. M. (1991) Creation and destruction of continental crust. *Geol. Rundsch.* **80**, 259–78; Rudnick, R. L. (1995) Making continental crust. *Nature* **378**, 571–8.

12. Jull, M. and Kelemen, P. B. (2001) On the conditions for lower crustal convective instability. *JGR* **106** (B4), 6423–46. However, delamination may occur in localized regions. Thus Gao, S. *et al.* (2004) Recycling lower continental crust in the North China craton. *Nature* **432**, 892–7, describe what appears to be an example of delamination in the North China craton. Late Jurassic high magnesian andesites, dacites and, significantly, adakites appear to have originated from interaction between eclogites and mantle peridotite. These have inherited Archean zircons and have strontium and neodymium isotopic compositions that overlap with those in eclogite xenoliths from the lower crust in the region. The authors suggest that these lavas are derived from lower crustal eclogite that foundered into the upper mantle and "interacted with peridotite" in the late Jurassic. This observation raises interesting questions. Is it a local episode or one that occurs on a large enough scale to influence bulk crustal compositions? One might have expected many such examples if delamination occurred on a large enough scale to remove most of the lower crust, as required in the felsic models of these authors. This forms a classic example of an inherent geological dilemma. Is this occurrence of local significance or can this observation be extrapolated to the whole crust? This problem is non-trivial as the entire crust has to be involved for these models to be viable. See also Gao, S. *et al.* (1998) How mafic is the lower continental crust? *EPSL* **161**, 101–17 for another example of what may be a local occurrence. In the andesite model, during the intra-crustal melting that

produced the granitic upper crust, europium was retained in the residual lower crust that became a sink containing excess Eu. In the felsic models, delamination of this material removed it into the mantle. The amount of Eu processed from the mantle into the crust over geological time constitutes perhaps 10–20% of the bulk Eu content in the Earth and so much of this would be returned to the mantle by delamination of the lower crust. In view of the difficulties of rehomogenization in the mantle, one might then expect that some of this signature of REE patterns with Eu/Eu* > 1 would be retained and that mantle-derived melts might display such an enrichment in Eu (a "positive Eu anomaly"). However there is essentially no evidence for such a signature; primary melts from the mantle invariably show little sign of either relative enrichment or depletion in Eu with Eu/Eu* = 1. In this context, Jackson, M. G. *et al.* (2007) The return of subducted continental crust in Samoan lavas. *Nature* **448**, 684–7, have identified the presence of upper continental crustal material with negative Eu anomalies in the EM2 lavas at Samoa. However, no record appears anywhere of lower crustal material, so the delamination model fails this test.

13. Arculus, R. J. (2003) Use and abuse of the terms calc-alkaline and calc-alkalic. *J. Petrology* **44**, 929–35.
14. Arculus, R. J. (2004) Evolution of arc magmas and their volatiles. *AGU Geophys. Monograph* **150**, 95–108; Davidson, J. P. and Arculus. R. J. (2006) The significance of Phanerozoic arc magmatism in generating continental crust, in *Evolution and Differentiation of the Continental Crust* (eds. M. Brown and T. Rushmer), Cambridge University Press, pp. 135–72.
15. e.g. Hildreth, W. and Moorbath, S. (1988) Crustal contributions to arc magmatism in the Andes of Central Chile. *Contrib. Mineral. Petrol.* **98**, 455–89.
16. Suyehiro, K. *et al.* (1996) Continental crust, crustal underplating and low-Q upper mantle beneath an oceanic island arc. *Science* **271**, 390–2; Taira, A. *et al.* (1998) Nature and growth rate of the northern Isu-Bonin (Ogasawara) arc crust and their implications for continental crust formation. *The Island Arc* **7**, 395–407. Takahashi, N. *et al.* (1998) Implications from the seismic crustal structure of the northern Izu-Bonin arc. *The Island Arc*, **7**, 383–94. Fliedner, M. M. and Klemperer, S. L. (1999) Structure of an island-arc: Wide-angle seismic studies in the eastern Aleutian Islands, Alaska. *JGR* **104**, 10,667–94. Ducea, M. N. (2002) Constraints on the bulk composition and root foundering rates of continental arcs: A California perspective. *JGR* **107**, doi: 10.1029/2001JB000643; The Tonga–Kermadec subduction zone forms another example, Crawford, W. C. *et al.* (2004) Tonga Ridge and Lau Basin crustal structure from seismic refraction data. *JGR* **108**, doi: 10.1029/2001JB001435; but see also Holbrook, W. S. *et al.* (1999) Structure and composition of the Aleutian island arc and implications for continental crustal growth. *Geology* **27**, 31–4. Further support for foundering of gabbroic and ultramafic basal sections of arcs comes from Behn, M. D. and Kelemen, P. D. (2006) Stability of arc lower crust: Insights from the Talkeetna arc section, south central Alaska and the seismic structure of modern arcs. *JGR* **111**, doi: 10.1029/2006JB00432.
17. Behn, M. D. *et al.* (2007) Trench-parallel anisotropy produced by foundering of arc lower crust. *Science* **317**, 108–11. These authors attribute the fast seismic shear-wave velocities that are parallel to the strike of the trench to "foundering of dense arc lower crust". See also Figure 1 from Williams, Q. and Revenaugh, J. (2005) Ancient subduction, mantle eclogite and the 300 km seismic discontinuity. *Geology* **33**, 1–4. These authors attribute the discontinuity to "the presence of eclogitic material at depth within Earth's mantle and the associated onset of stishovite formation (from coesite) at pressures between 8.5 and 11 GPa (80 to 115 kbar) within eclogite" (p. 2). Free SiO_2

forms from the basaltic compositions at depth due to the breakdown of feldspars to garnets and pyroxenes.

18. Arculus, R. J. (1999) Origins of the continental crust. *J. Proc. Royal Soc. NSW* **132**, 83–110. Davidson, J. P. and Arculus, R. J. (2006) The significance of Phanerozoic arc magmatism in generating continental crust, in *Evolution and Differentiation of the Continental Crust* (eds. M. Brown and T. Rushmer), Cambridge University Press, pp. 135–72. This concept receives support from the structure of some island arcs that contain deep layers of andesitic composition (P-wave velocities of 6–7 km/sec) such as the Izu-Ogaswara arc and the Tonga Ridge. Suyehiro, K. *et al.* (1996) Continental crust, crustal underplating and low-Q upper mantle beneath an oceanic island arc. *Science* **272**, 390–2. Crawford, W. C. *et al.* (2003) Tonga Ridge and Lau Basin crustal structure from seismic refraction data. *JGR* **108**, doi: 10.1029/2001JB001435.

19. Further support for an andesite model comes from a study of heat production in a vertical section of exposed arc in Japan: Furukawa, Y. and Shinjoe, H. (1997) Distribution of radiogenic heat generation in the arc's crust of the Hokkaido Island, Japan. *GRL* **24**, 1279–82. They estimated that the crustal radiogenic component of heat flow amounted to $25\,mW/m^2$ over 30 km of arc crust. Normalized to average continental crustal thickness of 41 km, this translates into $34\,mW/m^2$ which is virtually identical to that predicted by the "Andesite Model" of crustal composition.

20. Reymer, A. and Schubert, G. (1984) Phanerozoic addition rates to continental crust and crustal growth. *Tectonics* **3**, 63–77.

21. Von Huene, R. and Scholl, D. W. (1991) Observations at convergent margins concerning sediment subduction, subduction erosion and the growth of continental crust. *Rev. Geophys.* **29**, 279–316. An early proposal for continent shrinking can be found in Fyfe, W. S. (1976) Hydrosphere and continental crust: Growing or shrinking? *Geosci. Canada* **3**, 82–3.

22. Arculus, R. J. (2004) Evolution of arc magmas and their volatiles. *AGU Geophys. Monograph* **150**, 95–108; McLennan, S. M. (1988) Recycling of continental crust. *Pure Appl. Geophys.* **128**, 683–724; Plank, T. and Langmuir, C. H. (1998) The chemical composition of subducting sediment and its consequences for the crust and mantle. *Chem. Geol.* **145**, 325–94; Davidson, J. P. and Arculus R. J. (2006) The significance of Phanerozoic arc magmatism in generating continental crust, in *Evolution and Differentiation of the Continental Crust* (eds. M. Brown and T. Rushmer), Cambridge University Press, pp. 135–72 for an extended discussion of these issues.

23. e.g. Wyllie, P. J. (1977) Crustal anatexis: An experimental view. *Tectonophysics* **43**, 41–71; Clemens, J. D. (2006) Melting of the continental crust: Fluid regimes, melting reactions and source rock fertility, in *Evolution and Differentiation of the Continental Crust* (eds. M. Brown and T. Rushmer), Cambridge University Press, pp. 296–330; Brown, G. C. and Fyfe, W. S. (1970) The production of granitic melts during ultrametamorphism. *Contrib. Mineral. Petrol.* **28**, 310–18.

24. Chappell, B. W. (1984) Source rocks of I- and S-type granites in the Lachlan Fold Belt, southeastern Australia. *Phil. Trans. Royal Soc.* **A310**, 693–707. Shaw, S. E. and Flood, R. H. (1981) The New England batholith, eastern Australia: Geochemical variations in space and time. *JGR* **86**, 10,530–44; Flood, R. H. and Shaw, S. E. (1977) Two "S-type" granite suites with low initial $^{87}Sr/^{86}Sr$ ratios from the New England batholith, Australia. *Contrib. Mineral. Petrol.* **61**, 163–73. See summary of I and S types in Taylor, S. R. and McLennan, S. M. (1985) *The Continental Crust: Its Composition and Evolution*, Blackwell, pp. 218–24; Chappell, B. W. *et al.* (2004) *Trans. Royal Soc. Edinburgh* **95**, 124–38. See also Clemens, J. D. (2006) Melting of the continental crust: Fluid regimes, melting reactions and source rock fertility, in *Evolution and*

Differentiation of the Continental Crust (eds. M. Brown and T. Rushmer), Cambridge University Press, pp. 296–330 and Patino-Douce, A. E. and Beard, J. S. (1995) Dehydration-melting of biotite gneiss and quartz amphibolite from 3 to 15 kbar. *J. Petrol.* **37**, 707–38.

25. Shaw, S. E. and Flood, R. H. (1981) The New England batholith, eastern Australia: Geochemical variations in space and time. *JGR* **86**, 10,530–44; Kemp, A. J. S. *et al.* (2007) Magmatic and crustal differentiation history of granitic rocks from Hf–O isotopes in zircon. *Science* **315**, 980–3.

26. e.g. Abbott, D. H. and Isley, A. E. (2002) Extraterrestrial influences on mantle plume activity. *EPSL* **205**, 53–62; Albarède, F. (1998) The growth of continental crust. *Tectonophysics* **296**, 1–14; Stein, M. and Hofmann, A. W. (1994) Mantle plumes and episodic crustal growth. *Nature* **372**, 63–8; Ben-Avraham, Z. *et al.* (1981) Continental accretion and orogeny: From oceanic plateaux to allochthonous terranes. *Science* **213**, 47–54.

27. Rudnick, R. L. and Fountain, D. M. (1995) Nature and composition of the continental crust: A lower crustal perspective. *Rev. Geophys.* **33**, 267–309.

28. e.g. Albarède, F. (1998) The growth of continental crust. *Tectonophysics* **296**, 1–14.

29. e.g. Rudnick, R. L. (1992) Restites, Eu anomalies and the lower continental crust. *GCA* **56**, 963–70; Hill, R. I. (1993) Mantle plumes and continental tectonics. *Lithos* **30**, 193–206.

30. Wendlandt, E. *et al.* (1993) Nd and Sr isotope chronology of Colorado Plateau lithosphere: Implications for magmatic and tectonic underplating of the continental crust. *EPSL* **116**, 23–43.

31. Hill, R. I. *et al.* (1992) Mantle plumes and continental tectonics. *Science* **256**, 186–93.

32. Such models seem to have originated with the paper by Pakiser, L. C. and Robinson, R. (1966) Composition and evolution of the continental crust as suggested by seismic observations. *Tectonophysics* **3**, 547–57. Recent examples include Rudnick, R. L. and Fountain, D. M. (1995) Nature and composition of the continental crust: A lower crustal perspective. *Rev. Geophys.* **33**, 267–309; Wedepohl, K. H. (1995) The composition of the continental crust. *GCA* **59**, 1217–32; Condie, K. C. and Selverstone, J. (1999) The crust of the Colorado Plateau: New views of an old arc. *J. Geol.* **107**, 387–97; Rudnick, R. L. and Gao, S. (2003) Composition of the continental crust, in *Treatise on Geochemistry* (eds. H. D. Holland and K. K. Turekian), Elsevier, vol. 3, pp. 1–64.

33. McLennan, S. M. and Taylor, S. R. (1996) Heat flow and the chemical composition of the continental crust. *J. Geol.* **104**, 377–96.

34. See for example Rudnick, R. L. and Fountain, D. M. (1995) Nature and composition of the continental crust: A lower crustal perspective. *Rev. Geophys.* **33**, 267–309; Wedepohl, K. H. (1995) The composition of the continental crust. *GCA* **59**, 1217–32; Condie, K. C. and Selverstone, J. (1999) The crust of the Colorado Plateau: New views of an old arc. *J. Geol.* **107**, 387–97; Rudnick, R. L. and Gao, S. (2003) Composition of the continental crust, in *Treatise on Geochemistry* (eds. H. D. Holland and K. K. Turekian), Elsevier, vol. 3, pp. 1–64. The Rudnick and Gao estimate falls within the heat flow constraints but predicts relatively low mantle contributions ($< 13 \, \mathrm{mW/m^2}$). Rudnick and Fountain (1995) calculated that Archean crust on average is depleted in the most incompatible elements (Rb, Cs) by up to a factor of 3–4 and a factor of about 2 for heat-producing elements (HPE) compared to the overall bulk crust. In their evaluation of global continental heat flow, Jaupart and Mareschal (2003) – see Note 6 – suggested Post-Archean crust was enriched in HPE by a factor of about 1.7 over Archean crust. Our models are constrained by a variety of factors (crustal growth models, bimodal-suite compositions, heat flow) and predict relatively modest changes in composition;

the most incompatible elements (Rb, Cs) differ by a factor of about 1.6–1.7 (about 1.2 for HPE). Taylor and McLennan (1985, 1995) – see Note 7 – and McLennan, S. M. and Taylor, S. R. (1996) Heat flow and the chemical composition of the continental crust. *J. Geol.* **104**, 377–96. For a detailed discussion of secular changes in crustal composition see McLennan, S. M. *et al.* (2006) Composition, differentiation and evolution of continental crust: Constraints from sedimentary rocks and heat flow, in *Evolution and Differentiation of the Continental Crust* (eds. M. Brown and T. Rushmer), Cambridge University Press, pp. 92–134; Rapp, R. P. and Watson, E. B. (1995) Dehydration melting of metabasalt at 8–32 kbar: Implications for continental growth and crust-mantle recycling. *J. Petrology* **36**, 891–931.

35. e.g. Kay, R. W. and Kay, S. M. (1991) Creation and destruction of lower continental crust. *Geol. Rundsch.* **80**, 259–78; Albarède, F. (1998) The growth of continental crust. *Tectonophysics* **296**, 1–14.

36. Arculus, R. J. (1999) Origins of the continental crust. *J. Proc. Royal Soc. NSW* **132**, 83–110. Davidson, J. P. and Arculus, R. J. (2006) The significance of Phanerozoic arc magmatism in generating continental crust, in *Evolution and Differentiation of the Continental Crust* (eds. M. Brown and T. Rushmer), Cambridge University Press, pp. 135–72.

37. e.g. Moorbath, S. (1978) Age and isotopic evidence for the evolution of the continental crust. *Phil. Trans. Royal Soc.* **A288**, 401–13; Taylor, S. R. and McLennan, S. M. (1981) The composition and evolution of the continental crust: Rare earth element evidence from sedimentary rocks. *Phil. Trans. Royal Soc.* **A288**, 381–99.

38. Armstrong, R. L. (1981) Radiogenic isotopes: The case for crustal recycling on a near-steady-state no-continental-growth Earth. *Phil. Trans. Royal Soc.* **A288**, 443–72; (1991) The persistent myth of crustal growth. *Aust. J. Earth Sci.* **38**, 613–30; Fyfe, W. S. (1978) The evolution of the Earth's crust: Modern plate tectonics to ancient hot spot tectonics. *Chem. Geol.* **23**, 89–114; Bowring, S. A. and Housh, T. (1995) The Earth's early evolution. *Science* **269**, 1535–40; Vervoort, J. D. *et al.* (1996) Constraints on early Earth differentiation from hafnium and neodymium isotopes. *Nature* **379**, 624–7.

39. Valley, J. *et al.* (2005) 4.4 billion years of crustal maturation: Oxygen isotope ratios of magmatic zircon. *Contrib. Mineral. Petrol.* **150**, 561–80.

40. e.g. Kasting, J. F. and Holm, N. G. (1992) What determines the volume of the oceans. *EPSL* **109**, 507–15; Galer, S. J. (1991) Interrelationships between continental freeboard, tectonics and mantle temperature. *EPSL* **105**, 214–28; McLennan, S. M. and Taylor, S. R. (1983) Continental freeboard, sedimentation rates and growth of continental crust. *Nature* **306**, 169–72; Wise, D. U. (1974) Continental margins, freeboard, and the volumes of continents and oceans through time, in *The Geology of Continental Margins* (eds. C. A. Burke and C. L. Dragk), Springer-Verlag, pp. 45–58.

41. Armstrong, R. L. (1981) Radiogenic isotopes: The case for crustal recycling on a near-steady-state no-continental-growth Earth. *Phil. Trans. Royal Soc.* **A288**, 443–72; Armstrong, R. L. (1991) The persistent myth of crustal growth. *Aust. J. Earth Sci.* **38**, 613–30. But see McLennan, S. M. (1988) Recycling of the continental crust. *Pure Appl. Geophys.* **128**, 683–724. The EM2 example is described by Jackson, M. G. *et al.* (2007) The return of subducted continental crust in Samoan lavas. *Nature* **448**, 684–7.

42. e.g. McLennan, S. M. and Hemming, S. R. (1992) Samarium/neodymium elemental and isotopic systematics in sedimentary rocks. *GCA* **56**, 169–200.

43. Rogers, J. J. W. and Santosh, M. (2004) *Continents and Supercontinents*, Cambridge University Press; Anderson, D. L. (1994) Superplumes or supercontinents. *Geology* **22**, 39–42.

44. Unrug, R. (1992) The supercontinent cycle and Gondwanaland assembly: Component cratons and the timing of suturing events. *J. Geodynamics* **16**, 215–40; McLennan, S. M. and Taylor, S. R. (1991) Sedimentary rocks and crustal evolution: Tectonic setting and secular trends. *J. Geol.* **99**, 1–21.
45. e.g. Duncan, C. C. and Turcotte, D. L. (1994) On the breakup and coalescence of continents. *Geology* **22**, 103–6.
46. Coffin, M. F. and Eldholm, O. (1994) Large igneous provinces: Crustal structure, dimensions and external consequences. *Rev. Geophys.* **32**, 1–36; Korenaga, J. (2004) Mantle mixing and continental breakup magmatism. *EPSL* **218**, 463–73.
47. e.g. Storey, B. C. *et al.* (1992) *Magmatism and the Causes of Continental Breakup.* Geological Society of London Special Publication 68.
48. Wright, T. J. *et al.* (2006) Magma-maintained rift segmentation at continental rupture in the 2005 Afar dyking episode. *Nature* **442**, 291–4.
49. Christensen, N. I. and Mooney, W. D. (1995) Seismic velocity structure and composition of the continental crust: A global view. *JGR* **100**, 9761–88; Condie, K. C. and Selverstone, J. (1999) The crust of the Colorado Plateau: New views of an old arc. *J. Geol.* **107**, 387–97; McLennan, S. M. *et al.* (2006) Composition, differentiation and evolution of continental crust: Constraints from sedimentary rocks and heat flow, in *Evolution and Differentiation of the Continental Crust* (eds. M. Brown and T. Rushmer), Cambridge University Press, pp. 92–134; McLennan, S. M. and Taylor, S. R. (1996) Heat flow and the chemical composition of the continental crust. *J. Geol.* **104**, 377–96; Rudnick, R. L. and Fountain, D. M. (1995) Nature and composition of the continental crust: A lower crustal perspective. *Rev. Geophys.* **33**, 267–309; Rudnick, R. L. and Gao, S. (2003) Composition of the continental crust, in *Treatise on Geochemistry* (eds. H. D. Holland and K. K. Turekian), Elsevier, vol. 3, pp. 1–64; Shaw, D. M. *et al.* (1986) Composition of the Canadian Precambrian shield and the continental crust of the Earth, in *The Nature of the Continental Crust* (eds. J. B. Dawsan *et al.*), Geological Society of London Special Publication 24, pp. 275–82; Taylor, S. R. and McLennan, S. M. (1985) *The Continental Crust: Its Composition and Evolution*, Blackwell; Taylor, S. R. and McLennan, S. M. (1995) The geochemical evolution of the continental crust. *Rev. Geophys.* **33**, 241–65; Weaver, B. L. and Tarney, J. (1984) Empirical approach to estimate the composition of the continental crust. *Nature* **310**, 575–7; Wedepohl, K. H. (1991) Chemical composition and fractionation of the continental crust. *Geol. Rundsch.* **80**, 207–23; Wedepohl, K. H. (1995) The composition of the continental crust. *GCA* **59**, 1217–32.

13

Crusts on minor bodies

Where observation is concerned, chance favors only the prepared mind
(Louis Pasteur) [1]

13.1 Minor bodies in the Solar System

Apart from the major planets, a host of minor bodies are also in orbit around the Sun or around the planets. However, no commonly agreed definitions can be found for the bodies that formed by a variety of stochastic processes in the nebula. Furthermore the extra-solar planetary systems mostly do not resemble our own. This has led to sharp debates over what does or does not constitute a planet (Chapter 1, Ref. 2) and the matter has been resolved by adding qualifiers. So the Solar System consists of eight major planets, each one distinct and a huge variety of objects that include the rocky bodies that inhabit the asteroid belt, the satellites of the major planets and the multitude of small icy bodies in the Edgeworth–Kuiper Belt that are usually referred to as Trans-Neptunian Objects or TNOs [2] that possibly represent material from the primordial solar nebula. Perhaps the most interesting observation about the small bodies is that there is little uniformity. Dave Stevenson has noted that "the four giant planets exhibit a startling diversity of satellite systems" [3] while Brad Smith has remarked that "the sense of novelty would probably not have been greater if we had explored a different Solar System" [4].

Minor bodies are typically composed of various mixtures of the ice and rock components of the original nebula. Gases are strongly depleted. The asteroids, inner nebular objects sunwards of the "snow line" (Chapter 1), are dominated by rock. Only in the outer portions of the asteroid belt does water and some carbonaceous material appear as in the carbonaceous chondrites. The analyses of most meteorites inform us that the rock component is depleted, as are the inner planets, in volatile elements, although except in rare cases (e.g. the angrites), the depletion is mostly less than that observed in the terrestrial planets. The result is that there is much variation in meteorite, asteroid and rocky planetary compositions.

While the formation of the asteroids is related to high-energy events in the inner nebula, different conditions apply further out. Beyond the snow line, ices become stable so that the satellites of the giant planets (excluding the special case of Io), TNOs and Centaurs are composed of variable amounts of rock and ices.

We have various levels of information on the nature of the surfaces of the minor bodies. We are reasonably informed about the nature of the asteroid belt from the study of meteorites that are mostly derived therefrom. We likely have samples that originated from the asteroid 4 Vesta, while spacecraft have visited the surfaces of Eros and Itokawa. We are informed in much detail about the Galilean satellites of Jupiter. There are increasing data from Titan and the smaller satellites of Saturn such as Enceladus while there is a more hazy conception of the nature of the surface of Triton. Thus we are beginning to get insights into the processes that led to the origin and evolution of the asteroids and TNOs, both of which are refugees from the formation of the Solar System, as well as of the satellites of the giant planets.

The composition of the rocky asteroids has been affected, along with other inner nebular objects, by the early depletion of the volatile elements. Has this extended to the rocky component in the icy satellites? The processes that resulted in the depletion of volatile elements in the inner nebula seem unlikely to be operating at Jupiter at or beyond the snow line. So in the absence of definitive evidence, the assumption is made that the rocky component of the satellites of the giant planets and the TNOs is solar, that is equivalent to CI carbonaceous chondrites. A caveat is that collisions may induce fractionation, both between the ice and rock fractions (as may have happened with the Pluto–Charon pair and other TNOs) and within the rocky component.

As the temperature decreases further from the Sun, other ices appear such as ammonia, nitrogen and methane. The most significant geological effect of ammonia hydrates is to allow melting to occur at low temperatures. The $H_2O–NH_3$ system has a peritectic point (at 35% NH_3) at 173 K, with a density close to that of water ice. Such temperatures are readily reached by heating in the larger satellites, either tidally or by radioactivity. These ammonia–water liquids (density 0.946 g/cm^3 at 175 K) will easily rise to the surface of satellites whose typical densities are around 1.3 g/cm^3. Landforms resembling volcanic lava flows on Earth may appear, caused by semi-crystallized flows of ammonia–water mixtures that have erupted through the satellite crusts [5]. But many of these interpretations depend on uncertain radar or photogeological data so caution is needed.

Here we avoid the temptation to conduct a guided tour of the minor bodies of the Solar System and concentrate instead on the best understood examples of the crusts that have been formed on smaller bodies. These include the asteroid 4 Vesta and those of the Galilean satellites of Jupiter. We conclude with some comments on the surfaces of Titan and Triton.

However, a subtle distinction arises between Vesta (and other differentiated asteroids) and the Galilean satellites that relates to their formation and differentiation. Vesta is a type example of bodies in the inner Solar System that underwent melting and differentiation induced by short-lived radioactivity and accretional heating. The formation of the well-studied satellites of Jupiter offers an interesting contrast. They appear to have accreted at lower temperatures without initially undergoing melting and differentiation. So they grew in an environment very different to the hierarchical accretion of the terrestrial planets where sufficient energy to cause melting was supplied by impacts. Callisto, that is the extreme example, must have formed from rock and ice components without being heated sufficiently to cause early melting of ice.

13.2 Observational problems

The difficulties of observing the surfaces of remote small bodies have presented planetary scientists with many continuing problems, reinforcing the view that the surface of the Earth is not the best starting point from which to try to interpret the surfaces of satellites and asteroids, let alone those of other planets. Apart from photography from spacecraft, much of the information about the surfaces of the smaller bodies comes from studies of the reflectance spectra in the visible (> 0.5 micrometers [microns], hereafter μm) and near infrared (1–4 μm). These data provide information about the mineralogy of the surface but also reflect the effects of the solar wind, sputtering and micrometeorite bombardment of airless surfaces [6].

However there are some limitations of this technique. Only the upper few micrometers or the "very" surface are probed. Although water ice, with its characteristic absorption bands around 1.5, 2.0 and 3.0 μm, is expected to be a significant component of the surfaces of small bodies in the outer regions of the asteroid belt, it is often surprisingly difficult to detect.

There are various reasons. Melting induced by impacts, radioactivity or tidal flexing may result in an ice-rich crust, but impacts or sputtering may also ablate ices, leaving a darker surface of more refractory components or organic compounds. Larger impacts may penetrate beneath such surfaces to add clean ice to an older surface. Great complexity is to be expected.

The attempts to use reflectance spectroscopy to understand the surfaces of asteroids form an interesting story. At first the application of this technique appeared to be a relatively straightforward task. Meteorites were known to originate from the asteroid belt, so that it was a matter of comparing the laboratory spectra of ground-up meteorites with those obtained telescopically from asteroidal surfaces.

This approach was reinforced by the apparent success in identifying Vesta as the source of the eucritic class of basaltic meteorites, something that persuaded the asteroidal community that this was an approach that might be applied to the rest of the asteroids.

However, nature is subtle and a paradox soon arose. The eucrites are a comparatively rare class of meteorites. The most abundant meteorites to fall on the surface of the Earth are the H, L and LL classes of stony meteorites or chondrites, so common that they are referred to as the "ordinary chondrites" or OC. Orbital calculations of meteorites photographed during atmospheric entry and that were subsequently recovered pointed to their origin from within the main asteroid belt.

The identification of their parent bodies by comparing laboratory with asteroidal spectra, apparently successful in the case of the eucrites, proved elusive for the abundant chondrites. Despite extensive searches, no unequivocal candidate parent asteroids appeared among the thousands of possibilities. Although the abundant S-type asteroids, that dominate the inner part of the main asteroid belt, appeared to contain the appropriate mineralogy (metal, olivine, low-Ca pyroxene), the absorption bands in the asteroidal spectra, compared with the laboratory spectra from ground-up meteorite samples, were much weaker and redder. Clearly the processes operating on the pulverized surfaces of asteroids were more complex than simply grinding up rocks. Similar effects were observed when spectra from returned samples of the lunar regolith were compared with ground-up lunar rocks [7] (Fig 13.1). This was initially attributed to the formation of dark glass agglutinates by melting from impacts of meteorites into the lunar soils.

13.2.1 Nanophase iron

Renewed study of the lunar regolith revealed that it was not so much the production of glass agglutinates but the coating of grains with microscopic metallic iron particles that was the cause of the spectral changes. This so-called "nanophase iron" (npFe°) that is mostly between 4 and 30 nm in size, is formed by reduction due to sputtering and micrometeorite impacts followed by deposition from a vapor phase onto grain surfaces [8].

Thus the observed spectra of S-type asteroids are likely due to "space weathering" of ordinary chondrites and the paradox seems to have been solved, a cautionary tale that is worth relating at some length. As Clark Chapman has noted "we remain chastened in our expectations of the robustness of remote-sensing techniques applied to unreachable, or rarely reachable, objects in space. But when we rely on theory, imperfectly relevant laboratory simulations and indirect inference to determine the compositions of solid-surfaced solar system bodies, we must be wary that we could well be led astray" [8].

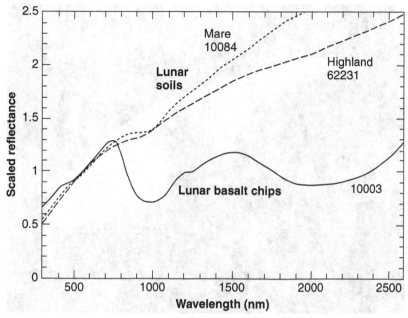

Fig. 13.1 Reflectance spectra scaled to unity at 560 nm showing the steeper higher wavelength spectrum of lunar mare (FeO 15.8%) and highland soils (Fe 4.2%) caused by "space weathering" compared to that of chips of mare basalt rock sample 10003 (FeO 19.8%). Adapted from Pieters, C. *et al.* (2000) *MPS* **35**, p. 1102, Fig. 1.

13.2.2 Eros and Itokawa

Some light has been shed on the problem of space weathering by the Near Earth Asteroid Rendezvous (NEAR)–Shoemaker mission to the peanut-shaped 433 Eros. This Mars-crossing Amor asteroid (13 × 13 × 33 km with a density of 2.67 g/cm^3) is the second largest of the near-Earth asteroids. The spacecraft carried X-ray and gamma-ray detectors and multispectral/near-infrared spectrometers and spent nearly a year orbiting the asteroid before landing on Eros [9] (Fig. 13.2).

Although Eros probably has the composition of an ordinary chondrite, the data from the NEAR Shoemaker mission were inconclusive. The Cr/Fe, Mn/Fe and Ni/Fe ratios match those of L or LL chondrites, but the problem is that the measured sulfur content is only about half that of ordinary chondrites. This led to Tim McCoy and others to question "the basic assumption that the regolith of Eros is indicative of the composition of the underlying bedrock" [10].

A touch-down on the 400 m S-type asteroid 25143 Itokawa was achieved by the Hayabusa spacecraft in 2005. It is unclear whether the mission, due to return in 2010, managed to collect samples. Itokawa is a Mars-crossing Apollo asteroid with a density of 1.95 g/cm^3. It is a loose rubble pile, covered with unconsolidated fine gravel. This tiny asteroid has a composition similar to that of LL chondrites and the

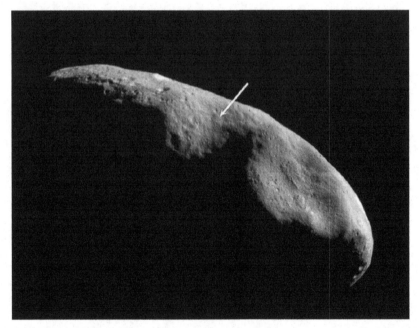

Fig. 13.2 The NEAR landing site on Eros. NASA PIA 03143.

Hayabusa spacecraft revealed that the surface is olivine-rich (70–80%), the other major mineral being pyroxene [11].

13.3 Vesta, a differentiated asteroid

4 Vesta is the third largest of the asteroids and is of particular interest as it differentiated, a process that occurred soon after T_{zero}. It thus forms a scale model of a terrestrial planet with a probable metallic core, an ultramafic mantle and a basaltic crust estimated to be around 10 km in thickness [12].

As revealed by reflectance spectroscopy, the surface is basaltic but with some regions richer in pyroxene and olivine. These are interpreted as regions where the mantle has been exposed by cratering. The surface indeed is heavily cratered: the south pole area is occupied by an enormous central peak crater, 460 km in diameter and 13 km deep formed by the impact of a 40 km diameter asteroid estimated to have occurred about 1 Gyr ago. This crater is close to the diameter of the asteroid and exposes ultramafic mantle rocks as well as providing a fresh regolith over much of Vesta. The impact produced many 10 km size fragments. As if riding on a conveyor belt, they are slowly drifting in the asteroid belt toward the 3/1 orbital resonance with Jupiter at 2.50 AU. It is probable that fragments produced by this impact that have reached that location are the source (the near-Earth "Vestoids") of the meteorites that are delivered to the Earth [13].

The consequence of all these observations is that Vesta is generally considered to be the ultimate source of the HED (howardite–eucrite–diogenite) class of meteorites. These include the eucrites, that are basalts, the diogenites (mostly orthopyroxene cumulates) that are derived from deeper in the mantle and the howardites that are brecciated mixtures of both [14]. Gabbroic eucrites with cumulate textures are thought to represent lower crustal samples, from a depth of around 8–10 km, while the basaltic, or non-cumulate eucrites are derived from lava flows at the surface.

The most interesting feature of Vesta, in the context of this study, is that the formation of an ultrabasic mantle and basaltic crust occurred on this planetesimal within a few million years of the earliest dates in the Solar System (4567 Myr) [15].

13.3.1 Evolution of Vesta

Vesta accumulated from dry planetesimals that were depleted in volatile elements. The asteroid melted during accretion or shortly afterwards, probably from heating due to ^{26}Al and ^{60}Fe perhaps assisted by impacts [16]. This interesting asteroid has preserved a differentiated crust over the history of the Solar System. Not unexpectedly, many models have been proposed to account for the petrogenetic evolution of Vesta without any consensus being reached. Several alternatives are current but effectively reduce to two [17].

The first proposes that the eucrites are partial melts from a mantle (sometimes assumed to be CI chondritic although here interpreted as depleted in volatile and siderophile elements). This requires that following melting, a metallic core formed that strongly depleted the mantle in siderophile elements and the mantle crystallized, forming cumulates. Partial melting of this mantle (possibly induced by heating from ^{26}Al) produced a 10 km thick secondary crust of basalt, at about 4560 Myr (the date from the oldest Pb–Pb ages of the eucrites) within a few Myr of T_{zero}. Basaltic volcanism then ceased, but the mantle and lower crust remained hot (>800 °C) for over 100 Myr. This model is in accord with conventional thinking, but requires that the differentiation of Vesta and subsequent partial melting occurred on very short timescales, within a few (< 5) million years of T_{zero}. Derivation by partial melting from the mantle is supported by the similarities between eucrites and lunar basalts (next section).

The other suggestion, first proposed by Brian Mason in 1962 [17] is that the eucrites are melts from an incompletely crystallized magma ocean, formed from the squeezing out of the eucrite fluid [18]. Major element compositions are expected to be similar in either model but residual melts would be expected to be more strongly enriched in incompatible elements, analogous to KREEP on the Moon. But no consensus has yet been reached.

Vesta is thus a type example of a differentiated planetesimal. There have been many claims for bodies thought to be analogues for conditions on the early Earth and Mars. Thus both Venus and Io are sometimes thought to represent such scenarios. However, the basaltic crust of Vesta may be a closer analogue for the earliest crust on Mars or on the Earth than the continuously heated crust of Io, or the present crust of Venus.

13.3.2 Eucrites and the Moon

The chemical composition of the basaltic eucrites bears a close resemblance to that of the low-Ti mare basalts from the Moon and this points to a similar evolution, despite a wide difference in their ages. This resemblance extends to high chromium abundances and to a similar depletion in siderophile and volatile elements. Thus the K/U ratios of the eucrites overlap the low values recorded both in the Moon and the angrite class of meteorites, of which Angra dos Res is the most famous example. So one asteroid at least underwent a sequence of volatile and siderophile element depletion that resembles that of the Moon. Such processes producing dry volatile-element-depleted "igneous" planetesimals (such as the Moon-forming Theia) that eventually accreted to form the terrestrial planets, may have been common in the inner nebula.

13.4 The Galilean satellites

The four major satellites of Jupiter are rightly famous for their diversity of crustal types that range from the volcanoes and mountains of the innermost, Io (the most volcanically active body in the Solar System), to the icy heavily cratered surface of the outermost, Callisto [19].

The Galilean satellites display a regular decrease in density outwards from the planet, with the proportion of water ice relative to rock in the bodies steadily increasing with distance [20]. The composition of the rock fraction remains unknown. Although it has been suggested that the composition of Io is best matched by L or LL chondrites, it seems more likely that the rock fraction composition is close to CI as discussed above.

The formation and differentiation of the Galilean satellites forms an interesting contrast to processes in the inner nebula that are dominated by hierarchical accretion with planetary-scale melting due to large impacts. The regular jovian satellites and those of the other giant planets, are generally thought to have formed in a nebula disk around Jupiter and are often referred to as a "miniature solar system". But like most facile comparisons, the analogy is not close and their formation did not resemble the hierarchical accretion of planetesimals, accompanied by impact-induced melting, that characterized the formation of the terrestrial planets. Instead, their accretion under cooler conditions recalls earlier models of planetary formation.

This difference is most dramatically shown by Callisto. It is only partly differentiated and must have accreted at relatively low temperatures without the ice fraction undergoing melting. Large impacts during accretion are thus unlikely and many of the large craters (e.g. Valhalla) are most likely due to later cometary impacts. The current model for the formation of the Galilean, and by inference, the other giant-planet satellite systems, calls for slow accretion of small (1 m) rock and ice particles in a late "gas-starved" circum-jovian nebula [21]. The temperature gradient from Jupiter controlled the ice/rock ratios, so that Io accreted rock only. Differentiation occurs through a combination of heating by later tidal resonances and long-lived radioactivity.

13.5 The extraordinary crust of Io

Io is differentiated. "Jupiter's moon, Io is a body replete with amazing geologic features, such as giant lava flows, high lava fountains, and tall mountains" [22]. The moment of inertia value is consistent with the presence of an iron core with a radius about 550 km or with an Fe–FeS core with a radius of about 900 km, that is overlain by a silicate mantle [23]. No impact craters have been observed and the surface is less than one million years old. There is substantial relief on Io. On a body about the size of the Moon, there are calderas 2 km deep and over 100 plateaus and mountains up to 17 km high. Five terrain types have been recognized. Layered plains, probably constructed of lava, lava flows of diverse colors and albedos, high mountains and "diffuse deposits" of material deposited from the volcanic plumes, present us with a bizarre landscape. No ice is present.

13.5.1 Volcanic activity

The surface of Io, as revealed by several close fly-bys of the Galileo spacecraft, is exotic. Hundreds of volcanic craters and calderas cover the surface of Io on a body that is only 1.5% that of the mass of the Earth. Over 60 volcanic centers are active on timescales from months to years. The giant plumes are probably driven by SO_2. These are up to 150–550 km in diameter and extend to 350 km high. They are long-lived on timescales of years. However new plumes also appear so that the surface is continually active.

Two types seem to be present: large "Pele-type" plumes that may reach 350 km seem to be driven by direct gas emission; the smaller "Prometheus-type" plumes that reach around 80 km seem to be due to remobilization of surface sulfur. Although a veneer of sulfur and sulfur compounds coats the surface, these probably form only a surficial coating on the lavas although some of the brighter areas may be lava flows of sulfur. However, the crust is mostly silicate, that alone has sufficient strength to support the rugged topography. The color of the surface is yellow with orange tints. Resurfacing rates are high, probably around 1 cm/yr. Even with the uncertainty of this estimate, several hundred cubic kilometers of lava per year are likely erupted, that is over an order of magnitude more than lava production on the Earth, on a body a little larger than the Moon. The heat flow of Io, a property difficult to calculate, appears to be about $2.2 \pm 0.9 \, W/m^2$, a value that is about 200 times greater than could be supplied by radiogenic heating [24].

Apart from the surficial coating of sulfur, there are three contenders for the silicate volcanism on Io: basaltic lavas, ultramafic lavas similar to the komatiite lavas of the Archean on Earth, or some type of refractory residual material from which the alkali elements, iron and perhaps silicon have been depleted during the $> 4 \, Gyr$ exposure of Io to the tidal embrace of Jupiter.

The ubiquitous coating of sulfur compounds makes it difficult to observe spectral features due to silicate. The lava temperatures appear to be between 1100 and 1500 K so that the lavas are basaltic or doubtfully ultrabasic [25]. Although ortho-pyroxene has been reported, suggestive of a high Mg content, the spectra are just as good a match for the Vesta basalts [26] so that the lavas are most likely basaltic on the basis of the temperature measurements, fluidity of the lava flows and landforms.

The suggestion that the lavas are some kind of "ceramic" residue is more easily dismissed. Although the partial vapor pressure of Fe, FeO and other components may exceed the ambient pressure of Io, it does not seem easy to lose these elements even from a body the size of Io. Much of the iron is in the core, for which there is excellent evidence [27]. After 4000 Myr of volcanic activity, the surface of Io remains coated with volatile sulfur, while a cloud of potassium and sodium ions around Io appears to have been maintained over this period. So it is not easy to lose elements as volatile as sulfur even from small bodies, something that must give pause to suggestions of volatile loss by such processes in the early Solar System [28].

Io orbits within six Jupiter radii of the giant and is subject to heating due to tidal stresses. The heat source for the widespread volcanic activity is tidal dissipation caused both by the orbital eccentricity combined with resonances with Europa and Ganymede. Europa is in a 2/1 resonance with Io while the orbit of Ganymede is in a 2/1 resonance with Europa. The effect of these resonances is to maintain the eccentricity of the satellite orbits against the tidal torques from Jupiter, that would circularize them [29].

Paterae are a form of volcanic caldera much larger than elsewhere and unique to Io. They are common, numbering about 50, are irregular in shape and have scalloped edges. The diameter of the paterae on Io average 41 km but they range to over 200 km in diameter. Often filled with lava flows, they do not contain shields or other volcanic constructional forms. However the paterae appear to be volcanic in origin. There is no resemblance to craters or basins produced by impacts nor do the paterae resemble the lunar maria [22].

13.5.2 Mountains

The mountains on Io are several kilometers high, with the highest reaching an altitude of 17 km, exceeding the height of Mt. Everest on the Earth. These mountains, however, are not volcanic but like Everest, owe their height to tectonic forces. No lavas issue from their summits although some are extruded from the bounding faults. There is a marked global anti-correlation between the distribution of these mountains and the volcanoes [30]. Their distribution is restricted to lower latitudes and does not extend to the poles where the heat flow is expected to be about an order of magnitude lower than at the equator [31] (Fig. 13.3).

No strong tectonic lineations are observed and the mountains mostly appear to be isolated uplifted and tilted crustal blocks, mimicking the Basin and Range Province in the western United States [32].

13.5.3 Nature of the crust

The basic question for this work is the nature of the crust. Probably it is between 20 and 30 km thick. Most of the mountains are steep sided and appear to be undergoing collapse. So we have a picture of a crust of fault blocks undergoing rapid growth and

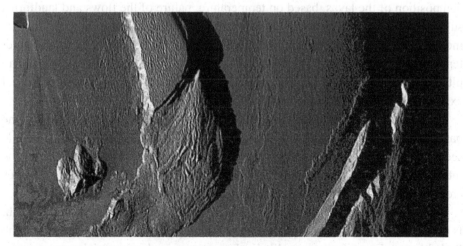

Fig. 13.3 Fault-block mountains on Io. The mountain just left of center is 4 km high. Area of view 210 × 110 km. NASA PIA 02520.

destruction on timescales of less than one million years, a figure consistent both with the observed youth of the surface and theoretical calculations [32].

The internal heat of Io is mostly dissipated by the volcanic conduits. Without this "heat-piping", the only crust that could be sustained on Io would be a thin crust, one or two kilometers thick, of cooled lava. It is the localization of heat loss through the volcanoes that enables a thicker crust to be built of lava flows and other volcanic deposits of low conductivity, enabling a cool crust about 20 or 30 km in thickness to accumulate. Heat loss through this crust by conduction is slow so the crust remains cool.

The tectonic processes responsible for such a landscape are conjectural, but a possible mechanism may be related to the rapid resurfacing of the satellite. As material is piled on top, this has a blanketing effect. Only the base is strongly heated, while the top continues to be loaded, pushing the colder material downwards. This continuing burial of older crustal material by the extremely rapid resurfacing rate leads to sinking with resulting development of compressive stresses, due to the shrinking radius of the layers. Heating at the base exacerbates these stresses. So faulting and the rise of tilted fault blocks occurs in this highly unstable situation. This combination of downward compression and thermal expansion of the lower-most crust seems adequate to account for this unstable block-faulted landscape and is consistent with the anti-correlation observed between mountains and volcanoes on Io [32].

Models for the development of the crust of Io thus envisage continuing burial and sinking of the crust back into the mantle, so recycling the crust [33]. Indeed recycling of crustal material is demanded by the rates of volcanic activity that must involve the entire mantle of Io on brief timescales. But the evidence for the composition of the lavas, based on temperature, nature of the flows and landforms suggests that they are basalts. Such compositions are traditionally formed by partial melting, but such a process should rapidly deplete the mantle of Io in the lower temperature components. Repeated recycling would result in a granitic crust and an ultrabasic mantle, something that might be confidently predicted from terrestrial experience. This has not happened on Io.

A model where the sinking basaltic crust is remelted and remixed to begin the basaltic cycle all over again seems likely. Although a completely molten mantle seems unlikely, a semi-molten magma ocean is generally assumed [33]. So the intuitive expectations from terrestrial experience of granites or refractory residues are overwhelmed by the enormous heat production on this Moon-sized body.

This is the only example of crust-to-mantle recycling other than on the Earth. Is this an analogue for conditions in the Hadean or Early Archean on the Earth? Probably not. Even if, as seems likely, there was an early terrestrial magma ocean that resulted from accretion of planetesimals, or from the Moon-forming collision, the energy source was not maintained. The surface of a terrestrial or lunar magma ocean, even if stirred by impacts, would be passive compared to the ongoing chaos on Io as there is no ongoing deep high-energy source provided on Io by the tidal stresses.

13.6 The thick icy crust of Europa

Europa probably has an Fe or FeS core that is overlain by a silicate mantle covered by a water ocean somewhere between 80 and 170 km thick with a frozen crust. The thickness of the water-ice crust probably exceeds 10 km, a figure based on crater morphology [34]. A crust this thick makes any exchange difficult between the underlying ocean and the surface.

The relief on the surface of Europa can reach 600 m. The satellite is crossed with fractures and is very lightly cratered, (Fig. 13.4) with only 28 primary craters with diameter greater than 4 km.

The age of the surface is probably about 60 Myr although there are wide variations in estimates among crater-counters. Tyre, a multiring structure 44 km in diameter with continuous ejecta extending out to about 100 km, is the largest impact feature. The crater morphology is distinctive. Craters are much shallower than those on Ganymede or Callisto, due to the impact of projectiles into a target that is weak at

Fig. 13.4 Chaotic surface at Conamara Chaos on Europa showing tilted and rotated fault blocks due to recent break-up and movement of pre-existing icy crust. Area shown is 30 × 50 km. NASA PIA 01403.

depth (an icy crust overlying an ocean). However, the impacts appear to have been into a solid icy crust, rather than into a thin icy shell [35].

Because the surface of Europa is so young, with a sparse population of large craters, this has revealed a disturbing fact. Most (95%) of the 17 000 small craters on the surface are secondaries from the few large primary craters. This brings into serious question the use of small craters to date planetary surfaces and is partly responsible for the variety of ages in the literature for this and other surfaces [36].

The orbit of Europa is in a 2/1 resonance with Io, a fact that keeps Europa in an eccentric orbit. Tidal heating from this participation in tidal resonances has supplied sufficient energy to keep the ocean liquid beneath the icy crust. Resurfacing of the crust must also occur due to tidal stresses. Moderate tidal heating not only keeps the ocean liquid, but probably causes melting at the base of the icy crust, inducing resurfacing and local faulting [37]. Possibly there are submarine hot springs on the ocean floor. This has led to the further speculation that thermophilic bacteria might live there, but the thick water-ice crust presents some serious difficulties in investigating this at the present time.

13.7 Two crusts on Ganymede

Ganymede has an Fe or FeS core about 820–900 km thick and the satellite has a strong magnetic field. The core is covered by a silicate mantle that in turn is overlain by a water-ice shell 890–920 km thick (about 50% of total mass) [38]. This shell possibly contains an ocean overlain by an icy crust, that is about 170 km thick. Although its neighbor, Callisto, is similar in composition to Ganymede, the surface terrains of these two bodies show striking differences that have always excited interest.

In contrast to the young smooth surface of Europa and the ancient heavily cratered surface of Callisto, the surface of Ganymede is divided into two distinct regions that indicate two distinct periods of crustal formation (Fig. 13.5). There is a dark heavily cratered older (perhaps 3.7–4.0 Gyr) terrain, that covers about 34% of the surface. This is rifted and intruded by a lower albedo, grooved terrain, covering 66% of the surface. This younger terrain is perhaps as old as 3.1–3.7 Gyr but possibly as young as 1 Gyr. This brighter terrain is strongly affected, on scales down to tens of meters, by tectonic forces that have destroyed crater rims. The resurfacing is so widespread that both tidal and convective forces must be involved.

The older crust is thought to be a relic from an earlier differentiation of Ganymede but when this occurred is problematical as the evidence from Callisto suggests cool accretion. The second younger period of planetary activity (differentiation?) was caused by the satellite passing through orbital resonances with Io and Europa [39] possibly assisted by radiogenic heating from the long-lived isotopes in the rocky

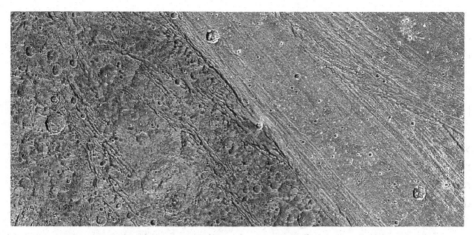

Fig. 13.5 The contrast between the dark and light terrains on Ganymede. The ancient dark terrain on the left is Nicholson Regio. The young bright terrain on the right is Harpagia Sulcus. Area shown is 97 × 213 km. NASA PIA 02577.

fraction. Extensional tectonics seem to have played a major role on Ganymede and the satellite may have expanded in diameter by up to 1%. Melting of high-pressure phases of ice deep in the mantle that expanded as the liquid crystallized in the lower-pressure surface regions may be one cause. The older crust appears to have been split with the intrusion of the lighter younger crust so that the satellite has been partly resurfaced.

The absence of flow fronts and other signs of "ice" volcanism has led to models in which fracturing and extension of an older crust revealed the underlying bright terrain, resulting in a sort of tectonic resurfacing. However the low-lying smooth terrains appear to be of uniform altitude, suggesting that an equipotential surface has been flooded.

Tectonic extension is of course responsible for the initial production of the grabens. The younger bright smooth terrains are 100–1000 m lower than the older terrains and appear to have formed by flooding of the shallow troughs. This model suggests that the grabens were flooded with low-viscosity water-ice "lavas" that accounts for the absence of volcanic features such as flow fronts [40]. In broader terms of crustal formation, this process is reminiscent of the flooding of the low-lying impact basins on the Moon with the low-viscosity iron-rich lunar basaltic lavas.

13.8 Callisto, an ancient crust

The moment of inertia value suggests that Callisto has a partly differentiated internal structure possibly with a rocky core, surrounded by a mixed rock–ice mantle and an icy outlying shell possibly up to 350 km thick [41]. The satellite is incompletely differentiated with a 20–40% separation of ice from rock, consistent with a relatively cool accretion and a slow differentiation. Callisto has a heavily cratered, ancient low-albedo surface without large-scale tectonic features and forms a great contrast to the other Galilean satellites, particularly to its neighbor, Ganymede, but "the geology of Callisto ... upon close examination reveals unexpected charm" [42].

Unlike the other Galilean satellites, Callisto has not subsequently undergone significant heating from tidal stresses due to participation in orbital resonances; such differentiation that has occurred is likely due to heating from long-lived radiogenic isotopes in the rock component. There is no sign of any resurfacing due to cryovolcanism.

However this somewhat ignored satellite has produced the usual set of surprises. At a scale of a few kilometers, a heavily cratered landscape presents itself but at a resolution of tens of meters, a bizarre landscape appears (Fig. 13.6). Major erosion of the cratered landscape has produced a field of isolated knobs, with intervening regions of debris removed from the crater walls. These resemble terrestrial rock glaciers or dry rock avalanches, possibly caused by sublimation as the surface is too cold for ice to deform plastically [43]. Whatever the process, massive mass wasting of the cratered landscape with lateral spreading of debris has occurred on a scale not seen on the other Galilean satellites [41]. Although degraded craters are common elsewhere, they mostly result from the impacts of smaller projectiles.

13.9 Sand dunes on Titan

The nature of the surface of Titan [44], that is the largest satellite of Saturn, has been much resolved by the Cassini mission and the successful landing of the Huygens probe [45]. The landing site was flat, solid and composed of water ice, dampened with methane. At first glance the surface resembled a basaltic martian plain. But the pebbles were centimeter-size water ice, eroded by methane floods [46]. The pebbles were coated with organic "goo" while the surface was mushy, like wet sand (probably ice grains wetted with methane) or a mix of fine-grained ice, methane and various phytochemical compounds resulting in a wet tar and clay mix [47].

The overall view of the landscape revealed a strangely familiar scene, that according to one observer, was like viewing England from a balloon [48]. Elevated rough areas have been dissected by stream valleys, with dark floors resulting from washed-in organic material. Dark low-lying regions on which the Huygens probe landed are also solid. River-like channels, 50–100 m deep, are cut into the brighter areas,

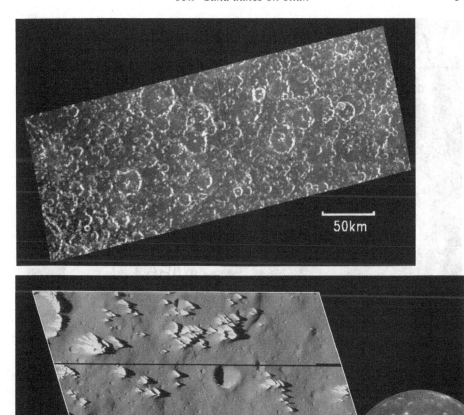

Fig. 13.6 (Top) Heavily cratered terrain on Callisto, antipodal to the Valhalla impact basin. Resolution is 330 m. NASA PIA 2593. (Bottom) Degraded terrain on Callisto south of the Asgard impact basin. The original cratered surface has been eroded or ablated to form isolated knobs, about 60–100 meters high. NASA PIA 03455.

apparently eroded by rains of methane, perhaps under "monsoonal" conditions or by flash floods (Fig. 13.7).

So in the distant reaches of the Solar System, Titan is aping both Earth processes and landscapes, surely an argument for inclusion in this book.

Radar observations from Cassini have shed more light. Icy "lava flows" appear as well as icy "volcanic" calderas and an extensive lava dome, Ganesha Macula that is 180 km wide and has young channels and flow features.

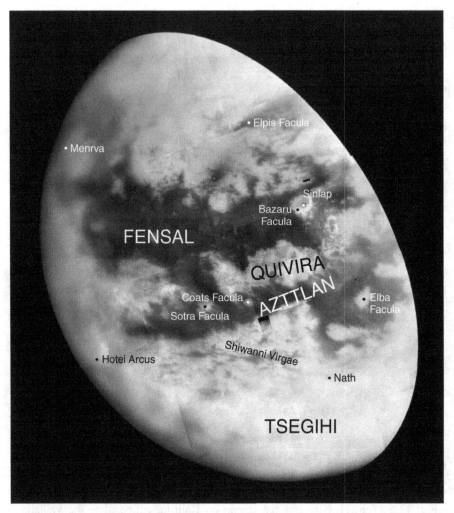

Fig. 13.7 The surface of Titan with named features, showing the contrast between the higher eroded regions and the darker low-lying regions. The origin of the names, mostly from mythology, can be found in Owen, T. C. *et al.* (2006) Titan: Nomenclature system and the very first names for one more world. *LPSC* **XXXVII**, Abst. 1082. NASA PIA 07753.

Small lakes (1–30 km in size) where organic compounds have pooled are observed and may form reservoirs for the methane present in the atmosphere. The gas needs to be renewed continually as it is destroyed by solar ultraviolet radiation. Titan thus appears to possess a meteorology dominated by a gas–liquid–solid cycle of methane in contrast to the water cycle on the Earth.

However, the surface may be mostly dry. Linear dune systems, similar to large-scale linear dune systems on Earth (e.g. the Namibia Sand Sea and Kalahari Desert in Africa and the Great Sandy Desert in Australia) have been observed in the equatorial regions of Titan. These, observed with a resolution of 300 m, are east–west oriented, with a spacing of one or two kilometers and lengths of tens of kilometers. Their presence implies processes to produce sand-sized particles, a lack of surface fluids and wind velocities of 0.5 m/s. These vast dune fields of dark, organic particles may indeed form a more permanent reservoir than the lakes for the organic compounds in the atmosphere [49].

Although the dense Titan atmosphere will prevent the formation of small craters, a very few large impact craters in the expected 20–100 km range have been identified. Both the 450 km basin (Menrva) and an 80 km crater (Sinlap) discovered so far have been subjected to infilling. The paucity of impact craters means that the surface is very young and must be being resurfaced probably as the result of cryovolcanism [50]. The large size of Titan and the high rock content may provide enough internally generated radiogenic heat to promote water–ammonia-ice volcanic activity.

13.10 Nitrogen ice on Triton

Triton, the single large satellite of Neptune, is differentiated with a metallic core, 600 km in radius, surrounded by a silicate mantle 350 km thick and is coated with a water-ice shell 400 km thick [51]. The surface is coated with nitrogen ice that gives it a blue color, along with H_2O, CO_2, CO, CH_4 and C_2H_6 ices [52]. This appearance of nitrogen ice both on Triton and Pluto contrasts with the surfaces of the other icy satellites, closer to the Sun, that are dominated by water ice. As the seasons change and the subsolar point moves, nitrogen ice sublimes and migrates. Geyser-like plumes that resemble terrestrial geysers, are in active eruption. Plume heights reach 8 km before encountering an inversion layer.

There is surface relief of about 1 km. Surface features that resemble lava flows, presumably composed of water ice–ammonia mixtures, are common and bear some resemblance to volcanic terrains on the Earth. The so-called "cantaloupe" terrain is probably the result of diapirism. Other regions resemble calderas [53]. Early heavily cratered terrains are absent. Impact craters are uncommon (the largest is Mazomba, 27 km in diameter). Thus the surface seems comparatively young (a few hundred Myr) [54]. Any early crustal history was probably destroyed by heating during the capture of Triton by Neptune [55]. Following this event, Triton probably remained hot enough for several hundred Myr to erase the early cratering record, so accounting for its youthful appearance (Fig. 13.8).

Some spectroscopic data exist for Pluto and Charon [56], but otherwise little is known about the crusts of bodies in the Edgeworth–Kuiper Belt that is the source of short-period comets and the Centaurs. The total population of TNOs is perhaps

Fig. 13.8 A computer image of the surface of Triton with Neptune in the background, showing a depression flooded by cryovolcanism. The origin of this "walled plain" is enigmatic, resembling neither a caldera nor a flooded impact basin. The surface relief, exaggerated here, is about 1 km and the field of view is about 500 km wide. NASA PIA 00344.

50 000. Many appear to have satellites and at least one, Eris, 3100 km in diameter, is larger than Pluto [57]. However, the total mass in the Edgeworth–Kuiper Belt is probably only about four Pluto-masses or about 0.01 Earth-masses, contrary to earlier estimates that much larger masses were present [58].

Synopsis

A major problem in studying the remote crusts of the minor bodies is observational. The use of remote-sensing techniques, particularly the interpretation of reflectance spectra, has been enhanced by the understanding of "space weathering" that is mainly due to the vapor deposition of "nanophase iron" during exposure of the surfaces to solar flares and micrometeorites.

Because we have samples, 4 Vesta provides us with the best-studied example among the asteroids of crustal development, that occurred within a few Myr of T_{zero}. It is a typical inner nebular body, forming from volatile depleted rock that melted and differentiated into a core and mantle, forming a basaltic crust shortly thereafter. The crustal basalts (eucrites) greatly resemble low-Ti lunar basalts but whether they are partial melts or residual fluids remains contentious.

In great contrast, the satellites of the giant planets and Trans-Neptunian Objects (TNOs) provide different examples of the diversity of crustal development and some interesting contrasts with the inner Solar System. Satellite formation around the giant planets occurred in a cooler planetary nebular environment than the energetic hierarchical accretion of the inner planets. In detail, the satellites display an astonishing variety of crusts, from the turbulent block-faulted and recycled crust of Io to the inert cratered crust of Callisto. In between, Europa has an icy crust overlying an ocean, while Ganymede exhibits two generations of crustal development.

Despite their cool accretion from meter-size particles, which recalls older models of planetary formation, the Galilean satellites have differentiated due to heating from tidal resonances and from the radioactive elements in their rocky component, that is probably of CI composition. Heating from tidal resonances provides the continuing energy source, accounting for the youthful crusts of Io and Europa.

Titan, like Ganymede larger than Mercury, exhibits another variant, with a youthful icy crust eroded into a terrestrial look-alike by methane monsoons. Triton, a first cousin of Pluto, provides an example of the operation of stochastic processes, being melted during capture and by possible radioactive heating and so presents another youthful surface. Like the Messenger mission to Mercury, the arrival of the NASA Pluto–Kuiper Belt mission is eagerly awaited.

Notes and references

1. Pasteur, L. (1854) Inaugural address, Faculty of Science, University of Lille.
2. Trans-Neptunian Objects (TNOs) are sometimes referred to as KBOs, or Kuiper Belt Objects. See Gladman, B. (2005) The Kuiper Belt and the solar system's comet disk. *Science* **307**, 71–5 for a review. For general information about asteroids, see *Asteroids I* (ed. T. Gehrels, 1979), *Asteroids II* (eds. R. P. Binzel *et al.*, 1989) and *Asteroids III* (eds. W. F. Bottke *et al.*, 2002), University of Arizona Press. A more general review of the asteroids appears in the *Encyclopedia of the Solar System*, 2nd edn. (2006), Academic Press, pp. 348–64. For detailed information about meteorites, see Papike, J. J. (1998) *Planetary Materials*, Reviews in Mineralogy 36; and *Meteorites and the Early Solar System II* (eds. D. S. Lauretta and H. Y. McSween, 2006), University of Arizona Press.

3. Stevenson, D. J. (1989) Implications of the giant planets for the formation and evolution of planetary systems, in *The Formation and Evolution of the Solar System* (eds. H. A. Weaver and L. Danly), Cambridge University Press, p. 82.

4. Smith, B. A. *et al.* (1979) The Galilean satellites of Jupiter – Voyager 2 imaging science results. *Science* **204**, p. 951. Some regularity indeed appears in that the satellites of Jupiter, Saturn and Uranus have similar mass fractions (about 10^{-4}) to their respective parents. This appears to be a consequence of their processes of formation. See Canup, R. M. and Ward, W. R. (2006) A common mass scaling for satellite systems of giant planets. *Nature* **441**, 834–9.

5. Croft, S. K. *et al.* (1988) Equation of state of ammonia–water liquid – Derivation and planetological implications. *Icarus* **73**, 279–93.

6. Hapke, B. (1993) *Theory of Reflectance and Emittance Spectroscopy.* Cambridge University Press; (2001) Space weathering from Mercury to the asteroid belt. *JGR* **106**, 10,039–73. See also Dominique, D. L. and Nelson, R. M. (eds, 2005) Special Issue on the Hapke Symposium. *Icarus* **173**. One micrometer (micron) = 10^4 Angstrom units.

7. Pieters, C. M. *et al.* (2000) Space weathering on airless bodies. Resolving a mystery with lunar samples. *MPS* **35**, 1101–7; Chapman, C. (2004) Space weathering on asteroid surfaces. *Ann. Rev. Earth Planet. Sci.* **32**, 539–67. See Hapke, B. (2001) Space weathering from Mercury to the asteroid belt. *JGR* **106** E5, 10,039–73. This paper has a good description of the distinction between glass production and vapor deposition. "Submicroscopic metallic iron" (SMFe) is a better term as some of the particles are hundreds of nanometers in diameter. However the term "nanophase iron" seems firmly entrenched.

8. Chapman, C. (2004) Space weathering on asteroid surfaces. *Ann. Rev. Earth Planet. Sci.* **32**, 563.

9. Trombka, J. I. *et al.* (2000) The elemental composition of asteroid 433 Eros: Results of the NEAR-Shoemaker X-ray spectrometer. *Science* **289**, 2101–5. The landing on Eros was remarkably successful although not initially planned as part of the mission but a "last-minute" decision after the spacecraft had been in orbit for a year.

10. Foley, C. N. *et al.* (2006) Minor element evidence that asteroid 433 Eros is a space weathered ordinary chondrite parent body. *Icarus* **184**, 338–43; McCoy, T. J. *et al.* (2001) The composition of 433 Eros: A mineralogical–chemical synthesis. *MPS* **36**, 1661–72.

11. Fujiwara, A. *et al.* (2006) The rubble-pile asteroid Itokawa as observed by Hayabusa. *Science* **312**, 1330–4; Miyamoto, H. *et al.* (2007) Regolith migration and sorting on asteroid Itokawa. *Science* **316**, 1011–14.

12. Keil, K. (2002) Geological history of Vesta: The "smallest terrestrial planet", in *Asteroids III* (eds. W. Bottke *et al.*), University of Arizona Press, pp. 573–84. Vesta is located at 2.36 AU (e = 0.097 and i = 7°). It has a density of 3.80 ± 0.6 g/cm^3 and an irregular shape with semi-axes $289 \times 280 \times 229$ (± 5) km with a mean diameter of about 520 km. The presence of a metallic core is inferred from the differentiation of Vesta into a silicate mantle and crust and the low abundances of siderophile elements in the eucrites.

13. Binzel, R. P. *et al.* (1997) Geologic mapping of Vesta from 1994 Hubble Space Telescope images. *Icarus* **128**, 95–103; Thomas, P. C. *et al.* (1997) Impact excavation on asteroid 4 Vesta: Hubble Space Telescope results. *Science* **277**, 1492–5.

14. Burbine, T. H. *et al.* (2002) Meteoritic parent bodies: Their number and identification, in *Asteroids III* (eds. W. F. Bottke *et al.*), University of Arizona Press, pp. 553–667. For a detailed description of the HED meteorites, see Mittlefehldt, D. W. *et al.* (1998)

Non-chondritic meteorites from asteroidal bodies, in *Planetary Materials* (ed. J. J. Papike), Reviews in Mineralogy 36, pp. 4.103–4.130.

15. Pb–Pb ages that give dates around 4560 Myr are available for a few of the eucrites. The most recent dating of the eucrites using the Mn–Cr system, gives a date of 4564.8 ± 0.9 Myr that is within error of the T_{zero} age of 4567 ± 2 Myr. However, this Mn–Cr age may be dating the fractionation of volatile Mn from refractory Cr in the early nebula, rather than an igneous event. The Sm–Nd data for the eucrites from this study define an age of 4464 ± 75 Myr but this age is effectively biased by the data for the cumulate eucrites and is interpreted to give the age for the cumulate eucrites, about 100 Myr after T_{zero}, while the age of the basaltic eucrites is close to that of the formation of the Solar System. See Carlson, R. W. and Lugmair, G. W. (2000) Timescales of planetesimal formation and differentiation based on extinct and extant radioisotopes, in *Origin of Earth and Moon* (eds. R. M. Canup and K. Righter), University of Arizona Press, pp. 25–44; Blichert-Toft, J. *et al.* (2002) [147]Sm-[143]Nd and [176]Lu-[176]Hf in eucrites and the differentiation of the HED parent body. *EPSL* **204**, 167–81. The cumulate eucrites generally yield ages about 100 Myr younger than the non-cumulate eucrites. This has led to the perception that there was an extended period of volcanism on this tiny body. However, an alternative explanation, supported by Ar–Ar dating is that there is only one episode of basaltic volcanism within a few million years of T_{zero}, followed by a prolonged period of heating to temperatures above 800 °C, that ended at around 4460 Myr due to excavation by cratering. This seems a more reasonable interpretation of the isotopic data. See Bogard, D. D. and Garrison, D. H. (2003) [39]Ar-[40]Ar ages of eucrites and the thermal history of asteroid 4 Vesta. *MPS* **38**, 669–710.

16. Evidence for this heat source comes from the presence of [26]Mg resulting from the decay of [26]Al with a half-life of 0.73 Myr. This evidence occurs in at least two eucrites, Piplia Kalan and Asuka 881394; Srinivasan, G. *et al.* (1999) [26]Al in eucrite Piplia Kalan: Plausible heat source and formation chronology. *Science* **284**, 1348–50. We note in passing that, according to the Hf–W isotopic data, metallic core formation on asteroids may have occurred less than 1.5 Myr after T_{zero} (4567 Myr) and so predate chondrule formation, well constrained to occur more than 2 Myr after T_{zero}. This may imply separate origins in different locations for the differentiated asteroids, the iron meteorites and the chondrules. Kleine, T. *et al.* (2005) Early core formation in asteroids and late accretion of chondrite parent bodies: Evidence from [182]Hf–[182]W in CAIs, metal rich chondrites and iron meteorites. *GCA* **69**, 5805–18; Bottke, W. F. *et al.* (2006) Iron meteorites as remnants of planetesimals formed in the terrestrial planet region. *Nature* **439**, 821–4.

17. e.g. Mason, B. H. (1962) *Meteorites*, John Wiley; Stolper, E. (1977) Experimental petrology of eucritic meteorites. *GCA* **41**, 587–681; Takeda, H. (1997) Mineralogical records of early planetary processes on the howardite, eucrite, diogenite parent body with reference to Vesta. *MPS* **32**, 841–53; Mittlefehldt, D. W. and Lindstrom, M. M. (2003) Geochemistry of eucrites: Genesis of basaltic eucrites and Hf and Ta as petrogenetic indicators of altered Antarctic eucrites. *GCA* **67**, 1911–35.

18. e.g. Righter, K. and Drake, M. J. (1997) A magma ocean on Vesta: Core formation and petrogenesis of eucrites and diogenites. *MPS* **32**, 929–44.

19. Discovered by Galileo in 1610, the presence of bodies in orbit about another planet was evidence that the Copernican System, rather than the Ptolemaic System, was the correct explanation for the arrangement of the planets. The four bodies are often erroneously regarded as a scale model of the Solar System.

20. The Galileo mission in particular has provided a flood of information: Schubert, G. *et al.* (2004) Interior composition, structure and dynamics of the Galilean satellites, in

Jupiter (eds. F. Bagenal *et al.*), Cambridge University Press, pp. 281–306; Schenk, P. M. *et al.* (2004) Ages and interiors: The cratering record of the Galilean satellites, in *Jupiter* (eds. F. Bagenal *et al.*), Cambridge University Press, pp. 427–56; Theilig, E. E. *et al.* (2003) Project Galileo: Farewell to the major moons of Jupiter. *Acta Astronautica* **53**, 329–52; Showman, A. P. and Malhotra, R. (1999) The Galilean satellites. *Science* **286**, 77–84; Belton, M. J. S. (ed., 1998) Special issue on remote sensing results of the Galileo Orbiter Mission. *Icarus* **135**, pp. 1–376; McKinnon, W. B. (1997) Galileo at Jupiter – meetings with remarkable moons. *Nature* **390**, 23–6; Kuskov, O. L. and Kronrod, V. A. (2001) Core sizes and internal structure of Earth's and Jupiter's satellites. *Icarus* **151**, 204–22.

21. For a full discussion of the problems of forming the Galilean satellites, see Canup, R. M. and Ward, W. R. (2002) Formation of the Galilean satellites: Conditions of accretion. *Astron. J.* **124**, 3404–23 and (2006) A common mass scaling for satellite systems of giant planets. *Nature* **441**, 834–9. See also Lunine, J. I. *et al.* (2005) The origin of Jupiter, in *Jupiter* (eds. F. Bagenal *et al.*), Cambridge University Press, pp. 19–34.

22. Radebaugh, J. *et al.* (2001) Paterae on Io: A new type of volcanic caldera? *JGR* **106**, 33,005–20.

23. Io's mean radius is 1821 ± 0.5 km and its density is 3.528 ± 0.006 g/cm^3. The moment of inertia of Io is 0.378 24 and the satellite is clearly differentiated. Anderson, J. D. *et al.* (2001) Io's gravity field and interior structure. *JGR* **106**, 32,963–9. It seems unlikely that Io has an intrinsic magnetic field. This implies the lack of a magnetic dynamo in the core, which accordingly must either be completely solid or liquid. The temperature is high enough that a core of Fe, FeS or rock would be molten. The temperature of an Fe–FeS eutectic is only about 1270 K at the 50 kbar (5 GPa) pressure of the core–mantle boundary. W. B. McKinnon (2006) personal communication; Keszthelyi, L. *et al.* (2004) A post-Galileo view of Io's interior. *Icarus* **169**, 271–86. There is a plethora of excellent papers: McKinnon, W. B. (2006) Formation and early evolution of Io, in *Io after Galileo* (eds. R. M. C. Lopez and J. R. Spencer), Springer Praxis Books, ch. 4, pp. 61–88; Turtle, E. *et al.* (eds., 2004) Special issue on Io after Galileo. *Icarus* **169**, pp. 1–286; McEwen, A. S. *et al.* (2004) The lithosphere and surface of Io, in *Jupiter* (eds. F. Bagenal *et al.*), Cambridge University Press, pp. 307–28.

24. Spencer, J. R. *et al.* (2002) A new determination of Io's heat flow using diurnal heat balance constraints. *LPSC* **XXXIII**, Abst. 1821.

25. Small areas may have temperatures indicating an ultramafic composition, e.g. the Pillan eruption in 1997 apparently reached a temperature of 1870 ± 25 K. Davies, A. G. *et al.* (2001) Thermal signature, eruption style, and eruption evolution at Pele and Pillan on Io. *JGR* **106**, 33,079–103. But eruption temperatures are mostly around 1100–1500 K indicating basaltic compositions (typical temperatures at basaltic Hawaiian volcanoes are 1423 K or 1150 °C). Davies, A. G. (2003) Volcanism on Io: Estimates of eruption parameters from Galileo NIMS data. *JGR* **108**, 5106–20. There is difficulty in measuring temperatures above 1450 K on Io so that caution is warranted. But measurements by the New Horizons mission to Pluto, that passed by Io in March 2007, yield magma temperatures in the range 1150 to 1335 K, consistent with basalt; Spencer, J. R. (2007) Io volcanism seen by New Horizons: A major eruption of the Tvashtar volcano. *Science* **318**, 240–3.

26. W. B. McKinnon (2006) personal communication. Terrestrial komatiites, with over 18% MgO, contain clinopyroxene, not orthopyroxene but perhaps the lavas on Io are different due to the much lower pressures in the Io mantle, another example of the difficulties in extrapolating from terrestrial petrology to that of other bodies.

27. McEwen, A. S. *et al.* (1998) High-temperature silicate volcanism on Jupiter's moon, Io. *Science* **288**, 1193–8. A similar mechanism of boiling off elements has been appealed to deplete the basaltic lavas on the Moon in the alkali elements. See O'Hara, M. J. (2000) Flood basalts, basalt floods or topless Bushvelds? Lunar petrogenesis revisited. *J. Petrol.* **41**, 1545–651; but the deficiency of the Moon in these volatile elements is endemic. It was accomplished prior to the accretion of the Moon and has nothing to do with the extrusion of the mare basalts hundreds of millions of years later as discussed by Taylor, S. R. (2001) Flood basalts. Basalt floods or topless Bushvelds? Lunar petrogenesis revisited: A critical comment. *J. Petrol.* **42**, 1219–20.

28. Williams, D. A. *et al.* (2000) The July 1997 eruption at Pillan Patera on Io: Implications for ultrabasic lava flow emplacement. *JGR* **106**, 33,105–19.

29. Probably the mantle of Io is partially molten, so that it represents a semi-crystalline magma ocean, that has remained in this state for 4 Gyr. Keszthelyi, L. *et al.* (2004) A post-Galileo view of Io's interior. *Icarus* **169**, 271–86.

30. McKinnon, W. B. *et al.* (2001) Chaos on Io: A model for formation of mountain blocks by crustal heating, melting and tilting. *Geology* **29**, 103–96. See his Fig. 4. Two broad zones of mountains appear around 70°W and 260°W while the volcanoes concentrate around 160°W and 340°W.

31. Lopes-Gautier, R. *et al.* (1999) Active volcanism on Io: Global distribution and variations in activity. *Icarus* **140**, 243–64.

32. Turtle, E. *et al.* (2001) Mountains on Io: High-resolution Galileo observations, initial interpretations and formation models. *JGR* **106**, 33,175–99; Schenk, P. *et al.* (2001) The mountains of Io: Global and geological perspectives from Voyager and Galileo. *JGR* **106**, 33,201–22; McKinnon, W. B. *et al.* (2001) Chaos on Io: A model for the formation of mountain blocks by crustal heating, melting and tilting. *Geology* **29**, 103–6.

33. Carr, M. H. (1998) Silicate volcanism on Io. *JGR* **91**, 3521–32; Carr, M. H. *et al.* (1998) Mountains and caldera on Io: Possible implications for lithospheric structure and magma generation. *Icarus* **135**, 146–65; McEwen, A. S. *et al.* (2004) The lithosphere and surface of Io, in *Jupiter* (eds. F. Bagenal *et al.*), Cambridge University Press, 307–28; Davies, A. G. (2003) Volcanism on Io: Estimates of eruption parameters from Galileo NIMS data. *JGR* **108**, 5106–20; Ojakangas, G. and Stevenson, D. (1986) Episodic volcanism of tidally heated satellites with special application to Io. *Icarus* **66**, 341–58; Keszthelyi, L. *et al.* (1999) Revisiting the hypothesis of a mushy global magma ocean on Io. *Icarus* **141**, 415–19.

34. Europa is mostly composed of rock ($d = 3.014$ g/cm^3; $r = 1560.7 \pm 0.7$ km) and the satellite is differentiated as shown by the moment of inertia value of 0.346. On account of its water-ice surface this satellite has the highest surface albedo (64%) of the Galilean satellites. Greeley, R. *et al.* (2004) Geology of Europa, in *Jupiter* (eds. F. Bagenal *et al.*), Cambridge University Press, pp. 329–62; Schenk, P. M. (2002) Thickness constraints on the icy shells of the galilean satellites from a comparison of crater shapes. *Nature* **417**, 419–21; McKinnon, W. B. (1997) Sighting the seas of Europa. *Nature* **386**, 765–6.

35. Moore, J. M. *et al.* (1998) Large impact features on Europa: Results of the Galileo Nominal Mission. *Icarus* **135**, 127–45; Moore, J. M. *et al.* (2001) Impact features on Europa: Results of the Galileo Europa Mission (GEM). *Icarus* **151**, 93–111.

36. Bierhaus, E. B. *et al.* (2005) Secondary craters on Europa and implications for cratered surfaces. *Nature* **437**, 1125–7. Thus many small craters on the Moon and other surfaces that have been used to date them could be secondaries, so that such surfaces are older than their apparent age. That this is a major problem is demonstrated by the finding that the 10 km crater Zunil on Mars has produced up to 10^7 secondaries ranging

from 10 m to 100 m in size. McEwen, A. S. *et al.* (2005) The rayed crater Zunil and interpretations of small impact craters on Mars. *Icarus* **176**, 351–81.

37. Showman, A. P. and Malhotra, R. (1997) Tidal evolution into the Laplace resonance and the resurfacing of Europa. *Icarus* **127**, 93–111.

38. Ganymede is bigger than Mercury and is the largest satellite in the Solar System ($r = 2634.1 \pm 0.3$ km; $d = 1.936$ g/cm^3). The moment of inertia of Ganymede is 0.3115, the lowest for any solid body in the Solar System. Accordingly this satellite is differentiated. Pappalardo, R. T. *et al.* (2004) Geology of Ganymede, in *Jupiter* (eds. F. Bagenal *et al.*), Cambridge University Press, pp. 363–96; Showman, A. P. and Malhotra, R. (1999) The Galilean satellites. *Science* **286**, 77–84.

39. Showman, A. P. *et al.* (1997) Coupled orbital and thermal evolution of Ganymede. *Icarus* **129**, 367–83.

40. Schenk, P. M. *et al.* (2001) Flooding of Ganymede's bright terrains by low-viscosity water-ice lavas. *Nature* **410**, 57–60.

41. Callisto has a radius of 2408 ± 0.3 km and density 1.839 g/cm^3. The moment of inertia value is 0.359, less than the value of 0.38 for an undifferentiated Callisto and indicates a partial separation of ice from rock. These data have resolved the former problem that Callisto appeared to be undifferentiated in apparent contrast to Ganymede. An induced dipole magnetic field appears to be caused by the presence of a salty ocean. Moore, J. M. *et al.* (2004) Callisto, in *Jupiter* (eds. F. Bagenal *et al.*), Cambridge University Press, pp. 397–426. Showman, A. P. and Malhotra, R. (1999) The Galilean satellites. *Science* **286**, 77–84; McKinnon, W. B. (1997) Galileo at Jupiter – meetings with remarkable moons. *Nature* **390**, 23–6.

42. Moore, J. M. *et al.* (2005) Callisto, in *Jupiter* (eds. F. Bagenal *et al.*), Cambridge University Press, p. 404.

43. Callisto has a very tenuous atmosphere of CO_2 so perhaps sublimation of CO_2 ice is responsible for this unique landscape. Bill McKinnon (2006) personal communication.

44. Titan is a little smaller than Ganymede ($r = 2575 \pm 25$ km; $d = 1.87$ g/cm^3), but is still larger than Mercury. Its orbit is eccentric (e $= 0.028$). The moment of inertia value is predicted to be low. A review of our knowledge about Titan before the Cassini–Huygens mission can be found in Taylor, F. W. and Coustenis, A. (1998) Titan in the solar system. *Planet. Space Sci.* **46**, 1085–97.

45. Lebreton, J.-P. *et al.* (2005) An overview of the descent and landing of the Huygens probe on Titan. *Nature* **438**, 758–64.

46. Hueso, R. and Sanchez-Lavega, A. (2006) Methane storms on Saturn's moon Titan. *Nature* **442**, 428–31.

47. Zarnecki, J. C. *et al.* (2005) A soft solid surface on Titan revealed by the Huygens Surface Science Package. *Nature* **438**, 792–5.

48. Bill McKinnon (2006) *New Scientist*, Jan. 14, p. 39.

49. Lorenz, R. D. *et al.* (2006) The sand seas of Titan: Cassini RADAR observations of longitudinal dunes. *Science* **312**, 724–7. W. B. McKinnon (2006) personal communication.

50. Elachi, C. *et al.* (2006) Titan radar mapper observations from Cassini's T$_3$ fly-by. *Nature* **441**, 709–13.

51. Triton ($r = 1352.6 \pm 2.4$ km), captured by Neptune into a retrograde orbit?, is of much interest as it is a first cousin of Pluto. Although Triton is smaller than the Moon ($r = 1738$ km), it is larger than Pluto ($r = 1175 \pm 25$ km) and 40% more massive. The density of Triton (2.064 ± 0.012 g/cm^3) is close to that of Pluto. Estimates for the moment of inertia are around 0.30–0.35, indicating that it is probably a highly differentiated body. The rock/ice ratio is about 0.72. McKinnon, W. B. *et al.* (1997) in

Pluto and Charon (eds. S. A. Stern and D. J. Tholen), University of Arizona Press, pp. 295–343; McKinnon, W. B. *et al.* (1995) Origin and evolution of Triton, in *Neptune and Triton* (ed. D. P. Cruikshank), University of Arizona Press, pp. 807–77.

52. Brown, R. H. and Cruikshank, D. P. (1997) Determination of the composition and state of ices in the outer solar system. *Ann. Rev. Earth Planet. Sci.* **25**, 243–77.
53. Croft, S. K. *et al.* (1995) in *Neptune and Triton* (ed. D. P. Cruikshank), University of Arizona Press, pp. 879–947.
54. Bill McKinnon personal communication.
55. McKinnon, W. B. (1989) Aspects of the possible capture of Triton. *Bull. Amer. Astron. Soc.* **21**, 916. Triton is significant as it is the largest survivor of the planetesimals that occupied the primitive solar nebula around 30 to 40 AU. It survived by being captured by Neptune while its smaller cousin, Pluto, escaped capture by being nudged into a 3/2 resonance with the ice giant along with a host of smaller Plutinos.
56. The mass of Pluto is only 18.5% of the mass of the Moon. Its orbit has an inclination of 17.2° with an eccentricity of 0.25 and so extends from 29.7 AU inside that of Neptune out to 49.5 AU, while Pluto is locked into a 3/2 resonance with Neptune. It has a density of 1.8–2.1 g/cm^3, about 73% rock. The basic reference is Stern, S. A. and Tholen, D. J. (eds., 1998) *Pluto and Charon*, University of Arizona Press. See also Brown, M. E. (2002) Pluto and Charon: Formation, seasons, composition. *Ann. Rev. Earth Planet. Sci.* **30**, 307–45. A popular account is given by Stern, S. A. and Mitton, J. (1988) *Pluto and Charon: Ice Worlds on the Ragged Edge of the Solar System*, Wiley. For the origin of the Pluto satellite systems, see Canup, R. M. (2005) A giant impact origin of Pluto–Charon. *Nature* **307**, 546–50; McKinnon, W. B. (1989) On the origin of the Pluto–Charon binary. *Astrophys. J.* **344**, L41–L44; and Stern, S. A. *et al.* (2006) A giant impact origin of Pluto's small moons and satellite multiplicity in the Kuiper Belt. *Nature* **439**, 946–8. Satellites may be common in the Kuiper Belt.
57. Bertoldi, F. *et al.* (2006) The trans-neptunian object UB$_{313}$ is larger than Pluto. *Nature* **439**, 563–4.
58. Luu, J. X. and Jewitt, D. C. (2002) Kuiper Belt objects: Relics from the accretion disk of the Sun. *Ann. Rev. Astron. Astrophys.* **40**, 63–101.

14

Reflections: the elusive patterns of planetary crusts

It is difficult to make predictions, especially about the future

(folklore) [1]

14.1 Too many variables

On commencing this study, we were hopeful of reaching some general conclusions about the origin and evolution of crusts, at least on the terrestrial planets. However a survey of the previous chapters reveals little that might assist one in predicting any of the details of crustal development. Crusts of many types are present but they are characterized by differences rather than similarities; there are more variables than there are planets. As with most aspects of planetology, reaching general conclusions or deriving some widely applicable principles remains elusive. Rather than the terrestrial planetary crusts representing points on a continuum of evolutionary style, crustal evolution is governed largely by stochastic processes that also influenced the origin and evolution of the planets themselves. So there are too many variables and too few outcomes to allow for any kind of statistical treatment, just as the accretion of the terrestrial planets, as we have seen in Chapter 1, is essentially a stochastic process, with outcomes impossible to predict.

Although it is possible to classify crusts on the terrestrial planets into "primary", "secondary" and "tertiary" (Section 1.5), this does not imply any logical or inevitable sequence of development. Thus both the primary anorthositic crust of the Moon and the tertiary continental crust of the Earth are unique (Fig. 14.1). So like many classifications, distinguishing the different types of crusts provides convenient pigeonholes but has little predictive power [2].

What factors indeed influence crustal origin, composition and development? Important variables appear to include planetary accretion, size, mass, composition, volatile-element depletion, early thermal history, oxidation state, water or ice content and, importantly, heat either from accretion or from short- or long-lived radioactivity.

352

	Primary crust	Secondary crust	Tertiary crust	Plate Tectonics
Mercury	X	?	–	–
Venus	–	X	–	–
Earth	–	X	X	X
Moon	X	X	–	–
Mars	X	X	–	–

Fig. 14.1 The diversity of types of crust on the terrestrial planets and the Moon.

Impact history is clearly important. Even if the Moon [3] is a special case, it provides the evidence from its pock-marked face of the validity of the planetesimal hypothesis (the accretion of the terrestrial planets from a multitude of smaller rocky bodies). Giant impacts may affect planetary and mantle compositions, sometimes dramatically as in the case of Mercury and may lead to the formation of a magma ocean, that seems to be a requirement for the production of a primary crust. Differentiation of a molten mantle resulting from impacts will affect the subsequent evolution and composition of secondary crusts. These comments about impact heating apply within the inner Solar System. The formation of the Galilean satellites occurred apparently from smaller particles in the circum-jovian nebula so that accretionary heating from major impacts did not occur.

Is planetary mass important? Here comparisons between the Earth and Moon are interesting, the Earth being 81 times more massive. Both were melted, particularly during the Moon-forming collision. However, the Moon formed a thick crust of anorthosite permeated by KREEP shortly following accretion, while the Earth developed a continental crust enriched in incompatible elements only very slowly and episodically. Smaller planets such as Mercury, Mars (and the Moon) cooled more rapidly so that crustal evolution ceased sooner; partial melting in the mantle shut down or, in the case of Mars dramatically diminished, thus inhibiting the development of secondary crusts. Even trivial differences in size may be crucial. The basalt to eclogite transition occurs at a depth of 65 km on Venus, probably too deep to initiate subduction and so allow for crustal recycling.

Is planetary composition important? Earth and Venus are nearly the same size and composition, but crustal development has proceeded very differently. Venus in contrast to the Earth, is a one-plate planet and appears to have undergone planetary-wide resurfacing with basalt perhaps every billion years or so. Here the difference seems to lie in minor variations in the water content, recalling the aphorism "no water, no granites, no oceans, no continents" [4]. Our assessment is that Venus accreted little water to begin with. However, whatever that history has been, the 500 ppm water content of the Earth, although trivial in comparison to the initial water content of the solar nebula, was sufficient to enable recycling and remelting of the oceanic crust, allow an asthenosphere to form and eventually to initiate the development of plate tectonics. As the study of Venus and the Earth shows, similarity does not constitute identity, a truism to philosophers now becoming obvious to students of planets.

Does the presence of a core or the amount of metal make a difference? Mercury and the Moon represent extremes. Mercury has a huge metallic core but the Moon has only a tiny one at best. Although major impacts are the cause of both these outcomes, it appears that the amount of metal, as perhaps might be expected, has little effect on crustal development from silicate mantles, so that both crusts may indeed be similar.

Is heliocentric distance important? This is a frequent trap for modelers. The high density of Mercury is particularly seductive. It is not due to proximity to the Sun, as is often imagined, but is the accidental outcome of an impact. The Moon has a low density for a similar reason. But Mercury and the Moon, far apart, may have similar crusts. No trends appear for the rocky planets. Mars is less dense and contains more volatile elements than either the Earth or Venus but both Venus and Mars have less water than the Earth. All seem due to the operation of random accretionary events. The Galilean satellites show that the amount of ice depends on distance from Jupiter but this trend is not repeated among the satellites of the other giant planets. Moreover Ganymede and Callisto are close in density and hence ice/rock ratio. However they have experienced a very different crustal evolution due to differences in their orbital history.

The effect of heliocentric distance is shown most dramatically by the behavior of "lavas" extruded at the surface of bodies in the Solar System even if such effects are complicated by secondary factors. On Venus, silicate lavas flow like water, but on the icy satellites of the outer planets, water ice produces landforms that mimic those formed by silicate lavas on Earth. Even in this example, the surface temperature of Venus is a consequence of a greenhouse effect that is specific to the planet; without the atmosphere the surface temperature would be about 11 °C [5].

The development of secondary crusts also reflects both planetary origin and evolution in significant ways (Fig. 14.2). Thus the basalts that form the lunar maria were derived from an interior from which the primary feldspathic highland crust was extracted, leaving behind the tell-tale signature of europium depletion. Martian basalts were also derived from depleted mantle sources, from which an LIL-enriched (e.g. with Th) primary crust was extracted very early, certainly before 3.5 Gyr. The mid-ocean ridge basalts (MORB) that dominate the oceanic

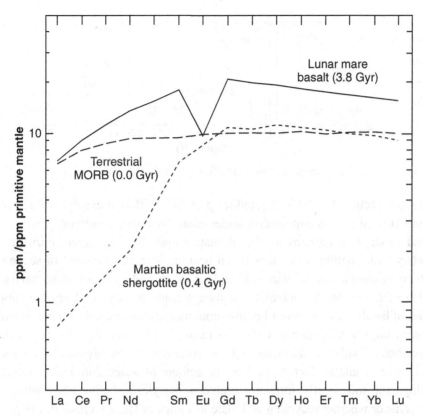

Fig. 14.2 Rare earth element patterns of secondary crusts from the previously depleted mantles of the Earth, Moon and Mars. The lunar mare basalts were derived from a mantle from which a primary crust of anorthosite was extracted. The pattern for terrestrial MORB reflects the long-term derivation of the tertiary continental crust. That for the martian basalts again reflects the extraction of a LIL-enriched primary crust prior to 3.5 Gyr.

crust on the Earth reflect instead the long-term extraction of the tertiary continental crust.

Growth rates of crusts likewise show different patterns in each planet. The Moon and Mercury (and possibly Mars) formed primary crusts. The early (pre-750 Myr) history of Venus is concealed as is that of the pre-200 Myr oceanic crust of the Earth, while the continental crust of our planet grew episodically (Fig. 14.3). The amount of crustal differentiation likewise varies among the planets. These differences are readily shown by their widely varying stratigraphic sequences.

This survey of the better documented smaller bodies of the Solar System reveals the same information as that derived from the study of the terrestrial planets. A diverse array of crusts presents itself, mostly unique to the particular body while the asteroid Vesta provides a sobering example of just how early such differentiation of

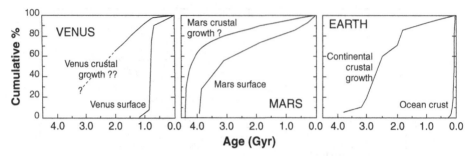

Fig. 14.3 Differing rates of crustal growth for Venus, Mars and the Earth.

bodies can occur. The Galilean satellites provide a different example of accretion in a gas-starved circum-Jupiter disk under relatively cooler conditions.

Heat production remains as the ultimate cause. Primary crusts form due to planetary-wide melting from accretional heating. The development of secondary crusts on the inner terrestrial planets depends on internal heat production, so that partial melting of the complex mineralogy of their silicate mantles produces the various species of basalt. Basalts appear because iron, magnesium and calcium are abundant elements: lavas with cerium or cesium as major constituents are not expected, here or elsewhere. Smaller bodies go through this process more quickly than larger bodies.

However, stochastic factors, such as the amount of water, dominate any further development and greatly influence convection styles, whether stagnant crustal lids persist or whether recycling and plate tectonics or tertiary crusts develop. The secular cooling of the Earth is the cause of the changes in crustal growth from the Early Archean to the emergence of plate tectonics in the Late Archean and for the shift in igneous compositions at the Archean–Proterozoic transition responsible for crustal growth.

Although heat, either due to accretion, built up from radioactive decay or from orbital resonances, is ultimately responsible, each planet, asteroid or satellite shows some peculiarities in the developments of its crusts. Our continental crust–oceanic crust dichotomy, temporal changes in composition and evolution and the history of the growth of the continental crust are largely the consequences of evolving plate-tectonic processes. Despite a persistent myth in the geological sciences of a primordial world-encircling crust of granite, the continental crust has in fact grown episodically throughout geological time, linked to the cooling of the Earth.

However, producing terrestrial continents is an inefficient process. It has taken the Earth over 4000 Myr to transform less than 0.5% of its volume into continental crust. Within this crust of overall andesitic composition, an upper granitic crust, totalling 0.2% of Earth volume has formed by intra-crustal melting, mostly during the past 3000 Myr. The Moon, in contrast, produced a primary crust of calcium-rich feldspar, that constitutes nearly 9% of lunar volume, within a few million years.

The continental crust of the Earth appears to be unique in the Solar System. In contrast to the other terrestrial planets, the significant parameter in the Earth is the occurrence of liquid water. Dry Mars and Venus produced differing varieties of basalt, but the presence of water, even at a minuscule cosmochemical abundance of only a few hundred ppm on the Earth, led to the formation of the continental crust. Water is the ultimate driver for plate tectonics, the establishment of an asthenosphere and subduction that enables the basaltic oceanic crust to be recycled through the mantle. The absence of subduction in the other rocky planets enables the barren basaltic plains to persist while the crusts of the smaller bodies are startlingly diverse.

So the one general conclusion that can be drawn in our planetary system is that the rocky planets commonly generate secondary basaltic crusts of diverse types due to partial melting in the mantles that are dominated by iron–magnesium silicates. The processes involved in the formation and evolution of these crusts are governed by the laws of geochemistry and the abundances of the elements. These are just as relevant on Venus or Vesta as on the Earth. But further generalizations have eluded us and crustal development, like the origin of the planets themselves, is driven by stochastic processes. So it should come as no surprise that their subsequent evolution, including the development of crusts, likewise follows differing paths on different planets. An analogy can be drawn with biology. While evolution acts as a guiding principle, it has no predictive power so that outcomes, such as ants or elephants, arise from stochastic processes involving the interplay of the environment with genetics and biochemistry.

Thus judging from the one example of a planetary system available for our close inspection, one can conclude that the development of plate tectonics and the formation of continents suitable for advanced species to live on, are indeed rare.

14.1.1 Sampling

In attempting to understand crustal development and derive crustal compositions, a major problem is that of sampling. Geochemists have often been pre-occupied with analytical problems in their endeavors to obtain increasingly precise data from samples rather than the more mundane problem of their geological relevance. Isotopic analyses of individual minerals are often difficult to place in a broader perspective. Planetary crusts are so huge and complex so that their investigation often recalls the parable of the blind men and the elephant. However, nature sometimes provides shortcuts. Thus we have a good understanding of our accessible upper continental crust because we have an efficient sampling of it through the erosion and deposition of sedimentary rocks, or less successfully from the herculean efforts of systematic sample collection. The lack of such adequate sampling means that the composition of the lower crust, the mantle and of the Earth itself, remain model-dependent. Because crusts concentrate the incompatible elements, sometimes to extreme levels, as in the continental crust of the Earth or in the lunar highland crust, there is a tendency to overestimate their abundances in compositional tables (e.g. Fig. 12.6).

When direct access to samples that can be studied in terrestrial laboratories is not available, further difficulties arise in the interpretation of the data. These problems, that were once so acute for the Moon, are now encountered with Mercury, Venus, the asteroids, the satellites of the giant planets and the objects in the Edgeworth–Kuiper Belt. Even the data from increasingly sophisticated spacecraft may be difficult to interpret, as with the Lunar surveyors, the Mars Viking and Pathfinder landers and MER rovers on Mars or the NEAR Shoemaker mission to Eros. Instruments designed for remote measurement are inevitably constrained in responding to unanticipated results.

Another problem is that of extrapolation from a limited number of samples such as occurs in our understanding of the lunar crust. Although the interpretation of the basaltic samples returned from the maria is relatively straightforward, the highland samples are complex breccias resulting from multiple impacts. Individual rock units of widely varying significance may be represented by less than gram-size pieces. Only the correlation of remote measurements from spacecraft with the ground truth from the returned lunar samples has enabled a general picture to emerge. The scattered outcrops of Archean rocks on the Earth form a final example. Mostly heavily deformed and altered through billions of years of geological processes, they have frustrated generations of geologists attempting to construct some overview of that remote epoch. Here we regard the averaging provided by sedimentary rocks as a key to understanding the overall crustal composition of the Archean crust as well as those of later times.

14.2 Earth-like planets elsewhere?

There is a surprisingly common perception that Earth-like planets will be found in extra-solar planetary systems although exactly what is meant by "Earth-like", "Earth-type", "Earth-mass", "terrestrial" or "rocky" planets is rarely made clear. Although the Earth is a unique planet in our system, the notion that the Earth represents a norm from which others differ to varying degrees, seems deeply entrenched. As George Basalla has remarked, "we are good at looking for things like ourselves" [6]. An interesting example occurs in a discussion of the possibility of the occurrence of "Earth-like" planets around 47 Ursa Majoris, a star that has two gas giant planets orbiting at 2.1 and 3.7 AU. The authors state that "we assume an Earth-like planet with plate tectonics, a crucial ingredient for our models" and they concluded by predicting the existence of such a body [7]. This indeed seems to be building the answer into the question and so make their calculations self-fulfilling. Another theoretical study is able to form Earth-mass planets in the habitable zone, despite the prior migration of giant planets through the inner nebula. Again the assumptions involved render their conclusions speculative rather than testable [8].

So how likely is the appearance of a clone of the Earth? Indeed no general principles that might be applicable to extra-solar Earth-like planets and the question of their habitability appear from our investigation. The tale told of crustal development in our system reiterates that obtained from a study of the formation of our planets: random processes dominate.

Stars differ mostly in mass and form dominantly from hydrogen and helium with a dash (from zero to a few percent) of "metals" (ices plus rock). But planets may form from any combination of gases, ices and rock. Even within this system of three components, there are complexities. The respective abundances of the ices (water, methane and ammonia are the most common) may vary, while the rock component may be that of the nebula or variously changed by processes, such as the loss of the more volatile elements, within the nebula [9]. Information from our own system reveals the obvious requirements for the formation of planets of metals, orbits of low eccentricity and the avoidance of giant planet migration into the inner nebula. The latter processes might have cannibalized the terrestrial planets, although perhaps others might form subsequently [8].

Collisional events may change the metal-to-silicate ratios and perhaps cause further fractionation of volatile components. The process that formed the Moon was probably responsible both for the obliquity of our planet and for its 24 hour rotation period, both properties affecting biological evolution. As Jacques Lascar and co-workers have noted, without the presence of the Moon, the obliquity of the Earth would be "chaotic with large variations reaching more than 50° in a few million years and even, in the long term, more than 85° in the absence of the Moon" and so "our satellite is a climate regulator for the Earth" [10].

Another effect of the giant Moon-forming collision was to remove any primitive atmosphere from the Earth. Then the presence of the Moon raises substantial tides in the Earth's oceans that also stimulate biological activity. So among questions without current answers, perhaps it is the presence of the Moon resulting from a random event, that is partly responsible for the habitability of this planet. The Moon is unique within the Solar System and the particular parameters of the Moon-forming collision make its duplication elsewhere improbable.

However, making these planets in our system is not just a matter of accumulating the rocky fraction from the primordial nebula. The inner nebula from which the rocky planets formed was also depleted in elements whose sole common property is volatility (Chapter 1). Ironically, this depletion, that occurred in the habitable zone around the Sun, included a large fraction of the biologically important elements, such as carbon, nitrogen, phosphorus and potassium. Of course, any consideration of extra-terrestrial biology must remain open to the possibility that life elsewhere need not conform to an "Earth-like template" but the Earth provides our only empirical example for such discussion [11].

The inner nebula was also bone dry, as shown by the anhydrous nature of the primary minerals of meteorites (olivine, pyroxene and plagioclase) so that the splash of water for our oceans had to come in randomly from the neighborhood of Jupiter. The operation of stochastic processes in building the terrestrial planets from differentiated planetesimals has also resulted in differences from the primordial abundances in the Solar System in planetary compositions even for the major elements (e.g. magnesium and aluminium relative to silicon in the Earth).

However, when nature got around to building two similar planets, it finished up with the Earth and Venus. These twins, unlike the earlier-formed Mars and Mercury, are close in mass, density, bulk composition and in the abundances of the heat-producing elements (potassium, uranium and thorium), but the geological histories of these "twin" planets have been wildly different as we have seen and Venus provides no haven for life.

Indeed a sober contemplation of our own Solar System, the only example available for close study, reveals just how difficult it is to make planets that resemble our own [12]. None of our planets nor their 160-odd satellites resemble one another except in the broadest outlines [13]. "Making planets is an inherently messy business. A growing planetary system resembles an overly energetic infant learning to eat cereal with a spoon. Some is consumed, but much of it ends up on the floor, walls and ceiling" as Steven Soter has remarked [14]. There is no sign of design, intelligent or otherwise. Planetary-forming processes in our system seem to be essentially accidental and the repetition of the particular sequence of events in another system seems as unlikely as multiple wins in a lottery. Uniqueness, it appears, is the common property of planets.

The results of all these factors are beginning to appear among the newly discovered planetary systems [15]. Our own eight vary from gas dominated (Jupiter), through ice and rock mixes (Uranus and Neptune) to rock and metal mixes among the terrestrial planets. Metal may dominate, as in Mercury, or rock if we include the Moon. But even the composition of the rock component may differ, not only from that of the original nebula, but also between the Moon, Earth and Mars, that are successively richer in volatile elements. The other 160 assorted satellites likewise fail to supply us with clones. None of this diversity encourages us to derive general principles for forming crusts. In summary, "the difficulty of producing clones of our present solar system makes (its) duplication as unlikely as finding an elephant on Mars" [16].

14.3 Planetary evolution and plate tectonics

To these problems of building planets are added those associated with their subsequent evolution, including the development of diverse crusts. In our study we find confirmation that the subsequent evolution of planets, most readily displayed by the development of crusts, likewise follows stochastic paths, often driven in different directions through minor differences in composition or mass. Amongst other differences, Venus and for that matter no other body in our system except the Earth, display any evidence of the operation of plate tectonics. It is a common error to suppose that rocky planets will inevitably develop plate tectonics [7]. But the existence of such a tectonic process depends on fine-tuning. Several requirements need to be met such as the formation of a weak asthenosphere, due to the presence of hydrous silicate melts below 100 km or so in the mantle, on which the plates can slide. This in turn depends on the abundance of water in the mantle (a few hundred ppm) coupled with a solubility minimum of water in aluminous orthopyroxene at below 100 km that allows hydrous melts to develop [17]. In the absence of such fine details (a mantle with aluminous orthopyroxene) the Earth might have developed a "stagnant lid" regime with a single plate.

However, apart from water, bulk planetary composition also influences mantle mineralogy and so the possibility of the development of an asthenosphere. Planetary size decides mantle pressures that constrain the conversion of crustal compositions to denser phases, as in the terrestrial basalt–eclogite transition that provides the slab-pull that drives our plate-tectonic system. Variations in planetary size, mineral composition or water content might inhibit or prevent the development of the whole process.

A combination of these factors results in plate tectonics, that is unique to the Earth in our Solar System. This process has the useful property both of building continents and of forming ore deposits useful for advanced civilizations and so enabling this discussion to take place. Indeed habitable planets that develop technical civilizations need a variety of exotic elements such as Ag, Au, Cu, Mo, Pb, REE, Sn, W and Zn. The concentration of these elements into ore deposits here is accomplished mostly by the operation of plate tectonics and the evolution of continents [18].

One such process for recycling and for concentrating rare elements on the Earth is hydrothermal activity beneath volcanoes, that occurs mostly in the vicinity of subduction zones. However, even this does not guarantee the formation of ore deposits. As any economic geologist will testify, it is only rarely that geological parameters conspire to precipitate the metal-rich fluids in sufficient quantity to form ore deposits suitable for mining. However, the long continuation of these accidental processes for over 3 Gyr has developed a continental crust rich in mineral deposits that have enabled the development of our complex technology [19].

This is in stark contrast to the basaltic crusts of Venus and Mars. A civilization developed on such planets might well have some difficulty in discovering the periodic table of the chemical elements. So the evolution of the continental crust through the operation of plate tectonics must be added to the many other parameters that have resulted in forming this habitable planet.

The problem of forming planets and crusts elsewhere would seem to depend on the repetition in detail of the essentially accidental processes of planetary accretion and subsequent geological evolution that have characterized the formation of planets in our Solar System. As it is very difficult here to form a clone of the Earth, it would not seem easy to make Earth-like habitable planets elsewhere. One must recall the difference between necessary and sufficient conditions for the emergence of life. Much more is needed than metals, orbits of low eccentricity within the habitable zone, water and even plate tectonics. None of the results from our study of the Solar System offer grounds for supposing that clones of the Earth exist elsewhere. Such accidental outcomes make questions of purpose or design irrelevant. When the biological and paleontological evidence of evolution is added to the evidence from the physical evolution of the Solar System that we have discussed here, the odds against finding examples of intelligent life forms on Earth-like planets elsewhere appear insuperable.

The concept that Earth-like planets (*sensu stricto*) are rare is not a conclusion that one might wish for, but like much in science, this is what can be read from the observations. Or, as Hervré Faye, writing in 1885, put it more elegantly "Nature has had to form a great number of worlds for that one habitable milieu to be produced ... by a fortunate concourse of favorable circumstances" [20].

Notes and references

1. Variously attributed to Yogi Berra, Niels Bohr, Winston Churchill, Albert Einstein, Enrico Fermi, Eugene Ionesco, Dan Quayle, Bernard Shaw and Mark Twain amongst others.
2. Taylor, S. R. (1982) Lunar and terrestrial crusts: A contrast in origin and evolution. *PEPI* **29**, 233–41.
3. Cameron, A. G. W. and Ward, W. R. (1976) The origin of the Moon. *LPSC* **VII**, 120; Stevenson, D. J. (1987) Origin of the Moon – the collision hypothesis. *Ann. Rev. Earth Planet Sci.* **15**, 271–315; Canup, R. M. (2004) Dynamics of lunar formation. *Ann. Rev. Astron. Astrophys.* **42**, 441–75.
4. Campbell, I. H. and Taylor, S. R. (1983) No water, no granites – No oceans, no continents. *GRL* **10**, 1061–4.
5. The surface temperatures for the terrestrial planets in the absence of atmospheres are: Venus 284 K, Earth 242 K, Mars 195 K. Kevin Zahnle (2007) personal communication.
6. Basalla, G. (2006) *Civilized Life in the Universe*, Oxford University Press, p. 201.
7. Cuntz, M. *et al.* (2003) On the plausibility of Earth-type habitable planets around 47 UMa. *Icarus* **162**, 217.

8. Raymond, S. N. *et al.* (2006) Exotic Earths: Forming habitable worlds with giant planet migration. *Science* **313**, 1413–16; Raymond, S. N. (2006) The search for other Earths: Limits on the giant planet orbits that allow habitable terrestrial planets to form. *Astrophys. J.* **643**, L131–4.

9. Variations in the relative abundances of the "metals" (rock and ices) may also occur according to the mix of elements produced from the differing nucleosynthetic processes, notably for oxygen, iron and for the heavier elements formed by the s- and r-processes. Even within a nebular disk, the mixing may be incomplete. e.g. Ranen, M. C. and Jacobsen, S. (2006) Barium isotopes in chondritic meteorites: Implications for planetary reservoir models. *Science* **314**, 809–12; see also Andreasen, R. and Sharma, M. (2006) Solar nebula heterogeneity in p-process samarium and neodymium isotopes. *Science* **314**, 806–9.

10. Lascar, J. *et al.* (1993) Stabilization of the Earth's obliquity by the Moon. *Nature* **361**, 615–17.

11. Knoll, A. N. *et al.* (2005) An astrobiological perspective on Meridiani Planum. *EPSL* **240**, 179–89.

12. The peculiar difficulties in constructing Earth-like planets have been considered in greater detail elsewhere. See Taylor, S. R. (1998) *Destiny or Chance*, Cambridge University Press; Taylor, S. R. (1999) On the difficulties of making Earth-like planets. *MPS* **34**, 317–29; Ward, P. D. and Brownlee, D. (1999) *Rare Earth*, Copernicus.

13. One common factor is that the satellites of Jupiter, Saturn and Uranus have similar mass fractions (about 10^{-4}) to their respective parents. This appears to result from a balance between the supply of material flowing in and the loss of earlier satellites to the growing planet. However this process does not result in uniform individuals and may limit the sizes of satellites of extra-solar planets. See Canup, R. M. and Ward, W. R. (2006) A common mass scaling for satellite systems of giant planets. *Nature* **441**, 834–9.

14. Soter, S. (2007) Are planetary systems filled to capacity? *Amer. Scientist* **95**, 421.

15. Two extra-solar planetary systems illustrate the diversity that is beginning to appear elsewhere. The system around the red dwarf star Gliese 876 contains three bodies. Two are gas giants with orbital periods of 30 and 60 days while the third, the smallest extra-solar planet found so far, is a planet of 6–8 Earth-masses with an orbital period of two days. What its crust is composed of is only for bold spirits to predict. In contrast, the Sun-like star, HD 67830 has three planets of 10, 12 and 18 Earth-masses (presumably rock–ice bodies analogous to Neptune and Uranus) with orbital periods of 9, 32 and 197 days.

16. Taylor, S. R. (1992) *Solar System Evolution: A New Perspective*, Cambridge University Press, p. xi.

17. Miederl, K. *et al.* (2007) Water solubility in aluminous orthopyroxene and the origin of the Earth's asthenosphere. *Science* **315**, 364–8. Orthopyroxene, nominally $MgSiO_3$, may contain around a few percent aluminium located in the M2 and T2 sites. Smyth, J. R. *et al.* (2007) Crystal chemistry of hydration in aluminous orthopyroxene. *Amer. Mineral.* **92**, 973–6.

18. An excellent summary of the formation of ore deposits and their relation to plate tectonics and continents is given by Groves, D. I. and Bierlein, F. P. (2007) Geodynamic settings of mineral deposit systems. *J. Geol. Soc. London* **164**, 19–30. Our current depletion rate of mineral and other resources exceeds their formation rates by many orders of magnitude.

19. A sobering assessment of the development of technology and of the limitations of complex societies can be found in Basalla, G. (2006) *Civilized Life in the Universe*, Oxford University Press.

20. Fayé, H. (1885) *Sur l'origine du Monde*, 2nd edn. Gauthier-Villars et fils, p. 299.

Author index

Subject index